CW00971059

DEBUGGING GAME HISTORY

Game Histories
edited by Henry Lowood and Raiford Guins

Debugging Game History: A Critical Lexicon, edited by Henry Lowood and Raiford Guins, 2016

Zones of Control: Perspectives on Wargaming, edited by Pat Harrigan and Matthew Kirschenbaum, 2016

DEBUGGING GAME HISTORY

A Critical Lexicon

edited by Henry Lowood and Raiford Guins

editorial assistant, A. C. Deger

The MIT Press
Cambridge, Massachusetts
London, England

© 2016 Massachusetts Institute of Technology

All rights reserved. No part of this book may be reproduced in any form by any electronic or mechanical means (including photocopying, recording, or information storage and retrieval) without permission in writing from the publisher.

This book was set in Gentium Plus and Futura Std by Toppan Best-set Premedia Limited. Printed and bound in the United States of America.

Library of Congress Cataloging-in-Publication Data

Names: Guins, Raiford, editor. | Lowood, Henry, editor.
Title: Debugging game history : a critical lexicon / edited by Henry Lowood and Raiford Guins.
Description: Cambridge, MA : MIT Press, [2016] | Series: Game histories | Includes bibliographical references and index.
Identifiers: LCCN 2015039706 | ISBN 9780262034197 (hardcover : alk. paper)
Subjects: LCSH: Video games—History. | Video games—Terminology.
Classification: LCC GV1469.3 .D43 2016 | DDC 794.809—dc23 LC record available at http://lccn.loc.gov/2015039706

10 9 8 7 6 5 4 3 2 1

CONTENTS

Series Foreword xi

Introduction: Why We Are Debugging xiii
Henry Lowood and Raiford Guins

1 Achievements 1
Mikael Jakobsson

2 Adventure 13
Nick Montfort

3 Amusement Arcade 21
Erkki Huhtamo

4 Artificial Intelligence 29
Rebecca E. Skinner

5 Character 37
Katherine Isbister

6 Classic Gaming 45
Melanie Swalwell

7 Code 53
Mark Sample

8 Console 63
Raiford Guins

9 Control 73
Peter Krapp

10 Controller 81
Steven E. Jones

11 Cooperative Play 89
Emma Witkowski

12 Culturalization 97
Kate Edwards

13 Demo 103
Michael Nitsche

14 Difficulty 109
Bobby Schweizer

15 Educational Games 119
Anastasia Salter

16 Embodiment 127
Don Ihde

17 Emulation 133
Jon Ippolito

18 Fun 143
David Thomas

19 Game Art 151
Mary Flanagan

20 Game Audio 159
William Gibbons

21 Game Balance 169
David Sirlin

22 Game Camera 177
Jennifer deWinter

23 Game Culture 187
Reneé H. Reynolds, Ken S. McAllister, and Judd Ethan Ruggill

24 Game Development 195
Katie Salen Tekinbaş

25 Game Engine 203
Henry Lowood

26 Game Glitch 211
Peter Krapp

27 Games as a Medium 221
Mary Flanagan

28 Genre 229
Mark J. P. Wolf

29 Identities 237
Carly A. Kocurek

30 Immersion 247
Brooke Belisle

31 Independent Games 259
John Sharp

32 Intellectual Property 269
Jas Purewal

33 Kriegsspiel 279
Matthew Kirschenbaum

34 Machinima 287
Jenna Ng

35 Mechanics 297
Miguel Sicart

36 Menu 305
Laine Nooney

37 Metagame 313
Stephanie Boluk and Patrick LeMieux

38 Modification 325
Hector Postigo

39 Narrative 335
Marie-Laure Ryan

40 Platform 343
Caetlin Benson-Allott

41 Playing 351
Jesper Juul

42 Perspective 359
Jacob Gaboury

43 Procedurality 369
Eric Kaltman

44 Role-Play 377
Esther MacCallum-Stewart

45 Save 385
Samuel Tobin

46 Simulation 393
David Myers

47 Toys 401
Jon-Paul C. Dyson

48 Walkthrough 409
James Newman

49 World Building 419
Marcelo Alejandro Aranda

Contributors 425

Index 437

SERIES FOREWORD

What might histories of games tell us not only about the games themselves but also about the people who play and design them? We think that the most interesting answers to this question will have two characteristics. First, the authors of game histories who tell us the most about games will ask big questions. For example, how do game play and design change? In what ways is such change inflected by societal, cultural, and other factors? How do games change when they move from one cultural or historical context to another? These kinds of questions forge connections to other areas of game studies, as well as to history, cultural studies, and technology studies.

The second characteristic we seek in "game-changing" histories is a wide-ranging mix of qualities partially described by terms such as *diversity*, *inclusiveness*, and *irony*. Histories with these qualities deliver interplay of intentions, users, technologies, materials, places, and markets. Asking big questions and answering them in creative and astute ways strikes us as the best way to reach the goal of not an isolated, general history of games but rather of a body of game histories that will connect game studies to scholarship in a wide array of fields. The first step, of course, is producing those histories.

Game Histories is a series of books that we hope will provide a home—or maybe a launch pad—for the growing international research community whose interest in game history rightly exceeds the celebratory and descriptive. In a line, the aim of the series is to help actualize critical historical study of games. Books in this series will exhibit acute attention to historiography and historical methodologies, while the series as a whole will encompass the wide-ranging subject matter we consider crucial for the relevance of historical game studies. We envisage an active series with output that will reshape how electronic and other kinds of games are understood, taught, and researched, as well as broaden the appeal of games for the allied fields such as history of computing, history of science and technology, design history, design culture, material culture studies, cultural and social history, media history, new media studies, and science and technology studies.

The Game Histories series will welcome but not be limited to contributions in the following areas:

- Multidisciplinary methodological and theoretical approaches to the historical study of games.
- Social and cultural histories of play, people, places, and institutions of gaming.
- Epochal and contextual studies of significant periods influential to and formative of games and game history.
- Historical biography of key actors instrumental in game design, development, technology, and industry.
- Games and legal history.
- Global political economy and the games industry (including indie games).
- Histories of technologies pertinent to the study of games.
- Histories of the intersections of games and other media, including such topics as game art, games and cinema, and games and literature.
- Game preservation, exhibition, and documentation, including the place of museums, libraries, and collectors in preparing game history.
- Material histories of game artifacts and ephemera.

Henry Lowood, Stanford University
Raiford Guins, Stony Brook University

INTRODUCTION: WHY WE ARE DEBUGGING

Henry Lowood and Raiford Guins

Five years ago, the journal *IEEE Annals of the History of Computing* published a special issue titled "History of Computer Games." The journal's editor in chief, Jeffrey Yost, made a bold and bracing claim in his introduction to this issue: "Little critical historical analysis has been written on computer games to date. Much of the existing literature is blindly celebratory, or merely descriptive rather than scholarly and analytical. Only a small number of scholars have undertaken rigorous analysis of computer games" (2009, 2). As the editor of that *IEEE Annals* issue, one of us (Lowood) intended primarily to expose scholarly work that would raise awareness of the opportunities presented by game history. The usual readers of that publication naturally would have been likely to see this potential in connection with the history of computing as well as within the broader history of science and technology. In 2009, when the issue was published, the number of scholarly historical writings on computer games was indeed quite small. Game history had not emerged from what Erkki Huhtamo has called the "era of chronicle" (2005, 4). Chronicles have many good qualities. They establish essential chronologies, and their writers often uncover and deploy extensive collections of historical documentation and other forms of evidence. Alas, chronicles are also limited in specific ways. As Huhtamo notes, such histories are usually descriptive rather than interpretive, offering fact-based linear accounts written in the tone of a collector, enthusiast, or journalist, with scant attention to contextualization, critical perspective, methodology, or historiography. This is a mode of writing history consumed with the "when" and "what" to the detriment of the "why" and "how."

The special issue of *IEEE Annals* might be considered a lightning rod for the historical study of games. On the one hand, it demonstrated that historical accounts can add new and valuable perspectives on games as well as contribute to the growing academic field of game studies. On the other hand, one might have wondered if there is a bit of risk involved. How do we encourage new historical work on games to stop digging and lean on its shovel by attending to related areas of scholarly research, from game studies generally to histories of media, technology, materiality, design, culture, and society?

So what is history if not a chronicle? We might turn to a number of historiographers and theorists to produce a lengthy response to this question. Space forces us to pick only one: Hayden White's influential writing on historiography provides a provocative way of thinking about historical writing. In a nutshell, White tells us that history is less about subject matter and source material than about *how historians write*. In his book *Figural Realism: Studies in the Mimesis Effect* (2000), he explains that it is only to the extent that events, persons, structures, and processes "are past or are effectively so treated that such entities can be studied historically; but it is not their pastness that makes them historical. They become historical only in the extent to which they are represented as subjects of a specifically historical kind of writing" (2). One thread through many of White's writings is the point that historians do not simply arrange events culled from sources in correct chronological order. Such arrangements constitute the kind of writing White identifies as "annals" or "chronicles." The authors of these texts compile lists of events, ones easily arranged on timelines to chart progress. This approach resonates, of course, with Huhtamo's thoughts on the "era of chronicle" in the history of games. The work of the historian begins with the ordering of these events in a quite different way. White points out in *The Content of the Form* that in historical writing "the events must be not only registered within the chronological framework of their original occurrence but narrated as well, that is to say, revealed as possessing a structure, an order of meaning, that they do not possess as mere sequence" (1987, 5). How do historians do this? They create narrative discourses out of sequential chronicles by making choices. These choices involve the form, effect, and message of their stories. White puts choices about form, for example, into categories such as argument, ideology, and emplotment. Suffice it to say that the short version of why historians make these choices is that the goal of historical writing is sense making through the structure of story elements, use of literary tropes, and emphasis on particular ideas. Let's call the resulting narrative structures *plots*. White then gives us the enactment of history as a form of story that goes beyond the chronology of events. We do not need to get into the details here of every aspect of White's analysis; we offer this brief discussion merely as one possible entry point into the history of games via an established historiographical framework.

We conceived *Debugging Game History* as a collection of essays to jump-start the critical historical study of games. This goal of *starting* rather than summarizing was motivated by our shared sense that the history of games we envision is today not much farther along than it was when the *IEEE Annals* special issue was published five years ago despite even more source materials being made available to researchers at museums and archives. For a variety of reasons, critical historical work has lagged behind other important areas of research as game studies has emerged in the past decade or two. One result of this general neglect is that few of the fundamental terms of game design and development, technology, and play have been studied in relation to their historical, etymological, or conceptual underpinnings. The historical contexts for the terms we have collected in this book are often inadequately documented or, even worse, misunderstood entirely. The important point is that the poorly developed state of game history has become a drag on rather than a driver in the field of game studies as a whole, particularly with respect to humanistic or critical methods of inquiry. Thus, the title of this book points to the need to *debug* the flawed historiography of video games. In using this term, we are, of course, nodding to the history of computing. Computer bugs are deeply embedded disruptions that cause a system to function in error or to fail. Like Grace Murray Hopper's work on the Harvard Mark II electromechanical computer in 1947, we must open up the machinery of the past and present to remove the moth jamming the writing of critical game history. And to draw from another resourceful allegorical debugging from the history of philosophy, this collection's close inspection of words that we often take for granted in the study of games can help "shew the fly out of the flybottle." This collection mixes methods enthusiastically, so it is appropriate that Hooper's intricate knowledge of machinery sits well alongside Ludwig Wittgenstein's insistence that the goal of critical thought is to clear away misconceptions.

Now, to return to White by way of paraphrase, how do game historians write? Indeed, we can step back from this question and ask, What are they writing about? Our response, to be frank, is that we do not have very many samples from which to answer these questions. Game history of Huhtamo's "chronicle era" has been devoted mainly to stories about the games industry[1] and generalist surveys premised on era-by-era coverage across the late twentieth century and into the twenty-first[2] or on the acceptance of the US "games industry crash" as a prime temporal marker for establishing a "pre" and "post" periodization. There is a shocking lack of social and cultural histories of games at present. And it is only by way of platform studies that game technology has come into critical examination, while media archaeology has emerged as one area that promises great expansion to the types of game histories in current circulation. On account of this lack of samples, *Debugging Game History* is not an encyclopedia presenting a digest of what is known but a collection of essays probing what it means to write critically about historical topics. In White's framework, we have asked our writers to deliver plots or stories that model how to think about topics of relevance to game history. Once more, with emphasis: these essays are *not* encyclopedia articles. We have not asked our writers to ruminate and digest. We prefer to describe their contributions as "takes" on historical topics opened up by focusing on selected, specific terms—terms that prove fascinating due to their historical etymology, changes over time, as well as inclusion within the lingua franca of game studies. Sobered by the thought that White's conception of historiography sets a high bar for critical narrative, we believe that basic terminology is a good place to begin to stretch out toward that goal. By describing this volume as a *critical lexicon*, we assert that the exploration of these key terms will deliver a foothold for historical studies of games.[3]

There is, however, at least one problem with our plan. If the history of games has been a relatively dormant field within game, media, and computing history, and if we cannot draw on a wealth of existing historical studies, where did we find so many authors to write our essays? This, of course, is the proverbial chicken-and-egg problem. Our solution reflects another goal that we shared as coeditors: the desire to model a conception of game studies that is *inclusive*. Our commitment to inclusive game studies is twofold. First, we wish to show that production of *historical* game studies must be open to a group of participants who embody and demonstrate diverse training, methods, and research interests in games. Second, we believe that historical *game* studies can have an impact on a wide range of disciplines beyond game studies. Both of us have experienced recent conferences and projects devoted to game history, from the International Conference on the History of Games held in Montreal in 2013 to the various "Debugging Game History" panels organized for the annual meeting of the Society of Cinema and Media Studies in 2012 and 2013 and papers given at the Society for the History of Technology's annual conferences in 2012, 2013, and 2015, along with the Annual Design History Society conference also in 2015. We have participated in New York University's conference "Pressing Restart: Community Discussions on Video Game Preservation," projects such as Preserving Virtual Worlds, and meetings on game preservation hosted by libraries and museums and have consulted on exhibitions of game history.[4] One of us (Guins) even took his interest in game history to the bottom of a 35-foot pit to excavate Atari's ewaste dumped in Alamogordo, New Mexico. We have been impressed and influenced by events and work that opened up game history to diverse perspectives. Our guiding vision for *Debugging Game History* has been that we will actualize the critical-historical study of games by encouraging work from scholars and professionals whose interests in game history are astute and serious, theoretically and methodologically multidisciplinary, and wide ranging in subject matter. We envisaged a book that will not only influence game studies generally but also reshape how electronic (and other kinds of) games are understood, taught, and researched in fields such as the history of computing, history of science and technology, cultural history, media history, design culture, design history, material cultural studies, new media studies, and science and technology studies. And we also hope that interested parties such as curators and archivists will benefit from this collection because both of us are especially committed to maintaining strong links connecting critical game historiography with documentation, preservation, and exhibition practices.

This last point brings us to our contributors. It is not our intention in this introduction to offer a detailed reader's guide or an annotated table of contents. Rather, we would like to call out a few points of emphasis regarding the mix of authors and topics. First, every article here is not a history or historical etymology of a term. Although we requested that authors write as historically as they felt was appropriate for their topics, we also welcomed takes that opened up terms in ways that would inform historical writing along various and not necessarily predictable lines. For example, David Sirlin's chapter on "Game Balance" is essentially an analysis of game design, Kate Edwards reveals the importance of "Culturalization" by considering its impact on game production and publication, and Jas Purewal explains "Intellectual Property" from a legal studies perspective. These essays are not historical per se; however, they provoke questions for historians to answer about how design and industry practices may have changed over time as well as about the historical contexts and impacts of those changes. Other chapters look at game history from the outside in. Examples include "World Building" (Aranda) and "Game Culture" (Reynolds, McAllister, and Ruggill), which are anchored in games as an option for historical narrative and environmental history, respectively; the essays "*Kriegsspiel*" (Kirschenbaum), "Amusement Arcade" (Huhtamo), and "Toys" (Dyson) take deep dives into the histories of play and simulation and provide broad context for the more recent development of digital and electronic games. And as the reader would expect, a number of essays tear right into the guts of games to examine formative components such as "Code" (Sample), "Game Audio" (Gibbons), and "Mechanics" (Sicart).

A second priority for us has been a stubborn insistence, as should be apparent by now, on a diversity of backgrounds among our authors. A quick and by no means exhaustive accounting would include game development, literary studies, new media studies, media archaeology, law, museum and library curatorship, game studies, cultural studies, software studies, history, film studies, art and art history, technology studies, and philosophy. As noted earlier, we believe that inclusivity benefits game studies and multiplies its capacity to touch other scholarly disciplines and areas of media production and consumption. If there is a cost, it is perhaps a blend of registers that may approach cacophony. We acknowledge that our writers have at times differed or even written past each other with regard to basic concepts, such as history, play, and game design. This is where our stubbornness plays a role. We insist that game history as a new field must encourage channels of debate and conversation that will open up both academic game studies and, optimistically, game culture itself. We don't strive simply for more game history but for better. We offer *Debugging Game History* to ignite a conversation across engagements with games and game studies by including a range of voices, even at the cost of sweet harmony. After all, as Roland Barthes once said of interdisciplinary endeavors, such work "is not a peaceful operation."

Before closing, we would like to thank our contributors for putting meat on the bones of our concept for the book. Their "takes" are invaluable to the project of the critical historical study of games. We are also grateful for the support of Doug Sery and the editorial team at MIT Press for their usual excellence in production and publication. We also wish to express our gratitude to our editorial assistant, Anne Deger, for managing the grunt work of compiling this volume. Finally, we are grateful for the support and patience of our spouses and kids, whom we continue to baffle and intrigue with our capacity for turning games and play into research, new projects, and conference travel.

Notes

1. Notable examples include *Commodore: Company on the Edge* (Bagnall 2010); *Zap: The Rise and Fall of Atari* (Cohen 1984); *Console Wars: Sega, Nintendo and the Battle That Defined a Generation* (Harris 2014); *The Ultimate History of Video Games* (Kent 2001); *Power-Up: How Japanese Video Games Gave the World an Extra Life* (Kohler 2004); *Masters of Doom* (Kushner 2004); *Service Games: The Rise and Fall of Sega* (Pettus 2013); *Super Mario: How Nintendo Conquered America* (Ryan 2011); *Game Over: How Nintendo Conquered the World* (Sheff 1999); and *Atari Inc.: Business Is Fun* (Vendel and Goldberg 2012).

2. Although detailed and insightful, *Joystick Nation* (Herz 1997), *Trigger Happy* (Poole 2000), and *High Score: The Illustrated History of Electronic Games* (Demaria and Wilson 2002) nonetheless furnish broad, sometimes nostalgic, historical accounts given primarily to business history or descriptive accounts of specific games. To further exemplify this all-too-common perspective, we might also consider more recent attempts at surveying the history of video games, such as *Phoenix: The Fall and Rise of Video Games* (Herman 2001), *Smartbomb* (Chaplin and Ruby 2006), *Replay: The History of Video Games* (Donovan 2010), and *The Golden Age of Video Games* (Dillon 2011). These books maintain a linear narrative through which readers are taken on a journey from the "founding fathers" (Russell, Baer, Bushnell) to the latest next-generation console on the market at the time of their writing. Such surveys and chronologies are not the exclusive property of journalism. Textbooks aimed at a game studies audience, such as the edited collection *The Video Game Explosion: A History from* Pong *to Playstation and Beyond* (Wolf 2008), restrict the subject of history within chapters that replicate and reaffirm reductive approaches to making sense of video game history.

3. We have followed the lead of Matthew Fuller's book *software studies\a lexicon* (2008) for the general structure and organization of our content, and Raymond Williams's classic text *Culture and Society: 1780–1950* (1958) has proven invaluable for this collection's intended goal of composing an etymology of "key terms" to better understand the historical formations of games. Like Williams, we are interested in the meanings and, in particular, the changing meanings of words as they continue to shape our understanding of culture. Ours is a more narrow focus but one that nonetheless shares common ground with Williams's insistence that his references are not studied at length "only to distinguish the meanings, but to relate them to their sources and

effects" (1958, 18). The terms collected here are examined for their commonality today so that we might gain insights into how such commonality occurred and at the same time, we hope, open up their past, which may prove anything but common.

4. Both of us served as reviewers and advisers to the Smithsonian National Museum of American History exhibition *Don't Watch Television ... Play It: The Beginning of Video Games* and worked with the New York Historical Society on its exhibit *Silicon City*, devoted to the history of computing in New York City and scheduled to launch in November 2015.

Works Cited

Bagnall, Brian. 2010. *Commodore: Company on the Edge*. Winnipeg, CA: Variant Press.

Chaplin, Heather, and Aaron Ruby. 2006. *Smartbomb*. Chapel Hill, NC: Algonquin Books.

Cohen, Scott. 1984. *Zap: The Rise and Fall of Atari*. New York: McGraw-Hill.

Demaria, Rusel, and Johnny L. Wilson. 2002. *High Score: The Illustrated History of Electronic Games*. Emeryville, CA: McGraw-Hill and Osborne.

Dillon, Roberto. 2011. *The Golden Age of Video Games*. Boca Raton, FL: A.K. Peters/CRC Press.

Donovan, Tristan. 2010. *Replay: The History of Video Games*. Lewes, UK: Yellow Ant.

Fuller, Matthew. 2008. *software studies\a lexicon*. Cambridge, MA: MIT Press.

Harris, Blake J. 2014. *Console Wars: Sega, Nintendo, and the Battle That Defined a Generation*. New York: HarperCollins.

Herman, Leonard. 2001. *Phoenix: The Fall and Rise of Video Games*. Springfield, NJ: Rolenta Press.

Herz, J. C. 1997. *Joystick Nation*. Boston, MA: Little Brown and Company.

Huhtamo, Erkki. 2005. "Slots of Fun, Slots of Trouble: An Archaeology of Arcade Gaming." In *Handbook of Computer Game Studies*, ed. Joost Raessens and Jeffrey Goldstein, 3–21. Cambridge, MA: MIT Press.

Kent, Steven L. 2001. *The Ultimate History of Video Games: From Pong to Pokemon—the Story behind the Craze That Touched Our Lives and Changed the World*. New York: Three Rivers Press.

Kohler, Chris. 2004. *Power-Up: How Japanese Video Games Gave the World an Extra Life*. New York: Brady Games.

Kushner, David. 2004. *Masters of Doom*. New York: Random House.

Pettus, Sam. 2013. *Service Games: The Rise and Fall of Sega.* CITY: CreateSpace Independent.

Poole, Steven. 2000. *Trigger Happy*. New York, NY: Arcade Publishing.

Ryan, Jeff. 2011. *Super Mario: How Nintendo Conquered America*. New York: Portfolio/Penguin.

Sheff, David. 1999. *Game Over: How Nintendo Conquered the World*. Wilton, CT: CyberActive.

Vendel, Curt, and Martin Goldberg. 2012. *Atari Inc.: Business Is Fun*. Carmel, NY: Syzygy Press.

White, Hayden. 2000. *Figural Realism: Studies in the Mimesis Effect*. Baltimore: Johns Hopkins University Press.

White, Hayden. 1987. *The Content of the Form: Narrative Discourse and Historical Representation*. Baltimore: Johns Hopkins University Press.

Williams, Raymond. 1958. *Culture and Society: 1780–1950*. London: Chatto and Windus.

Wolf, Mark J. P. 2008. *The Video Game Explosion: A History from Pong to Playstation and Beyond*. Westport, CT: Greenwood Press.

Yost, Jeffrey, R. 2009. "From the Editor's Desk." In "History of Computer Games," ed. Henry Lowood, special issue of *IEEE Annals of the History of Computing* 31 (3): 2.

1 ACHIEVEMENTS

Mikael Jakobsson

Achievements is a term coined by Microsoft in 2005 to describe a reward system for digital games that they introduced with the launch of the Xbox 360 console. The system quickly gained significant success, which led to the implementation of achievement systems on Valve's personal computer gaming platform Steam, Apple's iOS gaming network Game Center, and Sony's PlayStation Network (in this case called "trophies"). Although Nintendo has similar systems for certain games, it stands out as the notable exception to this trend. Achievements have also been implemented on many other gaming platforms, sometimes under different names such as "badges."

The term itself as well as the particulars of its contemporary form are relatively recent, but achievements build on previous systems of external awards for the accomplishment of objectives. Looking outside of video games, we will find that these kinds of systems go far back in time. After giving an overview of how achievement systems work, I examine the origin of contemporary achievement systems and trace the origin of the term. I use Microsoft's achievements as an example of how such systems work, including player reception and appropriation. Finally, I map out the history of reward systems that laid the ground for contemporary achievement systems.

Achievement Anatomy

Achievements are one of eight different types of reward systems in games, as categorized by Hao Wang and Chuen-Tsai Sun (2011). Different achievement systems vary in terms of the details of their implementation but share some general characteristics. Drawing on Markus Montola, Timo Nummenmaa, Andrés Lucero, and their colleagues (2009), Staffan Björk (2012), as well as my own work (Jakobsson 2011), we can describe achievements as persistent public reward systems separate from the rest of the game but connected to optional in-game challenges. The term *achievements* is somewhat misleading. The tasks that have to be completed to unlock achievements are often trivial, such as finishing a tutorial, using a certain weapon, or playing a game for a certain amount of time.

Juho Hamari and Veikko Eranti (2011) separate achievements into three necessary components: *signifier*, *completion logic*, and *reward*. The *signifier* is the part that is presented to the player. It normally consists of a name, an icon-size graphical representation, and a description that lets the player know the requirements for unlocking the achievement.

The *completion logic* defines the game state that will unlock the achievement. It can be triggered by a player action such as blowing up 20 so-called infected (zombies) in a single explosion in *Left 4 Dead* (Turtle Rock Studios/Valve South, 2008) for the Pyrotechnician achievement or by a system event such as 60 seconds passing without shooting or dying in *Geometry Wars: Retro Evolved* (Bizarre Creations, 2005) for the Pacifism achievement. Sometimes there are several triggers, such as the Scholar achievement in *Mass Effect* (BioWare, 2007), which requires the player to gather information about every alien race in the game, and certain triggers have to be activated a number of times, as in the case of the Seriously achievement in *Gears of War* (Epic Games, 2006), which requires the player to get 10,000 kills in multiplayer matches.

The *reward* is the persistent proof that an achievement has been unlocked. Rewards are aggregated somewhere outside of the actual game session, where they normally also are accessible to others. These rewards are typically only symbolic. They cannot be used for anything beyond public display of having completed the corresponding challenges.

Achievements at Microsoft

When achievements were first introduced, they were part of the extended player profiles for the Xbox 360. Although the online service Xbox Live had been around in a simpler form since 2002 (Takahashi 2006, 21), with the new console online connectivity moved from being an experimental feature to being a core aspect of the system. Player account information was expanded from a unique user name across all games (called a gamertag) and a friends list of up to a hundred other Xbox Live users to a public user profile (called a gamercard), including, among other player information, the five most recently acquired achievements ("Description of Xbox 360 Gamer Profiles" 2010).

In the Xbox Live achievement system, every retail game has approximately 50 achievements, and the smaller digital-only games have 12 achievements. Every achievement has a point value (called a gamerscore). The total gamerscore is 1,000 points for a retail title and 200 points for a downloadable game. Beyond that, there are very few guidelines for the developers to follow in creating achievements (Bland 2009; Greenberg 2007). There are, for instance, achievements just for pressing the start button and even for playing badly (dying or failing repeatedly).

Anyone can look up other players' achievement information on Xbox.com as long as they know the player's gamertag or directly on the console if they have the person in their friends list or in the recent players list. Players can adjust the privacy settings to modify who has access to this information. Players actively promote their identity and gaming history by customizing gamercards for use on websites, blogs, and forums as well as with automated signatures using sites such as MyGamerCard.net ("Achievement Unlocked" 2007), and there are many reasons for doing this. A substantial achievement record can add weight to a forum poster's opinions by showing his or her in-depth familiarity with the topic at hand. The information also says something about a person's gaming interests, which can work as a conversation starter between players who do not know each other and lead to players adding each other to their friends lists and playing together. When a player receives an achievement, there is a sound cue, and a notification pops up on the screen informing the player of the gamerscore value and name of the unlocked achievement. The notification can be disabled together with all other notifications, but the player still receives the achievement.

Achievements were first introduced to the public at the Electronic Entertainment Expo on May 16, 2005. J Allard, one of the driving forces behind the Xbox project at Microsoft, mentioned the feature as part of the official announcement of the Xbox 360. He described them as "sort of a record of everything you've accomplished across your library of games" ("Robbie Bach, J Allard, Peter Moore" 2005). The achievements system had a predecessor within the company in the MSN Games badges, so to locate the origin of the achievements concept within Microsoft, we have to go back a decade to the mid-1990s.

In 1995, Bill Gates wrote a memo entitled "The Internet Tidal Wave," in which he assigned the Internet "highest level of importance" to the whole company ("Letters of Note" 2011). A few months later the Microsoft Network (MSN), a collection of Internet sites and services, was launched. When Microsoft wanted to add an online gaming site to the network, it decided to buy a site called Electric Gravity, which after the acquisition was renamed several times, most notably as The Zone, before it became MSN Games ("Microsoft Online" 2004).

In 2005, around the same time the Xbox 360 was announced, the site went through an overhaul that introduced badges that the players could collect by playing the games on the site. There were four different types of badges. Three of them were awarded for reaching a certain high score, accumulated score, or accumulated number of hours in any given game. The fourth type was known as special achievement badges, which were awarded for completing challenges in particular games during a limited period of time. They had unique names and artwork, which make them significant precursors to the Xbox Live achievement system. On a systemic level, what was new with the badges was that players had to log in with their

MSN accounts, which made it possible to store player records on a central server rather than in the local browser cache, which in turn created the persistence required for a modern achievement system ("Countdown to Badges" 2014).

The introduction of the special achievement badges dates the first use of the word *achievement* to denote this phenomenon back to an announcement made in June 29, 2005, when the Gold Rush challenge for the game *Gold Fever* (Qwest) went live: "Between now and July 26, 2005, every point you earn in Qwest Gold Fever will be counted towards our first-ever special achievement badge ... the GOLD RUSH!" ("Special Achievement Badge Available" 2014).

When Microsoft Games was developing the Xbox Live Arcade service for the original Xbox, which focused on small downloadable games often aimed at more casual players, it worked with MSN Games and some of the latter's hit games, such as *Bejeweled* (PopCap, 2000) and *Zuma* (PopCap, 2003) were ported over. The experiences from developing the badge system were also useful to the Microsoft Games team in the process of designing the achievement system for the Xbox 360 ("Microsoft Online—from The Zone to Xbox Live" 2004).

Community Reception

When the Xbox 360 was released, the achievement system had a significant impact on player practices. The system was often cited as a competitive advantage for the Xbox 360 console and as affecting both sales and critical reception of separate game titles (Electronic Entertainment Design and Research 2007). A number of websites, online forums, and podcasts were created that focused solely on achievements. Some websites even offered achievement farming to those who were prepared to pay money to "level up" their gamertag. The most prevalent aspect of community engagement with achievements was sharing tips and tricks on how to earn particular achievements, but other topics such as how good the achievements in new games were, how gaming habits have changed since the introduction of achievements, and the organization of leagues and tournaments were also discussed (Jakobsson 2011).

At the same time, a vocal opposition to achievements emerged, consisting of players, journalists, and developers. On forums, in podcasts, and in chat channels, players showing an interest in collecting achievements were often criticized, pathologized, and referred to in derogatory terms. The criticism usually concerned how achievements were "addictive" and attracted attention away from the "game itself" (e.g., Hecker 2010).

In an attempt to reach beyond sweeping generalizations of how players engage with achievements, I conducted a two-year participatory study of the Xbox 360 gaming community. Three different groups emerged with respect to their approach to achievements in the analysis of the empirical material: *achievement casuals*, *achievement hunters*, and *achievement completists*.

It is important to note that these three categories are fuzzy and unstable. Players may inhabit traits from different categories at the same time, and their attitudes and play styles often change over time. The players' own perception of which category they belong to is often inconsistent with their actual gaming behavior, and they are often ambivalent regarding their play styles, noting that they wish they could act differently or that they do not understand their own behavior (Jakobsson 2011).

The first category takes its name from the "casual gamer" stereotype, which is used to denote players who are significantly less invested in gaming than their counterpart, the "hardcore gamer" (Juul 2009). The *achievement casuals* often pay little attention to the achievements until they have finished a game but want to continue playing it. Developers have always provided different ways to optionally extend the gaming experience, such as so-called new-game-plus modes where the player can start over from the beginning with abilities or items carried over from the first playthrough. With achievements, progression past the formal end of the game is made smoother. The story ends, but some achievement is usually within reach of being completed, which beckons the player to continue playing. In this regard, achievements work as scaffoldings that support players who wish to remain in the game world and continue adding to their play experience.

The term *achievement hunter* comes from the community itself and describes someone for whom the achievements have become more important than the games themselves. For these players, it is perfectly natural to play games just to collect the achievements. Old sports games and games aimed at children are, for instance, known for giving easy achievements while not always providing the most exciting gaming experience. *Achievement hunters* sometimes go through the achievement-unlocking process several times for the same game title (a process known as "doubling") because copies of the game from different regions sometimes give separate achievements. The way achievement hunters approach these games can be compared to playing massively multiplayer online games such as *World of Warcraft* (Blizzard Entertainment, 2004), where long-term goals of leveling up and gaining quest rewards can overshadow the emotional investment in moment-to-moment gameplay for many players. T. L. Taylor's statement about massively multiplayer online *powergamers*, who "[t]o outsiders ... look as if they are not playing for 'fun' at all" (2006, 71,) fits *achievement hunters* very well.

There is a connection between this category of players and the category "multiuser dungeon" (MUD) players whom Richard Bartle (1996) identifies as *achievers*. The similarity lies in the focus on outcomes of play activities, but whereas achievers are characterized by their focus on overcoming challenges, *achievement hunters* spend most of the time on fairly menial tasks. They always go for the achievements that they can get with the least expenditure of time and effort. They also exhibit traits of other player types. To be successful, achievement hunters need to create a network of friends to trade tricks, games, and game saves, which implies that they are *socializers* (Bartle 1996). But they also frequently engage in competition between each other, tracking who collects the most achievements overall or within a given time frame, which would associate them with *killers* in Bartle's terminology.

The term *completists* (or *completionists*) is used to describe collectors of everything from vinyl records (Plasketes 2008; Stump 1997) to memorabilia (Henderson 2007). Jenny Robb describes them as being "systematic in [their] approach, collecting everything, or an example of everything, that falls within a certain category" (2009, 249). Video game completists are different in that they are not collecting games but rather items and rewards in games such as unlockable character models (MacCallum-Stewart 2008). *Achievement completists* consider games to be unfinished until they have collected all the achievements. They can spend months playing a game that they have already grown tired of just to get that last elusive achievement (Jakobsson 2011). This willingness to submit to difficult challenges resonates with Bartle's (1996) *achiever* category. Similar to achievement hunting, the struggle to collect everything in a game cannot always be described as fun, but the overall experience can still be very valuable to the players (see Taylor 2006, 88–92). The introduction of achievement systems has made the rewards of the collection efforts more concrete and visible for achievement completists. Olli Sotamaa (2010) connects the external visibility of achievement collection to the concept of gaming capital (Consalvo 2007). The collection represents reputation within the community.

Medals, Badges, and Patches

In tracing the predecessors that have inspired the achievement systems of today, I have focused on symbolic awards that can be used to display some kind of accomplishment or milestone within an organization or in a competition. Achievement systems have also been compared to different types of reward systems, such as frequent-flyer miles (Bogost 2010), but the suggested similarities are in these cases related to the systems' psychological or social outcomes. In their construction, these systems consist of points and scores rather than of badges and patches, and, unlike achievements, they have more than symbolic value.

Decorations used to denote accomplishments within the military are known to have existed as far back as in the Old Kingdom of Egypt around 2000 BCE (David 1982). Modern military organizations usually have intricate systems of medals, badges, ribbons, and so on. Just as with video game achievements, these physical emblems do not always denote accomplishments. The US military decoration the Purple Heart is, for instance, awarded to soldiers wounded or killed in service. It may have been through military practices that medals and other decorations found their way to organized sports, where they now play a significant role, for instance in the Olympic Games.

Badges were originally a heraldic symbol worn as an identifying mark by a knight and his retainers (Gibson, Ostashewski, Flintoff, et al. 2013). The Boy Scouts adapted the concept for their Merit Badges. These badges are of particular interest in the history of accomplishment-based decorations. The Merit Badges have been an integral and important part of the Scouting program since the start of the movement in the United Kingdom in 1907 ("History of Merit Badges" 2014). The merit badges are meant to encourage Boy Scouts to explore topics that interest them, and they are awarded for a wide variety of activities in areas such as astronomy, scuba diving, and theater.

Although the Boy Scouts Merit Badges are meant to signal the pursuit of and adherence to constructive values, some systems work in the opposite direction. Some outlaw motorcycle gangs are said to use a skull-and-crossbones patch to signify that the wearer has killed or committed other acts of violence in service of the club. Sometimes these patches are worn by members who have not actually "earned" them because the patches still fill the intended function of intimidating and instilling fear in outsiders (Queen 2005).

The coercive power of external reward systems has long been used for purposes related to behavior modification (Matson and Boisjoli 2009). Already in 1859, N. H. Avendano y Carderera described a ticket or token that could be used to reward good behavior in children. Since then, systems similar to achievements have been used in schools and in treatment of mental disabilities (Gibson, Ostashewski, Flintoff, et al. 2013). With the recent rise of interest in gamification, or the use of game design elements in nongame contexts (Deterding, Dixon, Khaled, et al. 2011), the interest in using achievement systems in education has intensified, but some advise caution against the use of external reward systems in general (Kohn 1993) and of achievements in particular (Nicholson 2014).

In the early 1980s, the US video game publisher Activision started offering similar awards for gaming accomplishments. Mainly for the games it published for the Atari 2600 but also for some games for other platforms, Activision printed special challenges in the game manuals. If a player managed to beat the challenge, he or she could send a letter to Activision, normally with a photo of the TV screen included as proof, who in return would send a decorative fabric patch together with a form letter congratulating the player and welcoming him or her to "the club." The patches could then be ironed or sown onto a jacket to display the player's accomplishments in the schoolyard and elsewhere ("Activision" n.d.; "Activision Patch Gallery" n.d.; Thomasson 2003).

Activision in the United Kingdom also offered these rewards, calling them "badges" ("Master Gamers—Free Badges to Be Won" 1984), and the game developer and publisher Imagic closely emulated the concept with its award decals ("Imagic Experts Club" 1983).

The Boy Scouts took the concept of displaying symbols of accomplishments on their uniforms from the military, and Activision used the format of the Boy Scouts badges and

Figure 1.1
Activision patches.

introduced it to a video game context. The Activision system already structurally had the components of signifier, completion logic, and award. What Microsoft did was implement an existing system on a new technological platform. By utilizing the panoptic affordance of the Internet, this system has made a significant impact on modern gaming.

Works Cited

"Achievement Unlocked." 2007. *Edge* 171 (January): 52–57.

"Activision." n.d. *Wikipedia*. http://en.wikipedia.org/w/index.php?title=Activision&oldid=589750492. Accessed January 11, 2014.

"Activision Patch Gallery." n.d. *Atari Age*. http://www.atariage.com/2600/archives/activision_patches.html. Accessed January 11, 2014.

Avendano y Carderera, N. H. 1859. *Curso elemental de pedagogia*. 4th ed. Madrid: Imprenta de D Victoriano Hernando.

Bartle, Richard. 1996. "Hearts, Clubs, Diamonds, Spades: Players Who Suit MUDs." *Journal of MUD Research* 1 (1): 1–28.

Björk, Staffan. 2012. "Achievements." *Gameplay Design Pattern Wiki.* http://129.16.157.67:1337/mediawiki-1.22.0/index.php/Main_Pageindex.php?title=Achievements&oldid=17097. Accessed January 11, 2014.

Bland, Justin. 2009. "Achievements: Blessing or Curse?" *DarkZero*, January 14. http://darkzero.co.uk/game-articles/achievements-blessing-or-curse/. Accessed January 11, 2014.

Bogost, Ian. 2010. "Persuasive Games: Check-Ins Check Out." Gamasutra, February 10. http://www.gamasutra.com/view/feature/4269/persuasive_games_checkins_check_.php. Accessed January 12, 2014.

Consalvo, Mia. 2007. *Cheating: Gaining Advantage in Videogames.* Cambridge, MA: MIT Press.

"Countdown to Badges." 2014. MSN Games. http://games.ca.zone.msn.com/en/general/article/genkeepingscore.htm?dr=t. Accessed January 11, 2014.

David, Rosalie. 1982. *The Ancient Egyptians: Religious Beliefs and Practices.* London: Routledge & Kegan Paul.

"Description of Xbox 360 Gamer Profiles." 2010. Xbox.com. https://support.xbox.com/support/en/us/nxe/kb.aspx?ID=905882. Accessed August 22, 2010.

Deterding, Sebastian, Dan Dixon, Rilla Khaled, and Lennart Nacke. 2011. "From Game Design Elements to Gamefulness: Defining 'Gamification.'" In *Proceedings of the 15th International Academic MindTrek Conference: Envisioning Future Media Environments*, 9–15. MindTrek '11. New York: ACM.

Electronic Entertainment Design and Research. 2007. "EEDAR Study Shows More Achievements in Games Leads to Higher Review Scores." Press release. http://www.eedar.com/News/Article.aspx?id=9 Accessed August 22, 2010.

Gibson, David, Nathaniel Ostashewski, Kim Flintoff, Sheryl Grant, and Erin Knight. 2013. "Digital Badges in Education." *Education and Information Technologies* 20 (2): 1–8.

Greenberg, Aaron. 2007. "Addicted to Achievements?" *Gamerscore Blog.* http://gamerscoreblog.com/team/archive/2007/02/01/540575.aspx. Accessed August 25, 2010.

Hamari, Juho, and Veikko Eranti. 2011. "Framework for Designing and Evaluating Game Achievements." In *Proceedings of the DiGRA 2011 Conference: Think Design Play*, 122–134. Hilversum, Netherlands: Utrecht School of the Arts.

Hecker, Chris. 2010. "Achievements Considered Harmful?" Talk given at Game Developer Conference, March 9–13, San Franciso. http://chrishecker.com/Achievements_Considered_Harmful%3F. Accessed April 17, 2014.

Henderson, Stephen. 2007. "Fan Behaviour: A World of Difference." Paper presented at the EuroCHRIE Conference, October 19–21, Leeds, UK.

"History of Merit Badges (Boy Scouts of America)." 2014. *Wikipedia*. http://en.wikipedia.org/w/index.php?title=History_of_merit_badges_(Boy_Scouts_of_America)&oldid=575069296.

"Imagic Experts Club." 1983. *Numb Thumb News* 2: 6.

Jakobsson, Mikael. 2011. "The Achievement Machine: Understanding Xbox 360 Achievements in Gaming Practices." *Game Studies* 11 (1). http://gamestudies.org/1101/articles/jakobsson.

Juul, Jesper. 2009. *A Casual Revolution: Reinventing Video Games and Their Players*. Cambridge, MA: MIT Press.

Kohn, Alfie. 1993. *Punished by Rewards*. Boston: Houghton Mifflin.

"Letters of Note: The Internet Tidal Wave." 2011. July. http://www.lettersofnote.com/2011/07/internet-tidal-wave.html. Accessed January 11, 2014.

MacCallum-Stewart, Esther. 2008. "Real Boys Carry Girly Epics: Normalising Gender Bending in Online Games." *Eludamos (Göttingen)* 2 (1): 27–40.

"Master Gamers—Free Badges to Be Won." 1984. *Fun Club News*, Spring, 6.

Matson, Johnny L., and Jessica A. Boisjoli. 2009. "The Token Economy for Children with Intellectual Disability and/or Autism: A Review." *Research in Developmental Disabilities* 30 (2): 240–248.

"Microsoft Online—from The Zone to XBox Live." 2004. http://megagames.com/news/microsoft-online-zone-xbox-live. Accessed January 11, 2014.

Montola, Markus, Timo Nummenmaa, Andrés Lucero, Marion Boberg, and Hannu Korhonen. 2009. "Applying Game Achievement Systems to Enhance User Experience in a Photo Sharing Service." In *Proceedings of the 13th International MindTrek Conference: Everyday Life in the Ubiquitous Era*, 94–97. MindTrek '09. New York: ACM.

Nicholson, Scott. 2014. "A RECIPE for Meaningful Gamification." In *Gamification In Education and Business*, ed. Lincoln Wood and Torsten Reiners, 1–20. New York: Springer.

Plasketes, George. 2008. "Pimp My Records: The Deluxe Dilemma and Edition Condition: Bonus, Betrayal, or Download Backlash?" *Popular Music and Society* 31 (3): 389–393.

Queen, William. 2005. *Under and Alone: The True Story of the Undercover Agent Who Infiltrated America's Most Violent Outlaw Motorcycle Gang*. New York: Random House.

Robb, Jenny E. 2009. "Bill Blackbeard: The Collector Who Rescued the Comics." *Journal of American Culture* 32 (3): 244–256.

"Robbie Bach, J Allard, Peter Moore: Electronic Entertainment Expo (E3) 2005." 2005. Microsoft. http://www.microsoft.com/en-us/news/exec/rbach/05-16-05e3.aspx. Accessed January 11, 2014.

Sotamaa, Olli. 2010. "Achievement Unlocked: Rethinking Gaming Capital." *Games as Services*, 73–82.

"Special Achievement Badge Available: Gold Rush!" 2014. MSN Games. http://games.ca.zone.msn.com/en/jewelquest/article/explorer_jqstbadge_promo.htm Accessed January 3, 2014.

Stump, Paul. 1997. *Digital Gothic: A Critical Discography of Tangerine Dream*. Wembley, UK: SAF Publishing.

Takahashi, Dean. 2006. *The Xbox 360 Uncloaked: The Real Story behind Microsoft's Next-Generation Video Game Console*. Arlington, VA: Spiderworks.

Taylor, T. L. 2006. *Play between Worlds: Exploring Online Game Culture*. Cambridge, MA: MIT Press.

Thomasson, Michael. 2003. "Activision Patches." http://www.gooddealgames.com/articles/Activision_Patches.html. Accessed January 11, 2014.

Wang, Hao, and Chuen-Tsai Sun. 2011. "Game Reward Systems: Gaming Experiences and Social Meanings." In *Proceedings of the DiGRA 2011 Conference: Think Design Play*, 1–12. Hilversum, Netherlands: Hilversum, Netherlands: Utrecht School of the Arts.

2 ADVENTURE

Nick Montfort

When we hear the word *adventure* in the context of computer gaming, we may feel compelled to grab our brass lantern and elvish sword or to put on our robe and wizard hat, but we are in any case likely to think back to beloved adventure games of decades past. Adventure games are still being made, of course; the genre has actually been revived in innovative ways. Nevertheless, they are considered old school to many, more associated in the popular imagination with 8-bit computers and multimedia CD-ROMs than with achievements and downloadable content. Even if our purpose in considering these sort of games is not nostalgia, the term *adventure* particularly asks to be historicized and for its origins to be explained.

The Name of the Game

The name of this genre of games comes from a specific early game that is highly influential and in fact prototypical: *Adventure,* also known as *Colossal Cave* or *Colossal Cave Adventure* (Crowther, 1976), a text game in Fortran that was written by Will Crowther and then expanded into its canonical form by Don Woods. This game, available in its earliest form in 1976, was an early hit on the ARPANET (Advanced Research Projects Agency Network), the predecessor to the Internet, and, without using graphics or even requiring a screen (it could be played on a print terminal) it sketched out most of the important aspects of adventure games.

A genre name that derives from a particular work is unusual in video gaming and otherwise. The romantic comedy genre of movies certainly *could* have been called "Trouble in Paradise" films after the 1932 film that many consider to have defined the category, but of course it is not. What is sometimes referred to as the "coming of age" novel is not called the "Tom Jones" novel after Henry Fielding's early exemplar; instead, it is often known by the German term for novel of formation, *bildungsroman.* Although there are "Sim games," this name designates a list of official games whose titles begin with "Sim" and that were published by the same company, Maxis—more an example of branding success than of genre formation.

Very few genres, from tragicomedy through mockumentary to steampunk, actually take their name from a single work that exemplifies the genre. One of the few cases in which the naming of a genre is done in this way is the specialized literary genre "Robinsonade," indicating the type of colonial story of survival and prosperity exemplified by Daniel Defoe's novel *Robinson Crusoe* (1719). However, although you might expect to be understood if you walk into a retail computer game store and ask for an adventure game, you might have difficulty getting a response if you go into a large, commercial bookstore and ask to be directed to the Robinsonades.

A case might be made that the naming situation is similar with "maze games," which in the late 1970s and early 1980s were two-dimensional action games derived from the game *Maze* (often remembered as *Maze War*), written by Steve Colley at NASA in 1972–1973 and expanded beyond two players by Greg Thompson, Dave Lebling, and others at MIT in 1974. However, the chain of influence from *Maze* to, for instance, *K. C. Munchkin* (Averett, 1981) for the Magnavox Odyssey 2 is not really extremely clear. It is true that *adventure*, like *maze*, happens to be a nicely generic term (no pun intended), so there are semantic reasons why it has been used to define a category when we do not really have "DOOM," "Tetris," or "Dance Dance Revolution" genres. Nevertheless, the adventure genre's relationship to its early ancestor is important to remember because it helps to explain why adventure games seem mainly a thing of the past.

Graphical Adventures, beyond *Adventure*

Warren Robinett programmed the prototypical graphical adventure game, *Adventure*, for the Atari Video Computer System (VCS, later known as the Atari 2600) for release in 1978. Although it may be surprising to players familiar with this game, Robinett thought of his project as adapting the all-text game by Crowther and Woods for the graphical and joystick-controlled Atari VCS. The major early adventure game for a console system was thus named after and based on the text game *Adventure*, which was the major early adventure game overall. The repertoire of actions available to the player in the Atari VCS *Adventure* is not great, but this game has many of the aspects of quests that are found in more elaborate adventure games (Montfort and Bogost 2009). It also has, if the player is successful, a suitable beginning (in which the hero, equipped with nothing but his wits, sets forth), a middle (involving exploration and combat), and an end (when the chalice is brought back triumphantly to the gold castle)—although this storylike segmentation did not end up being exclusive to adventure games.

Although Crowther and Woods's *Adventure* lacked graphics, it already exhibited all of the major characteristics of the adventure game genre. In adventure games, a player *explores a challenging simulated and fictional world to understand this world and to exhibit this understanding through in-game actions.* By "simulated and fictional," I mean that the world is interactive, operable, and susceptible to the interactor's choices or comments and that the interactor is invited to approach it using some of the fictional imagination that the reader of a novel uses.

There are almost always characters in adventure games, and even when they seem curiously absent from the simulated world, as in the best-selling CD-ROM adventure *Myst* (Cyan Interactive, 1993), the environment offers important traces of past presences. Although some adventure games are constrained to a single narrative of quest or progress, others are quite open-ended, allowing different players to traverse them in different ways. In any case, they are typically high in *narrativity* as backstories are discovered, puzzles are overcome, and in-game actions are undertaken in ways that relate significantly to the game's fictional world.

Many versions of the text game *Adventure* exist, as do many games that are significantly inspired by it in certain ways—for instance, *Zork* (1977), originally a minicomputer game (by Timothy A. Anderson, Mark Blanc, Bruce Daniels, and David Lebling) developed at MIT that was then published by Infocom as a trilogy of games for minicomputers. *Zork* is discussed in some detail and the early scholarship on *Adventure* is reviewed in my book *Twisty Little Passages: An Approach to Interactive Fiction* (Montfort 2003). *Adventure* was the topic of what is now considered the first game studies dissertation (Buckles 1985). The most remarkable scholarship on the game is by Dennis Jerz (2007), who visited the Kentucky cave that is the basis for the original game and recovered Crowther's original source code from before Woods modified it, showing that game studies can both include the study of the game designer's relevant explorations and experiences and address the game at the level of code.

What Defines Adventure Games?

There have been other perspectives on the genre, certainly. Graham Nelson, developer of the Inform interactive-fiction programming languages and several major interactive fictions as well, famously wrote: "An adventure game is a crossword at war with a narrative" (1995). This vivid figure identifies well the ludic and story aspects present in adventure games. However, I find some important shortcomings in the statement.

In *Twisty Little Passages,* I worked to show that the two tendencies identified here are not opposed in interactive fiction (and I do not believe they are opposed in adventure gaming in general). I would no more oppose the two than say that "a poem is sound at war with sense" (Montfort 2003, 51). To mention crossword puzzles specifically when speaking of adventure games overall, including graphical ones, is probably misleading because adventure games generally do not need to have words in them. And, finally, although *narrative* does have a special place in adventure games, what is really essential to them, distinctive about them, and probably less controversial for "antinarratologists" to admit is not actually the narrative, a representation of one or more events, but rather the imaginative, invented world into which the player is invited. This world is *fiction*. To rephrase (rather extensively) Nelson's dictum, it seems more suitable to me to say, "An adventure game is a fiction to be solved."

I find that a useful, concise definition of player activity in this sort of game is consistent with the short quip I have already offered. In an adventure game, the player *explores a challenging simulated/fictional world to understand this world and to exhibit this understanding through in-game actions.* This idea of an adventure game as a solvable fiction is also consistent with my description of interactive fiction and its relationship to the riddle (Montfort 2003), but it also includes Jesper Juul's (2005) concept of digital games as generally having both constraining rules and fictional worlds as important aspects and Clara Fernández-Vara's (2009) discussion of adventure games as including text adventures, graphic adventures, and point-and-click games.

Fernández-Vara adds that adventure games always have a player-character of some sort (2009, 18), but, as she notes, some of them have more than one. Although not the norm, LucasArts games such as *Maniac Mansion* (1987), *Day of the Tentacle* (1993), and *Sam and Max Hit the Road* (1993) provide one set of examples, and interactive fiction provides others, including Michael Berlyn's game *Suspended: A Cryogenic Nightmare* (Infocom, 1987) and Sean Barrett's game *Heroes* (2001). Fernández-Vara also distinguishes some games that are often lumped with adventure games, such as those that allow for the exploration of environment and have puzzles to solve embedded in this environment. Examples of this sort of adventure game include *The 7th Guest* (Trilobyte, 1993) and *Professor Layton and the Curious Village* (Level-5, 2007) (2009, 30–31). Although these games include both major aspects of an adventure game, the two aspects are at best superficially integrated, and when a puzzle is solved, it neither gives the player insight into the story world nor demonstrates the player's understanding of that world. Thus, solving the puzzles does not mean solving the fiction.

The Case of the Curiously Incapacitated Adventurer

Adventure games often have strange fictional worlds and usually present extraordinary situations. For these reasons and others, the actions that player-characters usually undertake are bit odd, as described vividly in a USENET post from 1995:

> Imagine if you lived your life like an adventure game.
>
> Whenever you came to a location you had never been before, you would examine everything closely. ... Anything that was not bolted down, you'd grab it. You'd be some bizarre kleptomaniac. ... Whenever there was a problem that had you stumped, you would wander around all of the locations you'd already been to and reexamine them.
>
> If there was a locked door, you would just assume that you needed to get to the other side of it. And you'd try any inane solution. (Hermansen 1995)

The player-character's behavior is bizarre when framed in these real-life terms, but it is not so unusual for a person facing a consciously designed and constructed challenge—a puzzle or riddle, for example. If the designer of a game de-emphasizes puzzles and provides an environment that is worth exploring and interacting with for its own sake, the player might be more inclined to direct his or her character to "act natural." But because adventure games are not essentially open or explorable worlds but instead fictions to be solved, the impulse to approach the world as something to be solved is a core feature of the genre. The most successful adventure games are not ones that almost eliminate puzzles but ones where the activity of solving is made not only challenging but also enjoyable and meaningful.

To kleptomania and amnesia, one might add autism as a player-character trait, and, indeed, Espen Aarseth identified this neuroatypicality as a quality of the player-character in the game *Deadline* (Infocom, 1982) (1997, 115). Jeremy Douglass has taken up all of these ideas and argued that the player-character's unusual nature, often incorporating "disability and incapacity," is not incidental but rather central to interactive fiction and effective design. He finds that in interactive fiction frustrations that often stem from a player-character's being disabled or incapacitated "exist in order to be overcome" (2007, 229–237). This insight applies to adventure games in general.

Adventure's Hiatus

This brief essay has focused on the origins and broad outlines of the adventure game genre, but it is not the place for a full story of that genre or for describing or declaring a canon of games. More lengthy and thorough discussions can be found in several of the works already cited. Particularly of note for offering a broad consideration of adventure games is Anastasia Salter's work *Magical Books* (2014).

Crowther and Woods's game *Adventure* was not only the first game in the adventure game genre but also, to many people, the prototypical computer game. After that, Infocom, Sierra On-Line, LucasArts, and other adventure game makers were among the top companies in the nascent computer game business. Infocom was the preeminent publisher of text-based interactive fiction, including *Deadline* and *The Hitchhiker's Guide to the Galaxy* (1984); the company systematically explored popular literary genres as it developed fantasy, science fiction, mystery, humor, and other titles. Sierra On-Line made early text-and-graphics adventures as well as the *King's Quest* (1984) and related series. LucasArts games were characterized not only by the multiple player-characters mentioned earlier but also by being constructed to support a wry, witty style. There was a great deal of other important adventure game development during the 1980s inside and outside the United States.

The runaway success of the point-and-click adventure game *Myst* and the development of sequels meant that adventure gaming was going strong into the mid-1990s, even if text adventures were not being commercially developed. By the beginning of the twenty-first century, however, there were very few new offerings in the adventure game genre in the United States, and these new offerings were not among the best-loved and most-talked-about computer games. Adventure games (in the form of visual novels) are still extremely popular in Japan; both graphical and text-based adventure games are made by individuals and released for free, and a significant new indie development of adventure games is happening for several platforms. But in the United States, at least, there was a great decline in prominence, commercial success, and number of new releases of adventure games after the early heyday of adventure gaming.

Many have speculated about why this decline occurred. To these speculations, I would like to add a new one. In the early days of computing, whether one accessed a minicomputer at a university or brought home one of those rare and unusual microcomputers, being involved with computing meant exciting exploration, discovery, solution, and the development of understanding through these processes. Computers were not personal fashion items or common workplace fixtures: they were adventures in themselves. The ability to enter fictional worlds and to solve puzzles in these worlds connected with the experience of computing generally, which involved learning to format disks, copy files, program in BASIC (Beginner's All-Purpose Symbolic Instruction Code), communicate over a modem with bulletin board systems, and manipulate text and graphics. People often took notes on how to figure out aspects of computing, scribbling on paper as they made adventure game maps and taking notes on their progress through a game. For those who set out on a journey to understand a strange world, the computer, the simulated and fictional worlds of adventure games resonated.

Computing is now in the United States and many parts of the world ordinary and pervasive. However many enjoyable and worthwhile things we can do with computers, they are seldom considered to be an adventure. Even though adventure games may have built mainly on the excitement of computing circa 1980, perhaps today they might kindle excitement once again by showing how thoughtful examination, experimentation, and solution can be rewarding. Perhaps adventure games, in addition to any specific educational benefits they can provide, might also allow players to unfold understanding rather than offer the mere time-filling grind of other sorts of gaming. They might help to encourage computer users in general to be more adventurous.

Works Cited

Aarseth, Espen. 1997. *Cybertext: Perspectives on Ergodic Literature*. Baltimore: Johns Hopkins University Press.

Buckles, Mary Ann. 1985. "Interactive Fiction: The Computer Storygame *Adventure*." PhD diss., University of California, San Diego.

Douglass, Jeremy. 2007. "Command Lines: Aesthetics and Technique in Interactive Fiction and New Media." PhD diss., University of California, Santa Barbara.

Fernández-Vara, Clara. 2009. "The Tribulations of Adventure Games: Integrating Story into Simulation through Performance." PhD diss., Georgia Institute of Technology.

Hermansen, Erik. 1995. "Re: Gameplay Theory: Leaving Object Behind." Posted September 14. news://rec.games.int.fiction. Message ID 1995Sep8.233605.25903@news.cs.indiana.edu.

Jerz, Dennis G. 2007. "Somewhere Nearby Is Colossal Cave: Examining Will Crowther's Original 'Adventure' in Code and in Kentucky." *Digital Humanities Quarterly* 1, no. 2. http://www.digitalhumanities.org/dhq/vol/1/2/000009/000009.html.

Juul, Jesper. 2005. *Half-Real: Video Games between Real Rules and Fictional Worlds*. Cambridge, MA: MIT Press.

Montfort, Nick. 2003. *Twisty Little Passages: An Approach to Interactive Fiction*. Cambridge, MA: MIT Press.

Montfort, Nick, and Ian Bogost. 2009. *Racing the Beam: The Atari Video Computer System*. Cambridge, MA: MIT Press.

Nelson, Graham. 1995. "The Craft of Adventure: Five Articles on the Design of Adventure Games." January 19. http://www.ifarchive.org/if-archive/programming/general-discussion/Craft.Of.Adventure.txt.

Salter, Anastasia. 2014. *Magical Books from Adventure Games to Electronic Literature*. Iowa City: University of Iowa Press.

3 AMUSEMENT ARCADE

Erkki Huhtamo

Just as empires rise and fall, so do amusement arcades. Numerous examples thrive worldwide, particularly at theme parks, amusement piers, and other tourist venues, but there is a feeling that their golden age is over. The reasons why a cultural form that once was ubiquitous may be nearing its extinction are mainly sociocultural. Changes in pastime activities and in patterns of consumption, mobility, and media use work against the amusement arcade, which may survive mostly as a nostalgic showcase (it has always been such in Disneyland as part of a simulated America concocted by Walt Disney's imagination). Pocket game consoles and mobile phones point toward a "postarcadean era," where amusements travel with their users. Mobile lifestyles, FaceTime sessions while walking, rapid-fire tweets, and games that unfold on sweaty palms squeezing smartphones are tokens of a cultural transformation that has eclipsed the arcade.

The amusement arcade has not always been part of the cultural landscape. It appeared in the nineteenth century as part of the gradual onslaught of location-based entertainments. In earlier times, most popular amusements were seasonal and ambulatory. Traveling people would show up at fairs and other public gatherings (including religious festivals) to sell their wares, display curiosities, and perform tricks and stunts for money. Tents and market stalls attracted the curious, but when the event was over, they were dismantled and erected somewhere else. Only memories, stories, and hopes for the next visit lingered on.

Eighteenth-century cities had places where one could spend time in the company of friends and strangers, such as pleasure gardens and cafés. There were also theaters, operas, and churches. A trend that anticipated the amusement arcade appeared around 1800, manifesting itself in location-based spectacles such as the Phantasmagoria, the Panorama, the Cosmorama, and (a little later) the Diorama. Although itinerant shows continued to flourish, the newcomers offered optical and mechanical wonders in permanent venues for a paying public. Listed in guidebooks and city directories, they became fixtures of the urban environment. Their appearance coincided with the coming of the passage or arcade.[1] Introduced in Paris in the late eighteenth century, the arcade was a covered walkway "carved" into the structure of the city. Combining commerce with pastime, arcades invited anyone for a stroll to examine the shop windows lining them or to sit at a café table and observe the passing crowd. They became incubators of modernity where ideas were disseminated, products introduced, and fashions molded.

Relationships developed between the new spectacles and the arcades. Symptomatically, the entrances to the famous twin panorama rotundas along the Boulevard de Montmartre in Paris were located inside the Passage des Panoramas. In London, the British Diorama operated between 1828 and 1836 inside an arcade named the Royal (or Queen's) Bazaar at 73 Oxford Street. The most intriguing example of all was the Cosmorama, an "arcade within an arcade." The first one was opened by an Italian refugee named Abbé Gazzera in 1808 in Paris inside the Galerie vitrée (No. 231), a fashionable passage at Palais-Royal. Many others followed in cities such as London and New York (Huhtamo 2012). In the Cosmorama salon, the visitor encountered a series of magnifying lenses inserted into the surrounding walls. Behind each was an illuminated painting. Cosmorama was a gentrified version of the peepshow, an itinerant "raree show" encountered at fairs and marketplaces. Like the ones that street showmen exhibited in their showboxes, the Cosmorama pictures depicted geographic locations; the main difference was their more refined quality.

The new media spectacles were addressed to the middle and upper classes, as complaints from the lower classes about high entrance fees demonstrate. What makes the Cosmorama important for an archaeology of the amusement arcade is its arrangement as a dispositive—in other words, as an "experience engine" that purported to involve the visitors both physically and mentally. After paying the entrance fee, the visitors moved freely around the gallery, peeking into one lens after another. Spectator mobility was a feature that the Cosmorama shared with the Panorama, but the experience was different. Whereas the latter offered a single overwhelming wrap-around view shared by everyone on the same viewing platform, the encounters with the pictures in the Cosmorama were intermittent and semiprivate in nature. After each peek, experiences could be shared with others, who were waiting for their turn or just biding their time. Private immersion alternated with social interaction.

Mobile versions of the Cosmorama traveled to places where permanent establishments did not exist, rooting the idea of its apparatus firmly in the popular imagination. It is therefore not surprising that it later appeared in attractions such as the Kaiser-Panorama and the Phonograph and the Kinetoscope Parlor. The former, introduced by the German showman and entrepreneur August Fuhrmann (1844–1925) around 1880, centered on a large wooden cylinder with a row of stereoscopic eyepieces and seats around its perimeter. A program of hand-colored glass stereoviews, lasting for half an hour, was rotated by a clockwork mechanism past the peepers' eyes. In central Europe, the Kaiser-Panorama grew into an extensive network that at its peak had as many as 250 filials operating in permanent indoor venues in different countries. Although its arrangement harked back to the Cosmorama (whether consciously or not), the basic situation had been reversed by making the spectators stationary and the images mobile (instead of paintings, stereoscopic photographs were used).

The apparatus of the Phonograph Parlor was closer to that of the Cosmorama even though it was focused on the visitor's ears instead of on his or her eyes. Introduced in the United States in 1890, it was an effort to exploit commercially an improved version of Edison's Phonograph (1877), the first device to record and reproduce sounds. Descriptions of what took place at Phonograph Parlors are scarce, but surviving photographs allow us to decipher its apparatus. Automatic Phonographs, enclosed in floor-standing wooden cabinets, stood in rows next to the walls. Pairs of headphones were attached to each, inviting visitors to enjoy moments of semiprivate listening. The technological "miracle" that made it possible, the cylinder phonograph, could be seen running behind a glass panel, adding a visual dimension to the overall attraction. The visitors normally stood while listening to the brief tunes, although seats were provided in some parlors (especially later ones). Edison's automatic phonographs are usually described as coin-operated devices, but that was not always the case. As Ray Phillips (1997) has explained, at parlors the machines were usually operated by attendants, and single coin-operated specimens were placed at hotel lobbies, bars, and other public venues.

The next step was logical. Edison's mission for the team that created the Kinetoscope was to produce a device that would do for the eye what the Phonograph did for the ear. The result joined the long tradition of "peep media" (Huhtamo 2006). The Kinetoscope, introduced in 1894, presented moving images from a strip of celluloid film threaded between numerous spools inside the viewing cabinet. These images were peeped at through an eyepiece at the top of the cabinet; the placement of the eyepiece demonstrates that the Kinetoscope was not a children's toy. The first machines were often added to Phonograph Parlors as a new attraction.[2] Like the Automatic Phonograph, the Kinetoscope was powered by an electric motor. The user's role was fairly passive: pay and then listen or watch.[3] The situation did not change when the Kinetophone, a combination of the Automatic Phonograph and the Kinetoscope, was introduced; it did not change even when projected moving pictures were added—they were normally peeped at through a row of holes in the back wall of the parlor (exactly as in a Cosmorama).

A different solution was offered by a competing device, the Mutoscope, which enjoyed a long-lasting success, unlike Edison's parlor machines. It had three advantages. First, instead of fragile film, the frames had been copied on paper slips attached together as sturdy flip reels. Second, the "nickel-in-the-slot" operation was a standard feature of each machine. Third, the Mutoscope was hand-cranked, a seemingly minor feature that made all the difference. As abundant iconographic evidence suggests, the ability to negotiate the experience by turning a crank created a stronger bond for the viewer than Edison's automatic devices. The Mutoscope can be characterized as a "proto-interactive" device like scores of other coin-operated amusement machines that began appearing in growing numbers in technologically advanced countries such as England and the United States in the mid-1880s (Bueschel and Gronowski 1993; Pearson 1992; Costa 1988). Although many of these machines were simple dispensers of products from chocolate and cigars to postage stamps, others, such as strength testers, rifle ranges, and sporting games, introduced more engaging modes of interaction, emphasizing the user's physical or mental skills. At first, these machines were often installed at gathering places such as pubs and saloons, forging a link with the long trajectory of men-only amusements.

Freestanding Mutoscopes became common at amusement piers and found a home at amusement arcades in the company of great varieties of other slot machines. Such arcades were often known as "penny arcades" or "automatic vaudevilles."[4] They became familiar both in cities and at resorts from the 1890s onward. Penny arcades continued the trajectory from the passage and the Cosmorama, but as concentrations of curiosities they also incorporated features from dime museums.[5] From a sociopsychological perspective, they were a Janus-faced institution. In principle but not always in practice, they were open for anyone regardless of class, age, and gender (racial barriers were not breached until much later). An advertisement for W. S. Fremont's Amusement Parlor touted, "A GOOD PLACE FOR THE OLD, MIDDLE AGED AND YOUNG TO DRIVE AWAY CARE AND HAVE A GOOD TIME. COME AND ENJOY YOURSELF."[6] The arcade was open from 8:00 a.m. to 10:00 p.m. every day and charged no entrance fee—the cost of the visit depended on the number of pennies consumed.

Such arcades may have provided new freedom to spend one's spare time and to mix with people one would not otherwise encounter, but they could also be interpreted as a power structure through which the bourgeois society attempted to counter the political tensions created by the new "masses." These masses included a large component of uneducated and illiterate immigrants. Simplified mass entertainment was a way of calming down potentially troublesome layers of the population and at the same time to profit from them financially. Slot machines can be contrasted with the production machines that increasing numbers of workers used at factories and offices (Huhtamo 2005). The latter were "preprogrammed" to increase productivity, and their users were forced to perform monotonous tasks again and again. The slot machines that these same workers encountered in the evenings and weekends, however, returned the initiative to the user, although such empowerment was illusory: "preprogramming" dominated here as well.[7]

In spite of its democratic promise and widespread popularity, the penny arcade became a target for moral crusaders, who associated it with temptation and sin and saw it as a threat to the existing social order. This fear may have been in part caused by the vagueness of the phenomenon. Coin-operated machines became associated with illegal gambling, especially in the Prohibition-era United States. The authorities took action to suppress them by legislation and even by destroying them in ritual actions reported by the media (Fey [1983] 1997, 56, 104, 136–137).

In the first half of the twentieth century, coin-operated machines and amusement arcades remained ambiguous entities; their popularity was countered by their overly negative public reputation, which reflected general attitudes toward popular culture. When the post–World War II societies began emerging from the wartime economy, the coin-op industry had to shed its negative image and rebuild its audience. The challenges were met only gradually. Former soldiers may have been avid users of automatic entertainments, but the most important emerging user groups were male teenagers and young adults. In the United States, the new ideal of the tight nuclear family living in a suburban home heralded a desire for momentary escapes. Drive-in cinemas and restaurants; shopping malls; Disneyland (1955–); the habit of cruising in cars and on mopeds and scooters; rock 'n' roll concerts; and amusement arcades were all ways to fill this desire. When the Baby Boomers reached their late teens at the end of the 1950s and in the 1960s, they identified amusement centers and bars with their coin-operated machines as places where they could gather beyond the watchful eyes of their parents, momentarily escaping the domestic suburban panopticon.

The coin-operated machine that epitomized the era was pinball. Pinball machines already had a long history by this point; they had been popular during the Great Depression of the 1930s. Most had been games of luck; after a coin or token was inserted, the outcome depended on a combination of chance and machine operations. Such devices became vehicles of illegal gambling. The "bumbers" (metal posts with coiled springs for scoring points) and the tilt function (the freezing of the machine if it was shaken too hard) were invented in the 1930s, but the player's contribution remained limited. It was the introduction of "flippers" (push-button-operated bats that allowed the player to return the falling ball to the field) that contributed to the redefinition of pinball in 1947, turning it into a potential answer to the challenges the game industry and arcade business were facing.[8]

The flippers lengthened the gaming experience and enhanced its interactivity. Combined with the thrill of physically shaking the machine just enough to avoid the tilt, the rapid-fire push-button action created a strong bond between the machine and the gamer. Another innovation, the electronic counter, made it possible to break records and compete with others, and colorful backboard art and thematic designs set the mood and increased the machine's drawing power. The flippers helped the industry to disassociate the pinball from gambling by promoting it as a game of skill (extra games and fame, not money, were the reward). This was essential to keep the authorities from interfering with the business by legislation. The words *flipper* and *flipper game* were often adopted instead of the charged term *pinball*. The cult status of the game was legitimized by the Who's rock opera *Tommy* (1969). Its main character is a "deaf, dumb, and blind kid," an autistic boy who is unbeaten pinball master and excels when he "becomes part of the machine." Tommy was an (anti)hero for a generation oppressed by capitalism and the Vietnam War, desperately searching for outlets.

"By gosh, it's a Pinball," Steve Russell is said to have exclaimed about *Spacewar!* (Russell/ MIT, 1962), the pioneering video game he helped to create in the 1960s (qtd. in Levy 1984, 65).[9] It became an inspiration for some arcade video games. Such games were a far cry from the proto-interactive attractions of the late-nineteenth-century arcades. They offered highly interactive play that gradually became more and more immersive and "realistic." Simulator games such as *Daytona USA* (Sega, 1993–1994) allowed up to eight players to race against each other in real time, each seated in his or her own cabinet.

Like the pinball arcades of previous decades, video game arcades have often been considered the exclusive domain of adolescent males. In his self-published account of the phenomenon, arcade operator Adam Pratt has countered this idea, distinguishing different types of video game arcades for different users. There are traditional arcades, big and small, addressed to dedicated gamers. There are also chains of "Eatertainment" venues that "combine food and games into the same location" (2013, chap. 2). Family entertainment centers combine video gaming with other activities such as minigolf, laser tag, skating, and gokart, and the arcades in theme and amusement parks exist alongside high-tech attractions such as simulator rides and 3D movies. A metamorphosis of the game arcade into a family-oriented amusement had begun in the 1960s, particularly at shopping malls. The trajectory ran parallel with the trend that emphasized the home as the nexus of interactive media consumption. The early TV games, video game consoles, and personal computers for domestic users often offered versions of popular arcade games. A death match between the arcades and home game consumption was inevitable; so far the amusement arcade has not emerged as a winner, although it remains to be seen if it can be characterized as a loser.

Notes

1. From the Latin *arcus*, the medieval Latin *arcata*, and the French *arcade*, *arch*.

2. The Kinetoscope Company Parlor in Cleveland, Ohio, was located inside the imposing Arcade building, a prototypical shopping mall (Bueschel and Gronowski 1993, 80).

3. Both attendant and coin-operated versions of the Kinetoscope were produced.

4. Similar devices were also exhibited by traveling showmen at fairs and exhibition sideshows.

5. P. T. Barnum's American Museum in New York City was a "mega dime museum" and model for many later amusements. It contained a Cosmorama Room, as did many other similar establishments.

6. From circa 1902. Fremont's Amusement Parlor was at Coney Island, New York. See the ad in Buschel and Gronowski 1993, 43.

7. Therefore, I call these slot machines "proto-interactive" rather than "interactive."

8. Flippers were first used in the Gottlieb Company's game *Humpty Dumpty*, introduced in October 1947. The engineer Harry Mabs is said to be their inventor (Kurtz 1991, 52). Michael Colmar mentions Chicago Coin's game *Bermuda* (1947) as the company's first flipper game (1976, 84) but does not mention *Humpty Dumpty*.

9. Russell mentioned elsewhere "Doc" Smith's *Lensman* series of science fiction books as his main inspiration (Brand 1974, 55).

Works Cited

Brand, Stewart. 1974. *II Cybernetic Frontiers*. New York: Random House.

Bueschel, Richard M., and Steve Gronowski. 1993. *Arcade 1: Illustrated Historical Guide to Arcade Machines*. Vol. 1. Wheat Ridge, CO: Hoflin.

Colmar, Michael. 1976. *Pinball: An Illustrated History of the Pinball*. London: Pierrot.

Costa, Nic. 1988. *Automatic Pleasures: The History of the Coin Machine*. London: Kevin Francis.

Fey, Marshall. [1983] 1997. *Slot Machines: A Pictorial History of the First 100 Years*. Reno, NV: Liberty Belle Books.

Huhtamo, Erkki. 2012. "Toward a History of Peep Practice." In *A Companion to Early Cinema*, ed. André Gaudreault, Nicolas Dulac, and Santiago Hidalgo, 32–51. Chichester, UK: Wiley-Blackwell.

Huhtamo, Erkki. 2006. "The Pleasures of the Peephole: An Archaeological Exploration of Peep Media." In *Book of Imaginary Media: Excavating the Dream of the Ultimate Communication Medium*, ed. Eric Kluitenberg, 74–155. Rotterdam: NAi.

Huhtamo, Erkki. 2005. "Slots of Fun, Slots of Trouble: Toward an Archaeology of Arcade Gaming." In *Handbook of Computer Games Studies*, ed. Joost Raessens and Jeffrey Goldstein, 1–21. Cambridge, MA: MIT Press.

Kurtz, Bill. 1991. *Slot Machines and Coin-Op Games: A Collector's Guide to One-Armed Bandits and Amusement Machines*. Secaucus, NJ: Chartwell Books.

Levy, Stephen. 1984. *Hackers: Heroes of the Computer Revolution*. New York: Dell.

Pearson, Lynn F. 1992. *Amusement Machines*. Shire Album no. 285. Princes Risborough, UK: Shire.

Phillips, Ray. 1997. *Edison's Kinetoscope and Its Films: A History to 1896*. Trowbridge, UK: Flicks Book.

Pratt, Adam. 2013. *The Arcade Experience: A Look into Modern Arcade Games and Why They Still Matter*. [West Valley City, UT]: Adam Pratt.

4 ARTIFICIAL INTELLIGENCE

Rebecca E. Skinner

Artificial intelligence (AI), the "attempt to build intelligent machines without any prejudice toward making the system simple, biological, or humanoid" (Minsky 1968b, 7), began as a serious research field in the early 1950s and was named as such at the Dartmouth Summer Research Project on Artificial Intelligence (hereafter "Dartmouth Conference") in 1956. During the field's formation, its paradigmatic gedankenexperiments included IQ test type questions and logic problems. Yet before AI's expression and after the painstaking work on logical theorems, SAT-type word problems, and simple problem solving began, the abstract study of intelligence through games contributed to the concept of AI itself. Games offered the elucidation of search methods and a syntactically rich but semantically austere domain, so they were central to the foundation of and early research in AI.

Before AI: The Appeal of Automata

Games have a deeper provenance in the form of wargames and the study of probability as a way to win at cards and chess, both of which are beyond our ambit here (Crogan 2008; Wilson 2008). More closely proximate origins of AI are found in the encouragement of automata and chess. In the 1940s and 1950s, practically everyone who would later be involved with AI worked with automata or games, either as research work or as a leisure hobbyist activity. The study of automata, another way to create engaging abstract systems, closely preceded AI. Mechanical automata, including mechanisms that simulated homeostasis with water levels or vision with photoelectric cells, were a popular garage enterprise in the first several decades of the twentieth century (Walter 1970; Pylyshyn 1970, 60; Nemes 1962, 164). They were typically designed using governing principles such as tropism toward light or rest; in other cases, the automata acted as "turtles" that moved on wheels and turned in reaction to walls.[1] At Bell Labs, MIT student (and later professor) Claude Shannon built maze mice using materials filched on the sly from the lab, which frowned on such time wasters as computer games (McCorduck 1979; Grosch 1991). At Harvard, Marvin Minsky and fellow undergraduate Dean Edmonds built an immobile automaton with probabilistically triggered synaptic firings.[2] Whatever their motive force or mechanical methods, these mechanisms acted in a manner similar to cybernetics: they made their creators think about the nature of intelligence.

Before AI: Games No One Wins

Building automata forced the question of means by which to direct actions as well as of sensor input, processing, and output. Automata did not, however, require or lead to politics, even at the university level. Game theory was another sort of game entirely, played with clenched fists and white knuckles. John Von Neumann and Oskar Morgenstern's classic work *Theory of Games and Economic Behavior* ([1944] 2004), established the canonical standards of this field, not coincidentally just as the Cold War commenced.

Game theory addresses the calculation of the utility of an agent's behavior in contending with an opaque opponent or more than one opponent. Thus, game theory is applicable to microeconomics, the economics of information and financial markets, and the study of non-perfect markets. Game theory's implications extended past the merits of calculation and autonomous agency because games could provide measured and probabilistic calculations for the efficacy of strategies in markets and political struggles between nations. In a crucial respect, this development was pernicious because the rendering of nuclear war into a game also cauterized this type of war. It provided a way to write and speak extensively about the unspeakable and to think about the unthinkable at a level of abstraction at which the human significance of the game itself completely disappeared.[3]

Chess before AI

Despite its conceptual beauty, game theory is no more innocuous than a tiger. The same is not true of chess and cellular automata, which have had more civil(ian) uses. Game theory and the study of cellular automata made the concept of intelligent agents solving problems seem much more familiar and intellectually accessible. Investigating games implies such active rather than passive tasks as figuring out strategies, imagining offense and defense, considering long-term versus short-term moves, and the like. The field of game playing itself epitomized the transition from computer programming that saw human intelligence as a form of conditioned reflex productions to programming that emphasized cogitation. In the early 1950s, adherents of chess playing and checkers playing took up information-processing views of intelligence.

These enterprises were variously part-time or untrained activities in the cases of Claude Shannon and Arthur Samuel, who created chess- and checkers-playing programs, respectively, and explicit and evangelical in the case of Allen Newell, Clifford Shaw, and Herbert Simon.

It is said that chess was the drosophila of the AI field, and, indeed, it was not only the effect of the formation of AI but also its cause. The game's intense demands helped to give rise to AI because examination of the game repeatedly showed that chess posed too great a challenge to blind search, certainly, as well as to any form of search that did not try to embody problem-solving strategies. Asking such questions inevitably evoked the issue of *human* problem solving. The game appeared and reappeared in early AI research and continues to do so.

Claude Shannon Sets the Bar for Chess Analysis

The definitive presentation of the problems of chess, if not of its solutions, was and perhaps remains that of Claude Shannon's essay "A Chess-Playing Machine," which appeared in *Scientific American* in 1950. Shannon pointed out that electronic digital computers are powerful enough to carry out symbolic computing rather than solely to process numbers. He then indicated how the squares of a chessboard may be expressed in machine code (pseudocode, actually). He emphasized immediately that blind search is not an option: chess simply involves too much data and thus forces the thinker toward strategy. A machine engaged in chess using blind search would have to take on 10^{120} possible moves; Shannon suggested a numerical evaluation score to indicate how the player(s) are doing, with respect to certain obvious features of chess playing. Then he turned to Dutch psychologist Adrianus Dingeman de Groot's studies of the strategies of expert players (1946; English translation 1965) and suggested subprograms to test out various plays along several iterations. In *Philosophical Magazine* in 1949, Shannon had proposed in a more technical manner the form he expected problem solving to take. The finite and nondeterministic nature of chess dictates that the game may be construed as a branching tree, albeit always portrayed upside down, from which spring nodes (the positions) and branches (the moves).

The strategic portrayal of the sequences of a game may be pursued through such a graph, with the maximum benefit being chosen by the computer's side being played and the minimum benefit being chosen, naturally, by the opposing side. The alternating plays are referred to as mini–max because of their respectively differing objectives.[4] Shannon's papers unequivocally showed the inadequacy of the existing neurological or conditioned reflex model of information processing and suggested alternatives. Widely read and highly inspirational, they helped to instigate the introduction of cognitive modeling.[5]

Shannon's papers were a catalyst for various chess-playing programs in the mid-1950s. The work was carried out by a number of groups in different research centers, a grassroots movement of haut scientific intellect. These elaborated further concepts in strategies of game playing, all of which research underlined the feasibility of formal methods of game playing.

Throughout the 1950s and into the 1960s, practically every AI and pre-AI practitioner was an aficionado of computer chess playing as well. Alan Turing's proposal for a chess-playing program to run on the MADAM computer, worked out during 1947 and 1948, implemented the concept of a "dead" position, which is relatively stable for several moves and can thus be evaluated.[6] Chess-playing programs were also developed at Los Alamos, using an odd six-square board. Engineer Alex Bernstein, using IBM's New York Service Bureau 704 mainframe, also developed concepts of search, pruning, subroutines that generated possible moves, and state space evaluation (Newell, Shaw, and Simon 1958, 45). Bernstein's program was featured in *Scientific American*, the *New York Times*, and *Life* magazine (McCorduck 1979, 159). The publicity caused much irritation at IBM because this research was well prior to the era of enthusiasm for "exotic" computing applications.[7]

The Birth of AI Proper

The historically unique combination of digital computers, access to programming time, and nascent theories of formal computer languages and cognitive psychology allowed Newell and Simon to come up with complex information processing, as they called it, in the projects they started in 1952. During the next several years, Newell, Shaw, and Simon developed the Logic Theorist while working at the Rand Corporation and the Carnegie Institute of Technology, and John McCarthy and Minsky published articles and a conference volume on digital automata. By 1956, McCarthy, Minsky, Shannon, and IBM scientist Nathaniel Rochester had proposed, funded, and convened the Dartmouth Conference (McCarthy, Minsky, Rochester, et al. 1956). This event graced the field with the unifying title of a research program, yet its definition of AI was sufficiently loose that it did not impede the great variety of the founders' ongoing research programs.

Games Give Way to Schematic Problem Solving

However, the gravity of AI's formal program in universities edged out the lightness of automata built in the off hours. Games such as computer chess, which Newell and Simon as well as McCarthy at MIT continued to develop, were structured for the elucidation of the search process rather than for carefree enjoyment (Newell, Shaw, and Simon 1958). The same thing is true of the schematic logical theorems of Newell, Shaw, and Simon's Logic Theorist, succeeded by the General Problem Solver in the late 1950s. In the early 1960s, schematic forms of problem solving were given names such as "Monkey and Bananas" and "Cannibals and Missionaries," which were more qualitatively and semantically rich, although still well defined in terms of initial state and goal states (Abelson and Sussman 1985). Particularly in Newell and Simon's research, these problem-solving opportunities were used as a springboard for the study of human problem solving as cognitive psychology rather than for the development of games for their own sake.

The Watershed of *Spacewar!*

Soon after the formation of the field of AI, games were thus detached both from the deadly serious topic of nuclear warmanship as well as from AI research into fundamental concepts of knowledge representation. Games in AI's informal program, or "folk culture" of hacking, began in earnest late in the 1950s. This was simply the earliest time at which graduate students were able to get hold of unmonitored time on the computer. Soon after students at MIT's AI Lab were allowed time on the institute's Programmed Data Processor-1 computers, they invented the first computer game, and more chess-playing programs followed soon afterward (Peterson 2003; Levy 1984). At the end of the 1950s, the students who had been involved with the Tech Model Railroad Club decamped to computing and continued to create artifacts. *Spacewar!*—perhaps the most successful of these games—was so addictive that according to Minsky it had to be banned in Australia because it was draining mainframe power (1989, 179).[8]

Chess constitutes an unbroken thread in AI gaming, including the win by MIT AI Lab student Richard Greenblatt's MacHack program. Much later, in 1997, AI as chess would culminate in the triumph of Deep Blue against Gary Kasparov. Chess playing in AI turned into its own ongoing and eventually enormously successful program. However, it became relatively independent from AI as anthropomorphic problem solving simply because it could win by dint of its immense problem-solving hardware power and speed.

Computer Gaming Returns to Its Identity

Games played on time-shared systems are thus part of the canon of AI's hacking tradition. Moreover, the complementary technologies that made games not only engaging but absolutely enthralling came from AI as well. The first vision program was itself invented by the youthful Gerald Sussman during the AI Summer Vision Program in 1966 (Minsky 1968a; Sussman and Guzman 1966). However, by the early 1970s, games for games' sake, separate from AI, had reasserted themselves in commercial form. Nolan Bushnell's indefatigable entrepreneurial efforts were not dampened when he found that he could not succeed in commercializing *Spacewar!*. Instead, his new company, Atari, released *Pong* in 1972, and the rest is history (Wilson 2008, 109).

An endlessly creative cornucopia of games, intended for purely recreational purposes and greatly enriched by better visual representation, has followed, as studied elsewhere in this book. There is no doubt that wargames are being played out at this moment on the monitors of intelligence agencies of this nation and every other nation, but it is comforting that a great preponderance of the wars in game usage are confined to the kingdoms of cyberspace.

Notes

1. In this genre, Grey Walter's Machine speculatrix, Edmund C. Berkeley's Sqee, and Ross Ashby's homeostat or "Machina sopora" were embodied tropisms of living creatures (Walter 1953, 189; Berkeley 1949).

2. Jeremy Bernstein (1982) reports that this artifact subsequently disappeared.

3. Computers that controlled weapons systems and envisaged wars using weapons as a game have haunted the dreams of sci-fi movie screenwriters. As the WOPR (War Operation Plan Response) computer says in its final assessment after testing out a series of war scenarios in the film *WarGames* (John Badham, 1983), "The only winning move is not to play" (see Kaplan 1983).

4. Newell, Shaw, and Simon (1958) has an excellent explanation of the concept.

5. This article was apparently not connected to Caissac, Shannon's chess-playing machine (Lucky 1989).

6. See Newell, Shaw, and Simon (1958, 44–48), and Bowden (1953) regarding the Los Alamos physicists, who built a program on the MANIAC I (Mathematical Analyzer, Numerical Integrator, and Computer) that could beat beginners.

7. "So here was Bernstein getting what for Watson was unpleasant notoriety about his chess-playing machine and Arthur Samuel, reaping a harvest of publicity for his checkers playing program. Faced with these two very successful examples of homegrown AI—soon to be joined by a third, Herbert Gelernter's geometry theorem proving program—sales executives at IBM began to grow nervous lest the very machines they were trying to sell prove so psychologically threatening that customers would refuse to buy them. Thus they made a deliberate decision to defuse the potency of such programs by conducting a hard sell campaign picturing the computer as nothing more than a quick moron" (McCorduck 1979, 159).

8. Minsky discusses the origins and addictive popularity of *Spacewar!* in Minsky (1988). Also see Crogan (2008).

Works Cited

Abelson, Harold, and G. J. Sussman. 1985. *Structure and Interpretation of Computer Programs*. Cambridge, MA: MIT Press.

Berkeley, Edmund Callis. 1949. *Giant Brains; or, Machines That Think*. New York: Wiley.

Bernstein, Jeremy. 1982. *Science Observed: Essays out of My Mind*. New York: Basic Books.

Bowden, B. V., ed. 1953. *Faster Than Thought: A Symposium on Digital Computing Machines*. London: Pitman.

Crogan, Patrick. 2008. "Wargaming and Computer Games: Fun with the Future." In *The Pleasures of Computer Gaming: Essays on Cultural History, Theory, and Aesthetics*, ed. Melanie Swalwell and Jason Wilson, 147–166. Jefferson, NC: McFarland.

Grosch, Herbert R. 1991. *Computer: Bit Slices from a Life*. Novato, California: Underwood-Miller.

Kaplan, Fred. 1983. *The Wizards of Armageddon*. Stanford, CA: Stanford University Press.

Levy, Steven. 1984. *Hackers: Heroes of the Computer Revolution*. Garden City, NY: Anchor Press and Doubleday.

Lucky, Robert W. 1989. *Silicon Dreams: Information, Man, and Machine*. New York: St. Martins Press.

McCarthy, J., M. L. Minsky, N. Rochester, and C. E. Shannon. 1956. "A Proposal for the Dartmouth Summer Research Project on Artificial Intelligence." http://www-formal.stanford.edu/jmc/history/dartmouth/dartmouth.html.

McCorduck, Pamela. 1979. *Machines Who Think: A Personal Inquiry into the History and Prospects of Artificial Intelligence*. San Francisco: W. H. Freeman.

Minsky, Marvin. 1989. Interview by Arthur L. Norberg, audio tape(s) and transcript, Cambridge, MA, November 1. Charles Babbage Institute Interview. Oral History Interview no. 179: 1–60.

Minsky, Marvin. 1988. "Early MIT Artificial Intelligence Memos: An Introduction to the COMTEX Microfiche Edition." In *Readings from AI Magazine*. vol. 1–5. Ed. Robert Engelmore, 305–308. Menlo Park, CA: American Association for Artificial Intelligence.

Minsky, Marvin. 1968a. *"Remarks on Visual Display and Console Systems." AI Memo 162, July*. MIT Project MAC.

Minsky, Marvin, ed. 1968b. *Semantic Information Processing*. Cambridge, MA: MIT Press.

Nemes, Tihamér N. 1962. *Cybernetic Machines*. Trans. I. Foldes. New York: Gordon and Breach.

Newell, A., J. C. Shaw, and H. A. Simon. 1958. "Chess-Playing Programs and the Problem of Complexity." *IBM Journal of Research and Development* 2 (4): 320–335.

Peterson, T. F. 2003. *Nightwork: A History of Hacks and Pranks at MIT*. Cambridge, MA: MIT Press.

Pylyshyn, Zenon, ed. 1970. *Perspectives on the Computer Revolution*. Englewood Cliffs, NJ: Prentice-Hall.

Shannon, Claude. 1950. "A Chess-Playing Machine." *Scientific American* (February): 48–51.

Sussman, Gerald Jay, and Adolfo Guzman. 1966. "A Quick Look at Some of Our Programs." Summer Vision Project, MIT AI Lab, Internal Memo no. 102, July 20.

Von Neumann, John, and Oskar Morgenstern. 2004 [1944]. *Theory of Games and Economic Behavior, Sixtieth Anniversary Edition*. Princeton, NJ: Princeton University Press.

Walter, W. Grey. 1970. "Totems, Toys, and Tools." In *Perspectives on the Computer Revolution*, ed. Zenon Pylyshyn, 184–195. Englewood Cliffs, NJ: Prentice-Hall.

Walter, W. Grey. 1953. *The Living Brain*. New York: Norton.

Wilson, Jason. 2008. "Participation TV: Videogame Archaeology and New Media Art." In *The Pleasures of Computer Gaming: Essays on Cultural History, Theory, and Aesthetics*, ed. Melanie Swalwell and Jason Wilson, 94–117. Jefferson, NC: McFarland.

5 CHARACTER

Katherine Isbister

Digital games have traditionally invited the player into an alternate terrain that is likely to be populated by other, often imaginary beings. Although one would not expect a spreadsheet to harbor a mischievous antagonist—Microsoft's "Clippy" (Cozens 2001) being an unfortunate historical aberration—one would be quite disappointed to encounter a single-player first-person shooter entirely devoid of challenging Others. In many cases, a game without characters would seem characterless. The player expects challenge, camaraderie, perhaps even witty repartee. In a single-player game, *nonplayer characters* (characters puppeteered by the game's software) have evolved to meet this need.

Digital games have also originated and refined an innovative form of projection of oneself into the alternate terrain of the game world—the *avatar*, or *player-character*. The player sees and hears and moves in the manner of this someone else while in the game, using the digital body in place of his or her own, and being this someone else helps to guide gameplay and tether the player at many levels—emotional, cognitive, social, fantasy (Isbister 2006)—to who he or she is meant to be in the game. The first known reference to avatars among game designers was Chip Morningstar's use of the term when developing the *Habitat* game for Lucasfilm Games in 1986 (Morabito 1986). The word *avatar* itself is very old, deriving from Sanskrit and used in Hinduism to refer to a god or goddess who chooses to descend and appear on earth in some humanly comprehensible form. The movie *Tron* (Steven Lisberger, 1982) presents the descent of players into an arcade video game, to dwell among its code-based inhabitants, bringing the Sanskrit notion to digital life and framing well the reason this term was adopted by the game-development community.

In this chapter, I present the first known instances of avatars and nonplayer characters in digital games and trace the evolution and variations of both types of game character. I also discuss the ways that game characters have been tuned to suit our psychology to help maximize the emotional experience of gameplay.

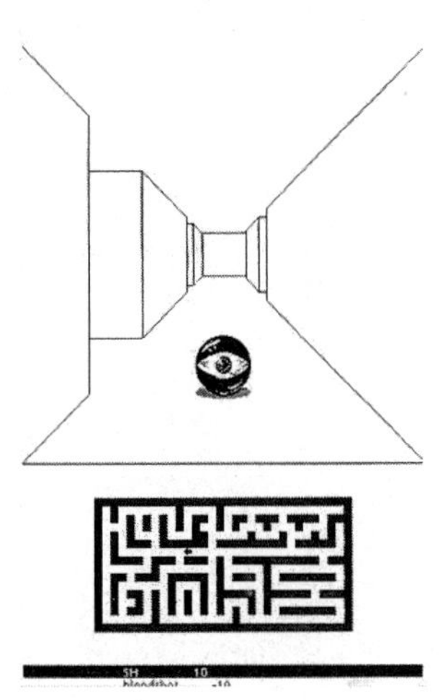

Figure 5.1
The *Maze War* (Colley, 1973–1974) eyeball avatar. Image acquired from *Wikipedia* by way of DigiBarn Computer Museum.

Origins

Avatar

The very first "organic"[1] avatar in a digital game seems to have been the eyeball in *Maze War* (Colley) (Damer 1997), a game that first appeared in 1973–1974 (DigiBarn 2004). *Maze War* was a networked multiplayer game and is considered the origin of the modern first-person shooter genre. Each player logged in and roamed a simple maze, looking to find and kill the other. The main view of the game was from the first-person perspective as if the player were moving through the halls of the maze herself. (Using this view, the player did not see a full view of her own avatar but instead saw the world as if she were traveling directly through it.) Each player was represented on-screen as an eyeball in this first-person view (see figure 5.1).

Beneath the main view of *Maze War* was a top-down view of the maze to help the player orient herself. In this view, the player was represented as a small arrow displaying her location within the maze. It is important to note that players saw only an eyeball when encountering another player, so even though each player was represented by an eyeball, the first-person view meant that the player was not constantly seeing her own eyeballian glory.[2]

Nonplayer Character

The first nonplayer character (NPC) in a digital game appeared earlier than *Maze War*. The honor goes to the dot representing a mouse in the *Mouse in the Maze* game created in 1959 by

MIT students for the TX-0 computer (International Center for the History of Electronic Games n.d.). The player would draw a maze using a light pen and then release the "mouse" to hunt through this maze for cheese. This game presages the simulation genre of computer games in which players create and modify virtual worlds and can then enjoy observing characters navigate and make choices within them.[3] Though the modern gamer might doubt that a dot can be a compelling NPC, earlier work done by psychologists Fritz Heider and Marianne Simmel (1944) demonstrates the incredible power of our minds to attribute agency and causality when viewing the interactions between simple geometric shapes. Graphic novelist Scott McCloud (1993) points out that abstraction in the visual representation of a character can actually increase the viewer's tendency to identify with and project herself into that character.

Evolution and Underlying Psychological Mechanisms

Avatars

The evolution of avatars in games over the past 40 years reflects the underlying technology's increasing capabilities. *Maze War*'s eyeball character was an elegant visual solution to the limits of vector-based graphics on a monotone screen. As computer-processing power, graphics-rendering capability, and input possibilities have grown, so too have the design options available for crafting avatars.

One innovation was the provision of multiple selves for the player at once—parties of characters. The player makes choices on behalf of several characters, managing the entire group as she travels through the game's challenges. This configuration of player characters arose from the conventions of nondigital role-playing games (RPGs) such as *Dungeons & Dragons* (Gygax and Arneson/Tactical Studies Rules, 1974), which were played among friends who then made up a party of adventurers. Digital RPGs designed for solo play re-created the party-based play experience by allowing the player to manage an entire party. *Wizardry: Proving Grounds of the Mad Overlord* (Sir-Tech, 1981) was the first example of a digital party-based game (Hallford and Hallford 2001). The game allowed players to compose a group of up to six characters, choosing their race (e.g., human, elf, dwarf), moral alignment (good, evil, neutral), and basic class (thief, mage, fighter, priest). Each class had different, complementary strengths to offer in combat. A party normally had to have uniform alignment, though there were workarounds. The player would suit up her party with armor and weapons and then head into the dungeon. She could choose which of her party would make a move against any monster the party encountered, alternating among the characters as needed depending on the situation. Based on experience, each of the party members would grow in abilities over time.

Having a cast of player-operated characters allows for a wider range of both gameplay actions and fantasy personas and projections (the player can be a powerful thief *and* a high-minded and mysterious mage). In addition, it enables the introduction of rich backstory about how party members get along and their history with one another. The *Final Fantasy* series (Square Enix, 1997) of RPGs in particular has evolved the narrative behind party members to a high art form, providing long animated sequences exploring their soap operatic struggles and triumphs, which then motivate player quests and aims.

Another innovation from more recent years is the increasing capacity to customize the appearance and capabilities of one's avatar. The best-known game to make customization of avatars a central feature is *The Sims* (EA Maxis). First released in 2000, *The Sims* offers players "living dollhouse"-style gameplay with extensive customization of avatars as well as of settings. A watershed moment in the evolution of customizable avatars was the game *Tiger Woods PGA Tour 2004* (EA Sports, 2003), which introduced a feature called "Game Face" (Gobel 2003): players could adjust their facial features and physical frame in great detail, including characteristics such as size of forearms, and then add accessories such as wristbands and glasses. As one reviewer put it, "The fact is that *Tiger Woods 2004* is the first golf game to allow you to re-generate yourself digitally, and the results can be quite remarkable" (Gobel 2003). The game encouraged the player to conduct an entire career as a digital golfer, which led the player to project herself more deeply onto the avatar (Isbister 2006) and increased investment in outcomes over the course of gameplay.

The introduction of camera and motion-based game inputs has also had an impact on how avatars are designed and how the player relates to them. The Sony EyeToy was an early example of this sort of input. Released in 2004, it repurposed basic webcam technology to insert a video image of the player moving about within the game world. The player's movements could result in game-state changes—for example, reaching a hand out to strike a ball or aiming a foot to kick a target. The initial novelty conceit of presenting the player's own video image within the game has been largely replaced in recent years by an avatar on-screen who is puppeteered through a player's movements. Game designers discovered that players did not necessarily want to see themselves and their own inaccurate attempts at maneuvers faithfully re-created. They wanted to see a fantasy version of themselves on-screen, performing such maneuvers successfully (Alexander 2011).

Avatars, whether they are piloted with a controller or with one's own limbs (or with some combination of the two, as in the Nintendo Wii system's "Mii" avatars), operate at multiple psychological levels for players. They serve as a kinesthetic, physical link between the player and the game world—I press this button, and I raise my weapon in the game. Human beings are able to adapt their proprioceptive sense (the feeling of the edges of the body in space) to accommodate tools such as hammers or even cars, and avatars act as a prosthetic physical self. Avatars also provide a conduit for problem-solving strategies, helping to frame a player's cognition. I know that my avatar can only run, hop, and duck, so I begin crafting action strategies that are delimited by this possibility space. An avatar also acts as a social proxy for the player—I can suddenly become a very powerful alpha male warrior just by picking up the controller, able to order minions around and lock horns with other powerful warriors with confidence. This leads to the final level at which avatars operate—fantasy. Avatars are typically (though not always) characters that hold fantasy appeal for players—people or creatures they are deeply interested in trying on for size to see how it feels to be such a thing. As game designer Tim Schafer once explained in a Game Development Conference talk, all games are wish fulfillments. He knew he was on the wrong track in thinking about a particular avatar when he realized that nobody else would want to be a psychic ostrich (!) (Kosak 2005).

NPCs

Most early digital game NPCs were provided as a challenge for the player in the form of enemies to shoot. A canonical early example is the two flying saucers in *Computer Space* (Nutting Associates), an arcade game released in 1971 (Edwards 2011). The saucers would fly in a zig-zag formation and shoot at the player's rocket ship.

Game designers evolved a more challenging form of enemy, the Boss, as a sort of crescendo to the end-game challenge for the player. A Boss is typically far larger and more powerful than the player's avatar and requires repeated attacks to conquer. It is usually placed at the end of a game level or at the end of the game itself, or it may be guarding some essential objective the player must reach. The first Boss in a digital game was in *dnd* (Whisenhunt and Wood, 1975), designed for the PLATO system. The game's objective was to retrieve an orb that was guarded by a very powerful enemy, the Golden Dragon. Defeating this dragon was the only way to have a chance at appearing on the game's high-scorer list (Universal Videogame List 2014).

Over time, the NPC's social repertoire has increased, moving from offering players a challenge to providing active support as well as motivation to continue. One recurring supporting role NPCs have played is loyal sidekick to the player character. A famous example is Floyd the robot of the Infocom interactive fiction game *Planetfall* (Infocom, 1983). When a student and I conducted a survey among gamers about moments in digital games that had made them cry (Isbister 2008), this NPC was mentioned again and again (spoiler alert!). Floyd travels with the player throughout the game and toward the end sacrifices himself so that the player can survive and succeed. Players reported being amazed at how strongly they felt about his loss and his loyalty. Well-designed NPCs can evoke a surprising amount of emotion.

Another common role for NPCs is someone to save—for example, a princess. The canonical example is Princess Peach in Nintendo's *Mario* game series. In the original game in the series (1985), she is kidnapped by the main villain in the game, Bowser, and must be rescued by Mario.

Though Princess Peach has occasionally had a more active role in the series over the years, for the most part she serves as the damsel in distress. Cultural critic Anita Sarkeesian (2013) points out that extensive reliance on this "damsel in distress" trope in digital games can be unsatisfying and demoralizing for female gamers.

NPCs unlock emotion by performing social roles that players know and understand from everyday life experience and from other media (Isbister 2006). They also exploit players' instincts in social situations—sizing up who is friend and who is foe and figuring out who is more powerful and who is less so. NPC designers heighten such effects by exaggerating these characteristics in NPCs—making a Boss enormous or making a damsel in distress delicate and helpless.

However, in recent years game designers have explored the use of contradictory role characteristics and expectations to create more nuanced feelings and even moral dilemmas. An example is the Little Sisters in *BioShock* (Irrational Games, 2007). The player must "harvest" material from these NPCs to achieve game objectives. She can harvest a little ADAM from a Little Sister harmlessly or kill the Little Sister to harvest much more. The Little Sisters look like small harmless children. As one game journalist described the experience,

> In my career as a gamer, I've racked up quite a virtual body count. And I've just taken down one more: Big Daddy, a hulking monster with a giant drill bit for an arm. I had to unload all my armor-piercing rounds into his body suit, and I barely escaped with my life. And now for my reward: a bioenhanced substance that will give me more superhuman powers. To claim it, all I've got to do is kill the thing the monster was protecting: a tiny little girl, known as a Little Sister, staring up at me with tear-filled eyes. Can I do it? Could you do it? She's not real. It's just a videogame. But that doesn't matter: I put her down and let her go free, forgoing my power upgrade so she can scamper away. It'll be that much harder to take down the next monster, but I feel better about myself. (Kohler 2007)

Conclusion

Characters have been a significant element of digital games since these games were first developed, and they are present in *almost* all game genres up to the present day.[4] As can be seen from the examples in this chapter, there are compelling reasons for this ubiquity. Characters ground players in what to expect and how to feel, providing them with challenge and emotional engagement that is readily mapped to deeply engrained patterns in the human psyche. They are a tremendous contribution games have made to the state of the art of media experience in everyday life.

Notes

I am grateful for the research assistance provided by Edward Melcer.

1. The earlier game *Spacewar!* (Russell/MIT, 1962) represented players as simply drawn spacecraft, but the game mentioned here has the first known organic/anthropomorphic avatar.

2. See http://www.youtube.com/watch?v=bVbRvVvKuHs for a video of the Xerox Alto version of this game to better understand how the game appeared in action.

3. See a demonstration of *Mouse in the Maze* at http://www.youtube.com/watch?v=ki95Z8Tx8go (accessed August 24, 2015).

4. The casual puzzle genre, for example, rarely includes characters of any sort, although a mobile-based puzzle game titled *Threes* (Sirvo, 2014) recently flouted this convention by using tiny faces on numbered puzzle blocks, which emote frequently both visually and verbally as the game is played. Most reviewers surprisingly appreciated the addition of these very minimal characters (e.g., "*Threes* Review" 2014; Webster 2014).

Works Cited

Alexander, Leigh. 2011. "PRACTICE: Inside Dance Central's Prototyping Process." Gamasutra, October 30. http://gamasutra.com/view/news/38204/PRACTICE_Inside_Dance_Centrals_Prototyping_Process.php. Accessed March 7, 2014.

Cozens, Claire. 2001. "Microsoft Cuts 'Mr. Clippy.'" *Guardian*, April 11. http://www.theguardian.com/media/2001/apr/11/advertising2. Accessed March 7, 2014.

Damer, Bruce F. 1997. *Avatars! Exploring and Building Virtual Worlds on the Internet.* Berkeley, CA: Peach Pit Press.

DigiBarn. 2004. "The *Maze War* 30 Year Retrospective at the DigiBarn." http://www.digibarn.com/history/04-VCF7-MazeWar/index.html. Accessed March 7, 2014.

Edwards, Benj. 2011. "Computer Space and the Dawn of the Arcade Video Game." *Technologizer*, December 11. http://technologizer.com/2011/12/11/computer-space-and-the-dawn-of-the-arcade-video-game/. Accessed March 7, 2014.

Gobel, Gord. 2003. "*Tiger Woods PGA Tour 2004* Review." *Gamespot*, October 13. http://www.gamespot.com/reviews/tiger-woods-pga-tour-2004-review/1900-6076746/. Accessed March 7, 2014.

Hallford, Neal and Jana Hallford. 2001. *Swords & Circuitry: A Designer's Guide to Computer Role Playing Games.* Roseville, CA: Prima Tech.

Heider, Fritz, and Marianne Simmel. 1944. "An Experimental Study of Apparent Behavior." *The American Journal of Psychology* 57 (2): 243–259.

International Center for the History of Electronic Games. n.d. *Video Game History Timeline.* http://www.museumofplay.org/icheg-game-history/timeline/. Accessed August 24, 2015.

Isbister, Katherine. 2008. "The Real Story on Characters and Emotions: Taking It to the Streets." Paper presented at the Game Developers Conference, San Francisco, February.

Isbister, Katherine. 2006. *Better Game Characters by Design: A Psychological Approach.* Boston: Morgan Kaufmann.

Kohler, Chris. 2007. "Creepy Moral Dilemmas Make *BioShock* a Sophisticated Shooter." *Wired*, August. http://archive.wired.com/gaming/gamingreviews/news/2007/08/bioshock_review. Accessed August 24, 2015.

Kosak, Dave. 2005. "Psychonaut Tim Schafer on Taking Risks." *Gamespy*, February 4. http://www.gamespy.com/articles/585/585524p1.html. Accessed March 7, 2014.

McCloud, Scott. 1993. *Understanding Comics: The Invisible Art.* New York: William Morrow Paperbacks.

Morabito, Margaret. 1986. "Enter the Online World of LucasFilm." *Run,* August, 24–28.

Sarkeesian, Anita. 2013. "Damsel in Distress (Part 1)." *Tropes vs Women*, March. http://www.feministfrequency.com/2013/03/damsel-in-distress-part-1/. Accessed March 7, 2014.

"*Threes* Review." 2014. *The Edge.* http://www.webcitation.org/6NaCdxGv4. Accessed archived version August 24, 2015.

Universal Videogame List. 2014. http://www.uvlist.net/game-160118. Accessed March 7, 2014.

Webster, A. 2014. "By the Numbers: 'Threes' Is Your New iPhone Addiction." *The Verge*, February 6. http://www.theverge.com/2014/2/6/5361708/threes-ipad-iphone-puzzle-game. Accessed March 7, 2014.

6 CLASSIC GAMING

Melanie Swalwell

The term *classic games* enjoys extended usage and is embraced in many popular histories and references to game history. For instance, Will Greenwald writes that the gaming festival Penny Arcade Expo (PAX) East, alongside the "biggest upcoming titles the gaming industry has to offer ... [includes] 'classic video games'" (2012); Troy Dreier draws the attention of "those who love to see the classic video games make a return" to *Centipede*'s appearance on the Wii console (2011); and Chris Kohler begins his book *Retro Gaming Hacks* by relating how impressed he was when he played a "classic Game Boy game, in all its 15-year-old monochrome glory ... on the Sony PSP [PlayStation Portable]," despite his fondness for "actual classic game hardware" (2006, xix). Its widespread use notwithstanding, the term *classic* remains problematic for many scholars of game history, in part because of the vagueness, nostalgia, or hyperbole with which it is frequently associated. Sometimes used interchangeably with the terms *old*, *vintage*, and *retro* or with the phrase "what was popular when I was a kid" (Herz 1997, 71), *classic* is often invoked with a celebratory rather than critical intent.

The term *classic* is difficult to define with precision. The *Oxford English Dictionary* (*OED Online*) gives many definitions, including "of acknowledged excellence or importance"; "of the first class, of the highest rank or importance; constituting an acknowledged standard or model; of enduring interest and value"; "archetypal; very typical of its kind, representative"; and "a work ... of acknowledged quality and enduring significance or popularity." The last of these definitions finds an extended use in denoting "something which is memorable and an outstanding example of its kind," a description that casts an exceedingly broad net (if memorability were taken as the sole criterion, then any game might be said to be a classic). Although these definitions refer to works, the *OED* also extends the term to the creator(s) of works "of acknowledged importance and quality." Usage in game history largely accords with these definitions. In addition, definitions extrapolated from classical antiquity may be meaningful ("of a timeless or unquestionable beauty" and "a simple, elegant style not greatly subject to changes in fashion"). Finally, the collectability of historic games also gives the common definition of the term *classic car* ("designating an older motor vehicle of acknowledged quality, especially of a type sought after by collectors") some relevance, though what is meant by this connection is not always clear.

To claim that a game is a "classic" is ultimately to make a judgment about its cultural status, value, or meanings. The term operates rhetorically to persuade one of the importance of the said game or the experience of the game. Given that digital games have historically been popular objects of low cultural status, we might pause to consider the cultural work that is done by making them into classics. What is at stake in endowing a game title or an artifact or a gameplay experience with classic status? Attempts to recognize digital games as possessing an enduring significance effectively refute the idea that games are ephemeral and recover the object in a way that will no doubt be attractive to some game historians. However, it also encourages the framing of game history through certain games and narratives rather than through others. In some sense, the descriptor *classic* functions to move the subject outside of historical time—"timeless," "transcend[ing] history" (Burnham 2003, 23)—putting its significance beyond question. Furthermore, who gets to decide what "constitutes an authentic and sanctioned canon of classic games and, by implication, the classic gaming experience" (Payne 2008, 56)? The making of classics involves making choices that are often unstated in the telling of game histories. Using the term *classic* to frame a history involves privileging certain narratives over others. There are enough books on game history now to see how this privileging plays out: "the" history is usually told either through a focus on classic (or "key" or "great") games or through a telling of the history of "the" industry and of the (typically) great men who founded it (Loguidice and Barton 2009; Demaria and Wilson 2004; Kent 2001; Gielens 2000). Brett Weiss's trilogy *Classic Home Video Games* (2011, 2009, 2007)—encyclopedic reference guides describing every official game for programmable consoles released in the United States—highlights another problem with the term. What sense does it make to claim classic status for every game for every console, and what do such "completionist" accounts deliver? What do they omit? Other approaches to game history are possible.

Asserting that a game or game experience is a "classic" may become a universalizing gesture that elides difference. To state an obvious point, game history did not unfold in the same manner everywhere, and the particularities of space, place, and time matter to historians. Given the great historic diversity of games and contexts for their play, an appreciation of sociocultural and geographic specificity is important to develop if other histories are to be told, for instance, from the "periphery" rather than from the "center." Especially prior to the development of the globalized gaming industry, a title considered a "classic" in one region may not have been available in another, which makes locality an important consideration. For instance, in the 1980s microcomputer systems that are well known in the West did not penetrate far behind the Iron Curtain in eastern Europe, and Vectrex, ColecoVision, and Intellivision consoles were not distributed in New Zealand. Assertions of classic status need to consider when, where, and for whom the subject is classic. Otherwise, the label simply naturalizes what it purports to study.

Several writers on game history have explicitly offered their rationales for what makes a classic game. J. C. Herz discusses the issue at some length in her book *Joystick Nation*, castigating those who are simply nostalgic for their youth and then argues that gaming "firsts" ought to be privileged. Herz considers "legitimate" the argument that

> many of these game consoles and arcade machines were the first of their kind. One can justifiably argue that the Atari 2600 (1977) is a classic in the way the Nintendo Entertainment System (1985) can never be, because Atari's machine was the first cartridge console to gain mass acceptance. Atari, Intellivision, and ColecoVision are all from a period when videogames were breaking into the mainstream and creating a culture of their own, during the first rise and before the first fall of the [North American] videogame industry. Comparing a *Pong* console to a Sega Master System is like comparing a '57 Chevy to, say, a '79 Mustang. One is from the period that created car culture. The other is simply a machine whose sentimental value will rise as its original owners wax nostalgic for their youth. (1997, 71).

Herz's analogy with car culture seems reasonable, and yet car enthusiasts have notorious difficulty in distinguishing between "classic," "vintage," and "antique" vehicles, despite institutional interests' attempts to create conformity for regulatory and insurance reasons. It's worth asking whether classic status should inhere only in "firsts"? Focusing only on firsts gives history a particular skew; historians may be interested in many firsts (or they may adopt a quite different approach, such as a history from below, a social history, a history of consumption, and so on). Furthermore, game design and gameplay concepts are often revisited in interesting ways in later games that draw inspiration from an earlier title. Herz's desire to reserve classic status for those games present at the "dawn of the era" leaves the problem of the "enduring significance" of nonfirsts—such as the Nintendo Entertainment System she mentions—unanswered.

Van Burnham is another who embraces the term *classic*, explaining her use of it as follows: "Classic videogames are determined by a number of factors—age, aesthetics, nostalgia—but ... I realize that modern games have the potential to be equally timeless. ... [I]t's important to appreciate videogames for their unique character—the balance they strike between the technology and the context they were created in. From that perspective, every game can transcend history and become the next generation of classics" (2003, 23).

Periodizing games and systems by the era of their manufacture—as is done in car culture—partially gets around the recognition of games that were nonfirsts as potentially having some claim to classic status. Not resolved, however, is the contradiction between considering games in the context in which they were created—as Burnham urges—and characterizing them as "timeless." Might it not be important to consider how games from other eras speak to our own? Or are they forever "time-capsuled," as Raiford Guins puts it (2014, 3)?

Some game historians self-consciously assert the classic status of games that are relatively contemporary. *Retro Gamer* magazine's "Future Classic" column reviews noncurrent releases (usually less than 10 years old), articulating why the reviewer thinks a particular game has classic qualities. For instance, a review of *The Warriors* (2005) is entitled "Discover Why We Feel Rockstar's Violent Brawler Will Stand the Test of Time" (Retro Gamer Team 2014). The "Future Classic" column seems to be an attempt to remember in the face of the rapid churn of new game titles by those who already identify as fans of retro games. However, as the "Future Classic" nomenclature recognizes, the business of making classics involves an indefinite deferral.

The term *classic gaming* refers to the practices of playing and collecting historic games, often early digital games, but also extends to coin-operated and electromechanical games. *Retro-gaming* is a perhaps more widely used synonym here, although the moniker *classic gaming* is used for various events and expos such as California Extreme and the Classic Gaming Expo, both in the United States. According to Elizabeth Guffey, the term *retro* acquired its current meaning of revival of the recent past with an often ironic or unsentimental detachment in the early 1970s, thereafter progressively coming into daily use. Guffey observes that whereas the term *classic* equates to "old" and an undefined time gone by, *retro* champions the revival of the recent past, in particular pop culture and other nonserious aspects of that past. Retro's focus on technological obsolescence is of relevance, with a retro sensibility sometimes pursuing this obsolescence for its own sake, for novelty and fashion, and other times out of "dissatisfaction with the present." Guffey asserts that although retro is characterized by "an admiration for the past," it entails a relation to the past that is often "non-historical" (2006, 9–21). Clearly, the game historian will want to tread a careful and conscious path around the more celebratory aspects of retro culture.

Retro-gaming has become a popular hobby, particularly over the past 10 to 15 years. Enthusiasts initially saw value in collecting games and related items that had been cast off because they were no longer new. Contra Guffey's accusations of irrelevance, their foresight has laid the foundation for some important collections of historic games and related artifacts. From a once obscure subculture, the hobby now boasts forums, festivals, expos, magazines, stores, social networks, and lively cultures of circulation. Commerce has been one beneficiary (and shaper) of this process: "old school" (sometimes spelled "skool") gaming has been anointed as "cool," so retro-gaming merchandise has entered the mainstream, commodifying the hobby (ironically, given the associations with rubbish and reuse) (Swalwell 2007).

Retro games have been revived with various derivatives—remakes and ports—of games, such as *Centipede* (Atari, 1981) on the Wii, and hacks that enabled Kohler to play Game Boy games on his Sony PSP, and, more recently, the remaking of 8-bit-era games as applications for the mobile phone screen. The reissuing of game compilations and the manufacture of plug-and-play "classic game consoles" such as the Atari Flashback have further fueled the revival of and markets for retro games in the twenty-first century. The Flashback is a remake of the Atari 2600 or Video Computer System console, sans cartridges but complete with faux wood paneling and controllers "exactly like the Atari 2600 joysticks of yore" (though some would disagree) (Harris 2005). Close attention to detail notwithstanding, the Flashback highlights a key consideration in studying retro gaming now—namely, that although the games might not have changed that much, the player has—"the games themselves have helped change us," as Sean Fenty puts it (2008, 30)—and so the experience of playing re-released Atari and licensed Activision titles now is distinct from the experience when they were newly released. As Ron Harris reflects, "The action games held up well, but adventure games like *Haunted House* and *Wizard* were mostly duds and offered little real suspense. Seriously. How scary can a blinking green square be anyway?" (2005).

Although retro-gaming can encompass derivatives made in the style of earlier games and games made playable via emulation, some collectors shun all such remakes in favor of "vintage" hardware—that is, from the "period ... [when the game] was made or produced" (*OED*). They want games in as close to their original state as possible, arguing in the case of arcade games that nothing less than original cathode ray tube (CRT) monitors and joysticks will do, especially where custom interfaces are involved. This view sets a demanding restoration standard comparable in some ways to the standard for vintage cars (Burnham, cited in Guins 2014, 261), though it is perhaps ultimately an impossible task, given bit rot and the closure of factories that produced CRT monitors. Ironically, as these objects age, they acquire not only cult value but perhaps also—for collectors and "game lovers"—an aura, that unique existence of a singular work of art thought not to adhere to mass-produced items (Swalwell 2013, 2007; Benjamin 1992).

The revival of retro style is frequently characterized by a longing for the past. Nostalgia has been the dominant mode of remembering early games for at least a decade now, and collectors and enthusiasts of retro games can have strong personal and emotional investments in game history. Some historians are concerned that game criticism might become mired in a nostalgia that "sinks ... efforts to create things that feel new" (Hilbert 2004, 57) and that inhibits criticality. It is worth noting that academics are not always immune to the nostalgic pull of their own histories with games. Indeed, acknowledging their own pasts as part of the "Nintendo Generation," Laurie Taylor and Zach Whalen call for a "nostalgic turn" in the field of game studies (2008, 4). C. Nadia Seremetakis reminds us that in Greek the term "*nostalghía* speaks to the sensory reception of history" in contrast to what she characterizes as "the American sense[,] [which] freezes the past in such a manner as to preclude it from any capacity for social transformation in the present." Nostalgia understood in the former sense, Seremetakis suggests, involves imagining the present in "a dynamic perceptual relationship" with the past (1996, 4). This distinction offers a different approach to those who either deride nostalgic sense memories as mere sentimentality or note the impossibility of the desired nostalgic return (Fenty 2008). Player memories, for instance, can reveal important perspectives on the social and material conditions of games' historic reception, matters that are difficult to collect, preserve, and display and that the game itself cannot relay (Stuckey, Swalwell, Ndalianis, et al. 2013).

Retro-gamer groups have developed some important historical and preservation endeavors of considerable significance to game historians. Some individuals and networks of enthusiasts possess a wealth of knowledge that renders them expert informants. Drawing on their members' knowledge, such networks have been responsible for important online collaborative archives that cover ephemera associated with games (the Arcade Flyer Archive), game magazines, and metadata on games (e.g., Killer List of Videogames). These groups have further developed emulators and contributed arcade game boards (e.g., Multiple Arcade Machine Emulator, or MAME), led to the development of preservation technology and standards (e.g., Kryoflux, the TZX file format), and built platform-specific sites with a remarkable collection of resources (e.g., World of Spectrum and Lemon64). These collectors and fans' visionary efforts deserve recognition. These individuals were involved in game history long before cultural institutions became interested in it, and their efforts have ensured that information, hardware, and software from the early years of gaming still exist.

Works Cited

Benjamin, Walter. 1992. "The Work of Art in the Age of Mechanical Reproduction." In *Illuminations*, ed. Hannah Arendt, trans. Harry Zohn, 219–253. London: Fontana.

Burnham, Van. 2003. *Supercade: A Visual History of the Videogame Age, 1971–1984*. Cambridge, MA: MIT Press.

Demaria, Rusel, and Johnny L. Wilson. 2004. *High Score! The Illustrated History of Electronic Games*. 2nd ed. Emeryville, CA: McGraw-Hill and Osborne.

Dreier, Troy. 2011. "Centipede Crawls to the Wii." *PCMag.com*, April 20. http://appscout.pcmag.com/video-games/268470-centipede-crawls-to-the-wii.

Fenty, Sean. 2008. "Why Old School Is 'Cool': A Brief Analysis of Classic Video Game Nostalgia." In *Play the Past: History and Nostalgia in Video Games*, ed. Zach Whalen and Laurie N. Taylor, 19–31. Nashville: Vanderbilt University Press.

Gielens, Jaro. 2000. *Electronic Plastic*. Berlin: Gestalten.

Greenwald, Will. 2012. "PAX East Gaming Expo Opens in Boston." *PCMag.com*, April 6. http://www.pcmag.com/article2/0,2817,2402688,00.asp.

Guffey, Elizabeth E. 2006. *Retro: The Culture of Revival*. London: Reaktion.

Guins, Raiford. 2014. *Game After: A Cultural Study of Video Game Afterlife*. Cambridge, MA: MIT Press.

Harris, Ron. 2005. "Gaming's Bygone Era Relived with Atari Gadget." *USA Today*, September 22. http://usatoday30.usatoday.com/tech/products/games/2005-09-22-atari-flashback_x.htm.

Herz, J. C. 1997. *Joystick Nation: How Videogames Gobbled Our Money, Won Our Hearts, and Rewired Our Minds*. London: Abacus.

Hilbert, Ernest. 2004. "Flying Off the Screen: Observations from the Golden Age of the American Video Game Arcade." In *Gamers: Writers, Artists, and Programmers on the Pleasures of Pixels*, edited by Shanna Compton, 57–69. New York: Soft Skull.

Kent, Steven L. 2001. *The Ultimate History of Video Games: From* Pong *to* Pokemon *and Beyond—The Story behind the Craze That Touched Our Lives and Changed the World*. New York: Three Rivers Press.

Kohler, Chris. 2006. *Retro Gaming Hacks*. Sebastopol, CA: O'Reilly Media.

Loguidice, Bill, and Matt Barton. 2009. *Vintage Games: An Insider Look at the History of* Grand Theft Auto, Super Mario, *and the Most Influential Games of All Time*. Burlington, MA: Focal Press and Elsevier.

Payne, Matthew Thomas. 2008. "Playing the Deja-New: 'Plug It in and Play TV Games' and the Cultural Politics of Classic Gaming." In *Play the Past: History and Nostalgia in Video Games*, ed. Zach Whalen and Laurie N. Taylor, 51–68. Nashville: Vanderbilt University Press.

Retro Gamer Team. 2014. "The Warriors." *Future Classic Column*, January 21. http://www.retrogamer.net/future_classics/21087/the-warriors/. Accessed January 29, 2014.

Seremetakis, C. Nadia. 1996. "The Memory of the Senses, Part I: Marks of the Transitory." In *The Senses Still: Perception and Memory as Material Culture in Modernity*, 1–18. Chicago: University of Chicago Press.

Stuckey, Helen, Melanie Swalwell, Angela Ndalianis, and Denise de Vries. 2013. "Remembrance of Games Past: The Popular Memory Archive." In *Proceedings of the 9th Australasian Conference on Interactive Entertainment: Matters of Life and Death.* New York: ACM. http://dl.acm.org/citation.cfm?doid=2513002.2513570.

Swalwell, Melanie. 2013. "Moving on from the Original Experience: Games History, Preservation, and Presentation." In *Proceedings of DiGRA 2013: DeFragging Game Studies*, Atlanta: DiGRA. http://www.digra.org/digital-library/publications/moving-on-from-the-original-experience-games-history-preservation-and-presentation/.

Swalwell, Melanie. 2007. "The Remembering and the Forgetting of Early Digital Games: From Novelty to Detritus and Back Again." *Journal of Visual Culture* 6, no. 2: 255–273. doi:. http://vcu.sagepub.com/cgi/doi/10.1177/1470412907078568.10.1177/1470412907078568

Taylor, Laurie N., and Zach Whalen. 2008. "Playing the Past: An Introduction." In *Play the Past: History and Nostalgia in Video Games*, ed. Zach Whalen and Laurie N. Taylor, 1–15. Nashville: Vanderbilt University Press.

Weiss, Brett. 2011. *Classic Home Video Games, 1989–1990: A Complete Guide to Sega Genesis, Neo Geo, and TurboGrafx-16 Games*. Jefferson, NC: McFarland.

Weiss, Brett. 2009. *Classic Home Video Games, 1985–1988*. Jefferson, NC: McFarland.

Weiss, Brett. 2007. *Classic Home Video Games, 1972–1984*. Jefferson, NC: McFarland.

7 CODE

Mark Sample

Code is a set of instructions to a computer. Code is more or less readable by both humans and machines under certain conditions and for different purposes. Programmers write code; machines enact it. In between these two stages, the original code—called the source code—often needs to be compiled. The compiler itself is a computer program, and it translates the lines of legible commands, conditionals, variables, functions, and loops into machine language, a sequence of numbers in hexadecimal or binary notation that the computer can understand. These numbers essentially tell the computer when to add and by how much. This is what even the most sophisticated software comes down to—combining ones and zeroes to manipulate the logic of a circuit. Video games are no different in this regard from word processors, spreadsheets, email programs, or web browsers. All start out as source code, and they eventually become executable programs. If the program is a video game, it runs on a gaming platform, which might be anything from a handheld Tamagotchi to the latest console system in the living room. Or a smartphone or a personal computer or even a calculator.

Code may appear arcane to nonprogrammers or be utterly inscrutable in its compiled form. Because of these occult properties, code is often privileged over other aspects of software design. As Wendy Hui Kyong Chun (2008) puts it, there is a tendency to fetishize source code. The attention that game designer Jordan Mechner received when he rediscovered and then released his original Apple II assembly code for *Prince of Persia* (Brøderbund, 1989) is a testament to this fetishization (Mastrapa 2012). The availability of the source code for Gabriele Cirulli's game *2048* (2014) has likewise spawned dozens of so-called clones (Ballard 2014), as if code alone—and not gameplay, mechanics, and a compelling balance of ease and difficulty—were all that is required for a hit video game. The term *clone* itself suggests there is something genetically essential to game code, the software equivalent of DNA. Understood this way, game code implies a vast, hidden universe under the surface of the game, difficult to crack, a cryptic mystery. When it comes to games, it's code all the way down.

Indeed, one of the earliest video games was about code. *Darwin* (1961) was a "game between computer programs as programs" (Aleph Null 1972, 93). Created by Victor A. Vyssotsky, Robert Morris Sr., and M. Douglas McIlroy at Bell Telephone Laboratories, *Darwin* set small computer programs in competition to take over each other's virtual environment. Each program attempted to PROBE and CLAIM memory blocks in an "arena" on an IBM 7090. Memory blocks occupied by an opponent could be eliminated with a KILL command. The first program to fill up the available memory in the arena won the game, a ludic inversion of the troublesome memory-leak problem that can plague sloppy code. *Darwin* was only the first in a long tradition of games that thematized code either overtly, as in the cult classic *Core War* (D. G. Jones and A. K. Dewdney, 1984) or, more subtly, as in the procedurally based game *Minecraft* (Mojang, 2011).

A more recent though similarly obscure video game also appears to insist on the primacy of code. Yet a close look at the game challenges the very idea of code as a stable property of video games. *The Naked Game* is a Flash version of the first commercially successful arcade game *Pong* (Atari, 1972). Released in 2008 by the now defunct company RetroDev Games, *The Naked Game* re-created the tennis-for-two play of *Pong*, except both of the dueling opponents were computer-controlled artificial intelligences. If the computer is playing itself, then what does the player do? Quite simply, the player plays the code.

In a rare kind of eversion, making available to the player what is usually kept below the surface of a game, *The Naked Game* appears to expose the ActionScript code that powers the game (see figure 7.1). The game's initial state variables appear to the right of the retro black and green *Pong* playing field, a kind of head-up display (HUD) that emulates how the computer itself sees the game. Meanwhile, the physics algorithms are displayed below the playing field, seventeen lines of code "governing the mechanics of the game," as the developer puts it. These lines of code are what the player plays, clicking on a line to gray it out, making the game engine ignore it. The effect of a line's removal occurs immediately, in real time, as the two AIs continue playing each other. The player can compound the effect by toggling multiple lines of code.

For example, clicking the line of code `Ball._x+=balldx;` removes that line from the program's algorithm, resulting in a ball that no longer moves along the *x* axis. The ball will now move only along the *y* axis, up and down in a straight line on the screen. Clicking the line again reinstates the code, normalizing gameplay. Turning off other lines of code leads to similarly unsettling effects. Eliminate line 18, the line that begins `if(Ball._y < PLAYFIELD_MINY)`, and suddenly the top and bottom "walls" of the game no longer matter, and the ball ricochets out of the playing field, right down into the code itself. Toggling some code a few lines down transforms either player's paddle into a phantom, which the ball simply passes through.

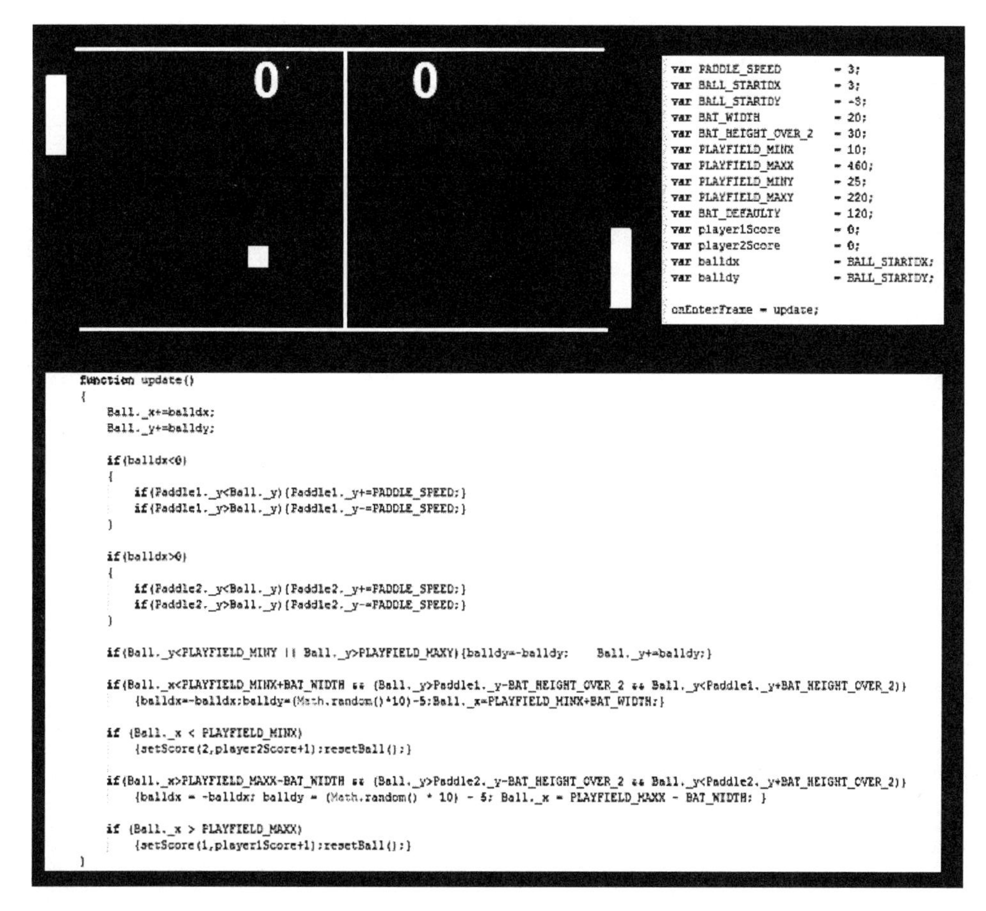

Figure 7.1

The Naked Game (RetroDev Games, 2008). At http://wayback.archive.org/web/20140405064152/http://www.retrodev.co.uk/MiscGames/NakedGame/TheNakedGame.html.

The Naked Game is what Ian Bogost might call a "throwaway," a game meant to be played once and then tossed aside (2011, 96). After all, it takes only a few minutes to explore the possibility space of *The Naked Game.* The game is also a throwaway in the sense that it is no longer readily available because the developer's website has long since disappeared—though a playable version of the game lives on in the Internet Archive. But even discarded or forgotten games can reveal a great deal about video games and in this case about video game code. *The Naked Game* clearly emphasizes the role of computer code in making video games what they are: rule-based interactive systems. Like *Darwin, The Naked Game* is in some sense a game about

Table 7.1

Playable Code	ActionScript Code
`function update()`	`function update()`
`{`	`{`
`Ball._x+=balldx;`	`if(LinesActive[1])`
`Ball._y+=balldy;`	`{`
	`Ball._x = Ball._x + balldx;`
	`}`
	`if(LinesActive[2])`
	`{`
	`Ball._y = Ball._y + balldy;`
	`}`

programming, turning the kind of glitch that can mar the seamless play of a video game into the game itself.

The Naked Game is also instructive in determining what counts or does not count as code. When a player toggles a line of code on the screen, has the code inside the game actually changed? Is the code on the surface of the game in fact the code inside the game? Decompiling NakedGame.swf—that is, unpacking the executable Flash file into lines of source code—provides an answer.[1] Table 7.1 compares the first lines of playable code to the corresponding ActionScript code in NakedGame.swf.

The program operates by scanning the 17 lines of playable code (the `LinesActive` array in the ActionScript), determining whether the line is active or not. If the matching line is active—that is, not clicked by the player—then the program updates the variable in that line (for example, `Ball._x = Ball._x + balldx;`). In other words, toggling off a line of code does not change the code of the game so much as it causes a variable to default to its previous value. The playable code is not code; rather, it is an interface to the variables of the game.

An important conclusion about code can be drawn from the example of *The Naked Game*: code is not a monolithic entity. What is commonly referred to as code includes a number of software—not to mention epistemological—dimensions:

1. Code is frequently interpreted to mean the rules of the game (Galloway 2006, 35; Manovich 2001, 222). This conflation crystallizes in *The Naked Game*, for when the player changes the "code," what is being changed is really the rules.
2. Code can be a loaded signifier aimed at human readers or players instead of at a machine. The onscreen code of *The Naked Game* is evocative of code, not literally code. Mez Breeze's or John Cayley's codework provides provocative examples of nonfunctioning pseudo-code outside the realm of video games (Raley 2002).

3. Functional code might also include what Mark Marino calls "extrafunctional significance" (2006). Extrafunctional significance means the code participates in a system of signs beyond those executable by the machine. This extrafunctional significance might be found in a developer's choice of a variable name, such as `BAT_WIDTH`, which hints at *The Naked Game*'s British origins, likely referencing a cricket bat, or the more troubling "FeministWhore," discovered in the code of the zombie survival horror game *Dead Island* (Techland, 2011) (John 2011). Extrafunctional significance might also occur in the comments in the code, meant to be read only by other programmers on the development team (Sample 2013).
4. Code can also include nonexecutable digital assets such as plain text, images, movies, texture maps, avatars, sounds, and music. Decompiling NakedGame.swf reveals a host of assets (see figure 7.2) that have been bundled in the package, including sprites of the paddles and ball as well as the audio file. Flash's packaging of assets in an executable file is relatively straightforward. Other video games may use other methods, including compressing assets in a separate file, such as WAD ("works as designed") or VPK (Valve Package) files, or as stand-alone files referenced by the main program. In all of these cases, the media assets are not strictly speaking instructions to the computer, but they are composed of code—ones and zeroes.
5. Finally, code may also refer to reconfigurations of the game the player makes outside of gameplay. These reconfigurations include modifications, hacks, and cheat codes (Galloway 2006, 13). The famous Konami Code on the Nintendo Entertainment System game controller (up, up, down, down, left, right, left, right, B, A) is not code on the level of source code, but it nevertheless plays on the occult mystique of code, changing the properties of the game through an arcane process (Consalvo 2007, 29).

Taken together, these five supplements to the definition of code teach an important lesson: *it is a mistake to think about code as always and only instructions to a computer.*

This situation is complicated by an exigency of modern software development: software rarely speaks directly to hardware; it speaks instead to a container piece of software within which it resides, what is known as a virtual machine. For example, Flash functions as a virtual machine within a browser. The case of *The Naked Game* raises an existential question that applies to nearly all modern video games: What counts as the code of a game? If code isn't in the playable code on the screen, is it the NakedGame.swf file, the web browser, the Flash extension needed to run the game? Or is it in the operating system of the computer itself? Faced with these propositions, one might readily see *The Naked Game* as circumstantial evidence that games are indeed code all the way down.

Yet this is not the case, and *The Naked Game* helps to explain why. Recall that *The Naked Game* is, as the developers put it, a "primitive version of 'Pong.'" It's unclear whether "primitive" is

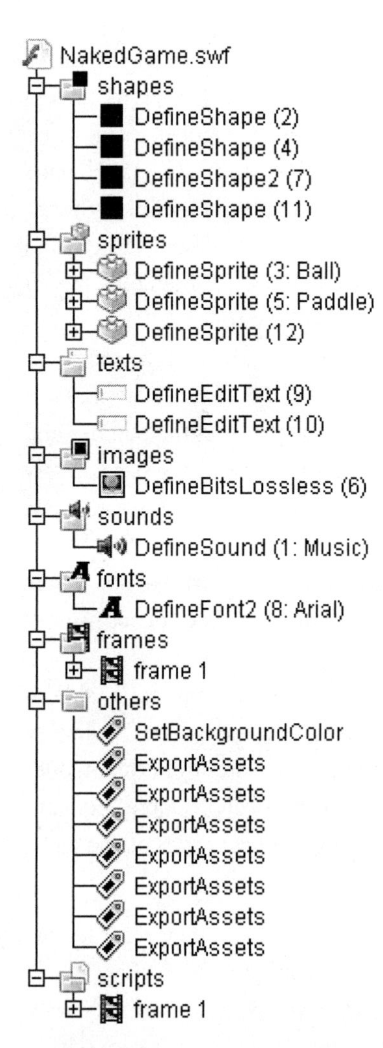

Figure 7.2
The assets of NakedGame.swf.

meant to be ironic, but it surely must be taken as so, for here is a mostly forgotten detail about *Pong*, a legendary game that by all accounts launched the video game industry: *Pong* had no code. Charles Babbage's theoretical Analytical Engine had more of a code base than *Pong*. In fact, *Pong* was so "primitive" that it had no microprocessor. Atari engineer Al Alcorn designed *Pong* using transistor-to-transistor logic, in which hardwired transistors completed the logic, performing the calculations necessary for the game (Lowood 2009). *Pong* is just one of many games that required no code; the game was preceded by a half-century's worth of electromechanical arcade games that required no code. Pinball machines, shooting games, racing games—all without code (Wolf 2008, 35). If video games can exist without code, then what more ought to be said about code and video games?

There are at least three answers to this question, and they all stem from the way code arises from and yields to certain physical exteriorities and material conditions. To begin with, code occupies space, and not merely space as a computer understands it—in RAM (random-access memory) and ROM (read-only memory). Code is usually displayed on a screen, but it can be printed, and sometime it must be. The prototypical video game *Spacewar!* (Russell/MIT, 1962) was punched into the equivalent of twenty-seven pages of paper tape, the storage medium for the Programmed Data Processor-1 computer (Levy 2010, 63). The 80-column IBM punch card was another version of printed code, in use well past the heyday of batch processing (Lubar 1992). Early handheld electronic games powered by repurposed calculator chips, such as a Mattel's *Football* (1977), relied on instructions punched into these IBM cards and assembled on a mainframe computer (Lesser 2007).

Code of course occupies digital space as well. The final code for Mattel's *Football* occupied 512 bytes of ROM, or half a kilobyte. That's roughly one-eighth the size of Mattel's contemporaneous games on the Atari Video Computer System (VCS), which used cartridges with 4 kilobytes of ROM (Montfort 2006). As video games became more data and process intensive, their underlying code grew as well.[2] Modern video games are such massive projects that large chunks of code are frequently reused (Bogost 2006, 62). These shared codebases are called "game engines," such as id Software's Quake engine (1996) and Epic's Unreal Engine (1998). Consisting of millions of lines of code themselves, these game engines manage the rendering and physics of a game's virtual world, ostensibly freeing the developers to focus on gameplay and narrative.

Code often costs money. Sometimes it is given away for free (to wit, the Quake engine is open source and free, whereas the development of commercial games with Epic's Unreal Engine requires a monthly subscription). Either way, code counts as intellectual property and is copyrightable. Copyright provides the copyright holder with exclusive legal rights pertaining to the copying or distribution of the copyrighted object, a status conferred on software by US copyright law in 1980 (Kelty 2008, 183). Software is also frequently subject to End User License Agreements (EULAs), legal contracts players agree to when they open a software package or install a game. A typical EULA, such as the license agreement for *Grand Theft Auto 5* (2013) from Rockstar Games, requires the player to agree not to "reverse engineer, decompile, disassemble, display, perform, prepare derivative works based on, or otherwise modify the Software, in whole or in part" (Rockstar Games 2013).[3] Game developers occasionally release the source code of their games under alternative copyright licenses, which allow players to do precisely what Rockstar disallows. Quake, for example, is now under a GNU Public License

(GPL), which gives users permission to do anything they like with the code, as long as they make the resulting source code free under another GPL (Carmack 1999).

Finally, one more dimension of code must be acknowledged. It is the greatest material condition of all: code is made by people. Development teams range from a single person designing a game bound for newgrounds.com to small studios designing an indie game for Steam to massive multinational companies such as Ubisoft and EA that dominate the video game industry. In the video game industry, most of these people are male, and there is roughly a $9,000 pay gap between men and women programmers (Dyer-Witheford and De Peuter 2009, 54–59). During the "crunch time" before a game's release, the developers may work 50 to 60 hours a week, a working condition made possible perhaps by the relatively young age of the workforce. The average worker in the industry is 31 years old, and fewer than a quarter of those surveyed have children. Burnout is common in the industry; on average, programmers will spend less than four years at their current jobs (Weststar and Legault Téluq 2012).

The work of the men and women who write the code that powers the games recapitulates one of the defining features of code. A few exceptions aside, code is invisible, as are the people who write it. The legendary computer scientist Alan Kay once called *SimCity* (Maxis, 1989) a "pernicious ... black box," full of assumptions and "somewhat arbitrary knowledge" that cannot be questioned or challenged (2007). Kay may be correct that code comes to us in a black box, but it is not a box that can't be opened. It is the task of historians, archivists, teachers, students, and players to examine that box, to make its contents visible, and to continue to complicate our understanding of code, where it comes from, what it does, and what it means.

Notes

1. Not all compiled code can be so easily decompiled, but in the case of Flash a program such as the JPEXS Free Flash Decompiler makes decompilation a relatively trivial task. See http://www.free-decompiler.com/flash/.

2. For an overview of the concepts "data intensive" and "process intensive," see Crawford (2003), 89–90.

3. This EULA suggests an alternative way of defining code. Code is that which can be reverse engineered, decompiled, disassembled, displayed, performed, used as the base for derivative works, or otherwise modified.

Works Cited

Aleph Null. 1972. "Computer Recreations." *Software, Practice, & Experience* 2 (1): 93–96. doi:.10.1002/spe.4380020110

Ballard, Ed. 2014. "Want to Stay Anonymous? Don't Make a Hit Computer Game." *Wall Street Journal Blogs—Digits*, March 18. http://blogs.wsj.com/digits/2014/03/18/want-to-stay-anonymous-dont-make-a-hit-computer-game/.

Bogost, Ian. 2011. *How to Do Things with Videogames*. Minneapolis: University of Minnesota Press.

Bogost, Ian. 2006. *Unit Operations: An Approach to Videogame Criticism*. Cambridge, MA: MIT Press.

Carmack, John. 1999. "Id-Software/Quake." GitHub. https://github.com/id-Software/Quake.

Chun, Wendy Hui Kyong. 2008. "On 'Sourcery,' or Code as Fetish." *Configurations* 16 (3): 299–324. doi:.10.1353/con.0.0064

Consalvo, Mia. 2007. *Cheating: Gaining Advantage in Videogames*. Cambridge, MA: MIT Press.

Crawford, Chris. 2003. *Chris Crawford on Game Design*. San Francisco: New Riders.

Dyer-Witheford, Nick, and Greig De Peuter. 2009. *Games of Empire: Global Capitalism and Video Games*. Minneapolis: University of Minnesota Press.

Galloway, Alexander R. 2006. *Gaming: Essays on Algorithmic Culture*. Electronic Mediations no. 18. Minneapolis: University of Minnesota Press.

John, Tracey. 2011. "Misogyny in Code Is Still Misogyny." *Tracey Writes Stuff*, September 8. http://traceyjohn.blogspot.com/2011/09/misogyny-in-code-is-still-misogyny.html.

Kay, Alan. 2007. Email to Don Hopkins. November 10. http://www.facebook.com/topic.php?uid=6321982708&topic=3486.

Kelty, Christopher M. 2008. *Two Bits: The Cultural Significance of Free Software*. Chapel Hill, NC: Duke University Press. http://twobits.net/

Lesser, Mark. 2007. Interview by Scott Stilphen. Digital Press. http://www.digitpress.com/library/interviews/interview_mark_lesser.html.

Levy, Steven. 2010. *Hackers: Heroes of the Computer Revolution*. Sebastopol, CA: O'Reilly Media.

Lowood, Henry. 2009. "Videogames in Computer Space: The Complex History of *Pong*." *IEEE Annals of the History of Computing* 31 (3): 5–19.

Lubar, Steven. 1992. "'Do Not Fold, Spindle, or Mutilate': A Cultural History of the Punch Card." *Journal of American Culture* 15 (4): 43–55.

Manovich, Lev. 2001. *The Language of New Media*. Cambridge, MA: MIT Press.

Marino, Mark C. 2006. "Critical Code Studies." *Electronic Book Review*, December 4. http://www.electronicbookreview.com/thread/electropoetics/codology.

Mastrapa, Gus. 2012. "The Geeks Who Saved *Prince of Persia*'s Source Code from Digital Death." *Wired*, April 20. http://www.wired.com/2012/04/prince-of-persia-source-code/.

Montfort, Nick. 2006. "Combat in Context." *Game Studies* 6 (1). http://gamestudies.org/0601/articles/montfort.

Raley, Rita. 2002. "Interferences: [Net.Writing] and the Practice of Codework." *Electronic Book Review*, September 8. http://www.electronicbookreview.com/thread/electropoetics/net.writing.

Rockstar Games. 2013. "Rockstar Games End User License Agreement." October 1. http://www.rockstargames.com/eula.

Sample, Mark L. 2013. "Criminal Code: Procedural Logic and Rhetorical Excess in Videogames." *Digital Humanities* 7 (1). http://digitalhumanities.org/dhq/vol/7/1/000153/000153.html.

Weststar, Johanna, and Marie-Josée Legault Téluq. 2012. "More Than the Numbers: Independent Analysis of the IGDA 2009 Quality of Life Survey," 8–9. October. http://gameqol.org/wp-content/uploads/2012/12/Quality%20of%20Life%20in%20the%20Videogame%20Industry%20-%202009%20IGDA%20Survey%20Analysis.pdf.

Wolf, Mark J. P. 2008. *The Video Game Explosion : A History from* PONG *to PlayStation and Beyond*. Westport, CT: Greenwood Press.

8 CONSOLE

Raiford Guins

I. console, *n.*

The *Oxford English Dictionary* (*OED Online*) informs its readers that the roots of the term *console* are French from the sixteenth and seventeenth centuries. A console is a masonry and/or architectural "bracket" or "shouldering" element for "bear[ing] up" roofs and support for figures, busts, vessels, and other interior ornaments. In the contemporary context, bracketing used to brace a fireplace mantel would be an example of this type of support. With such definitions, the architectural form "console" serves *to* and *as* support. By the nineteenth century, the supporting element *itself* was regarded as an ornament, a decorative piece in its intended function of bolstering other objects. Our contemporary fireplace bracket not only supports a wooden mantel but complements its interior surroundings. The console is both functional and aesthetic, lending a "supporting atmosphere" to those objects whose weight rests upon a lone console or upon a series of consoles.

The term *console* and the compounds *console cabinet* or *console model* achieved a more familiar usage in the United States in the early to mid–twentieth century when joined with the then new media of the gramophone, radio tuners, television receivers, and later magnetic tape–based consumer technologies such as videotape or videocassette recorders in the late 1970s. As we already know from the history of television, the TV console was never a mere "cabinet" for its electronic medium but was regarded as an object itself, an attractive piece of furniture designed to interplay with existing twentieth-century interiors or as a social and informative object within public spaces (Boddy 2004; McCarthy 2001; Spigel 1992).

The *OED* offers a third definition that adds another semantic layer to the etymology of the term *console*: "a desk, cabinet, or the like, incorporating switches, dials, etc., for the control of electrical or other apparatus; a control panel; a switchboard." In the 1940s, the term denoted not only containment, encasement, or the enclosure of electronics, as in "a cabinet for," but also "the control *of* electrical or other apparatus." A console contains while providing a means to control. In engineering, the console is where "central controllers" (both nonhuman and human, I should add) are grouped into a "console unit," an element within a larger entity. In the 1960s, the word *computer* commingled with *console* to form yet another compound, *computer console*, with the operator of a computer being a "console operator." The console coincides with a computer's (or other machine's) "control panel" as a material means whereby instructions or commands are entered into a system by an assortment of input devices, including a variety of keyboard types. By this definition of *console*, MIT's game *Spacewar!* built on a Digital Equipment Corporations' Programmed Data Processor-1 in 1962 could be regarded as a "computer console *game*," thus further muddying the waters of hard-and-fast distinctions between a computer game and a console game.

Of course, another specialized compound term soon replaced *computer console* in the lexicon of game history. I refer to the common nomenclature *game console* (even though our game consoles are computational devices and have long since ceased merely to afford the playing of games). According to the *OED*'s entry reference from 1976, a "games console" (not the more familiar "game") is "a new variety of electronic game that can be hooked up to the family TV set. Most recent development is a small console with replaceable cartridges to provide a large number of game selections." It is interesting to note that despite the architectural etymology of the term *console* (and the compounds *console cabinet* or *computer console* in particular), the word did not also apply to coin-operated arcade video games. *Arcade cabinet* or simply *cabinet*, not *console cabinet* or *arcade console*, became the preferred term for the large upright wooden structure containing electronic and mechanical parts launched by Nutting Associates, INC., in 1971 with Nolan Bushnell and Ted Dabney's game *Computer Space*. On the original promotional flyer for the coin-op arcade video game, the phrase "BEAUTIFUL SPACE-AGE CABINET" precedes the actual description of the game's objectives. Like *console*, the word *arcade* also coincidentally shares an architectural etymology ("a vaulted place" or "arch"). Such lineage goes unacknowledged in the *OED's* rather slim entry for *arcade game*: "a (mechanical or electronic) game of a type orig. popularized in amusement arcades." The first reference to the term *arcade game* is from 1977, five years after *Pong* (Atari, 1972) altered the face of the "amusement arcade" as well as store fronts.

The requirement that a "games console" "be hooked up to the family TV set" may have helped establish *console* rather than *cabinet* as the dominant descriptor to differentiate between "electronic games" containing televisions *within* their closed cabinet physical structure/interface and "games consoles" connected *to* televisions. Such an emphasis on the precise utilization of a display medium rather than on the physical location of a game device—either within the private sphere of the home, which warrants the title "home console" and "family TV set," or in diverse public settings for arcade video games, such as amusement arcades or operator routes across quotidian spaces—also helps to draw a technological as well as historical differentiation between the two game types. Although both types of electronic games possess cathode ray tube monitors, a "games console" from the 1970s to mid-1980s required a manual radio frequency (RF) switch (a.k.a., TV/GAME switch) joined to the back of a television via thin U-shaped metal connectors to alternate between a "games console" video/audio input and TV broadcast/cable/antenna signals. These little switch boxes—common until televisions started to incorporate additional signal input/output jacks into their physical design—were externally added (often just dangling) to a TV home "receiver," whereas the "arcade game contains its own TV monitor," thus not requiring the manual signal switch (Buchsbaum and Mauro 1979, 5). In this instance, the term *game*, as in "arcade game," refers to a self-contained unit, whereas the phrase "games console" treats the word *games* as a prefix to reference a specific type of console device.

Another reason the etymologies of the terms *console* and *cabinet* are seldom paired may be related to the existence of separate industries: coin-operated amusement manufacturers and television manufacturers with "console product" divisions. To illustrate, a games company such as Atari (even when incorporated into Warner Communications as Atari, Inc.) maintained two product lines: a coin-operated division and consumer electronics video games division. Within the company's description for the Securities and Exchange Commission (July 8, 1976) such a distinction is presented as follows: "Coin-operated games are self-contained units that Atari sells to independent distributors, who, for the most part, in turn sell the games to operators for placement in amusement parlors, arcades, taverns, and other public places. Consumer games are designed to be used in conjunction with home television sets and are sold by Atari, either directly or through sales representatives, to retail merchants" (Atari Inc. 1976). These descriptions demonstrate, at least according to Atari's model, that "self-contained units" are markedly different entities from those designed to work "in conjunction with home televisions sets," as previously noted (Buchsbaum and Mauro 1979, 5). Equally, the market of "independent distributors" and "operators" rather than "retail merchants" for consumer games is also a key determinant of differences between a games console and an arcade game.

It is also noteworthy that the *OED*'s first reference in the definition of *games console* is from 1976. The year 1976 marks the release of the first interchangeable read-only memory (ROM) cartridge-based, programmable "games console," the Fairchild Channel F. The inclusion of the line "[the] *most recent development* is a small console with replaceable cartridges to provide a large number of game selections" (my italics) registers but does not account for previous candidates for the title *games console*. To what did we refer all of those "ball and paddle games" or "*Pong*-in-a-chip" devices adorning North American coffee tables throughout the mid-1970s? Such devices elude the *OED*'s definition of *console*. The development of replaceable cartridges may have proved a catalyst for the *OED*'s 1976 entry for *games console*— four years after the release of the Magnavox Odyssey, which is recorded as "the world's first commercial home video game console" in the *Wikipedia* entry for this term ("Video Game Console" n.d.). And it is now clear why the *OED* opted for the term *games console* rather than *game console* given a console's ability to play multiple games. In the *OED's* next reference in its entry for *games console* from 1983, the terms *units*, *system*, and *consoles* all appear in the same single sentence, thus adding further confusion to the term's meaning: "The basic Atari system may be the strongest-selling units, but all buyers are looking for exceptionally strong growth from more sophisticated games and consoles." How have these three terms been used in the history of games?

II. System versus Console

Since I have already mentioned the Fairchild Channel F, it is important to share its full title, found on its packaging and instructions manual: "The Fairchild Video Entertainment *System*" (my italics). The game console (figure 8.1) is one component part among others. But what about noncartridge-based "games"? How were they described to potential consumers given that their playable games were hardwired? The Magnavox Odyssey entered the market in 1972 as an "electronic game center," and the device at the center was announced as the "master control unit"—the "heart" of the Odyssey (figure 8.2). Odyssey inventor Ralph H. Baer's US patent application for Sanders Associates Inc. included registered titles like "Television Gaming Apparatus and Method" (1972), "Television Game and Training Apparatus" (1973), and "Preprogrammed Television Gaming System" (1975). The various devices running ping-pong/

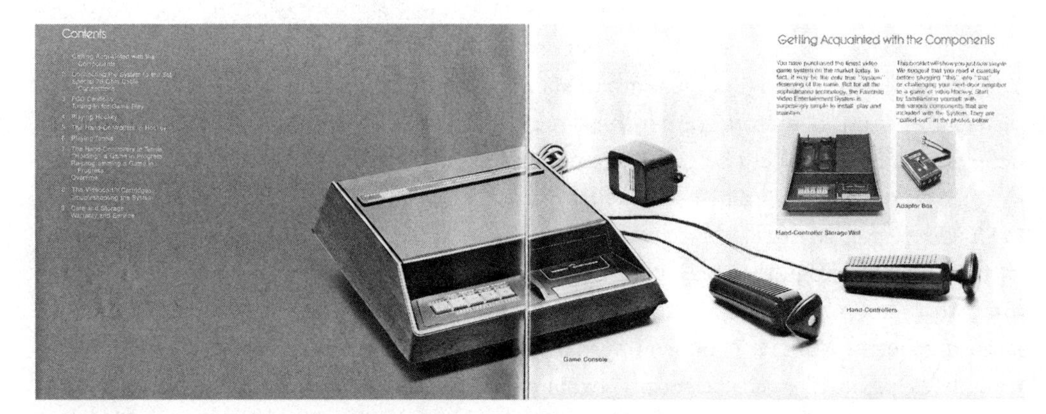

Figure 8.1
Instruction manual from the Fairchild Video Entertainment System or Channel F, showing its "Game Console" component.

INSTALLATION

GETTING TO KNOW YOUR ODYSSEY

Your master control unit . . . is the heart of Odyssey. It contains all the electronics necessary to play the many games provided. It performs much like a television station sending signals to your television receiver.

Batteries . . . are used to power your Odyssey. Remove the lid on the battery compartment by sliding the fasteners in the direction the arrows point. Place the six C cell batteries (included with your Odyssey) in the battery holder, as shown.

A channel switch . . . is provided inside the battery compartment to select either Channel 3 or Channel 4. This switch is normally set at the factory to the Channel 3 position. If a television station is operating on Channel 3 in your area, move the switch to the Channel 4 position. For future reference, indicate here the channel being used for Odyssey: ____

The external power jack . . . may be used to connect an optional 9-volt external power supply adapter to your unit instead of using batteries. This adapter may be purchased at your local dealer.

CONNECTING THE PLAYER CONTROL CABLES AND THE GAME CORD

Plug the Player Control unit cables into the Master Control Unit as illustrated. The two Player Control Units are identical; however, the Player Control Unit connected to the PLAYER 1 socket will be called Player Control Unit Number 1 and the one connected to the PLAYER 2 socket will be called Player Control Unit Number 2.

The GAME CORD is used to connect the Master Control Unit to the ANTENNA-GAME SWITCH. Insert one end of the GAME CORD into the socket marked GAME CORD on the Master Control Unit. The other end plugs into the socket on the top of the ANTENNA-GAME SWITCH marked GAME CORD.

4

Figure 8.2
Instruction manual from the Magnavox Odyssey illustrating the installation of its "master control unit."

tennis simulations for the "home TV" were promoted via the descriptors *video sports games* and *home video game*. The term *game*, too, was used not just to refer to the game or games hardwired within a device *but to the device itself*. The stickiest name of all was "TV-Games" given the aforementioned definitive (and physical) connection between gaming device and television. "The History of Video Games" chapter in *TV Games—Just The Beginning* (1977) by Steve D. Bristow, Atari's vice president of engineering, distinguishes the "consumer game" (or "consumer version") from the "coin-operated game," with further reference to "TV game," and "PONG game" is given as a specific example of this game type. In fact, it was not uncommon for companies to promote their consumer game devices via brand names, as in the Coleco Telstar line and Magnavox's numeric series, or based on the actual games offered on a particular device, such as Atari's *Video Pinball* and *Stunt Cycle* (both 1977) and Telstar's *Combat* (1977).

Such names can be found in William J. Hawkin's article "TV Games Turn Your Set into a Sports Arena," published in *Popular Science* in 1976. Hawkin wrote, "Tennis, in one form or another, is basic to all *units*" (88). He employed the term *unit* or *control unit* (as in the *OED*'s *console unit*) when referring to hardwired, dedicated *Pong* devices. Walter H. Buchsbaum and Robert Mauro utilize similar descriptors in their diagram for a "basic TV game," while referring to pong-based devices upon which "master controls" allow players to "select the type of game we want to play and choose its key parameters" (1979, 6). Hawkin concludes by briefly mentioning the soon-to-be-released Fairchild Channel F: "Ordinarily, *the system* plays either hockey or tennis. But you can change games by inserting into *the game console* a special cartridge that Fairchild calls Videocart" (1976, 90).

Can we surmise that *games console* emerged as a preferred term with the rise of interchangeable ROM programs? The first programmable system to utilize ROM cartridges did launch with the term *game console* (singular, unlike the *OED*'s preference for the plural form *games console*) included in its instructional documents. Len Buckwalter (1977) also favors the term *game console* when tackling the question "What is a video game?" "The heart of most videogames," he writes, "is a small console placed a comfortable viewing distance from the TV set." And at the time of publication, only the Channel F and RCA Studio II offered a console, unlike "the pongs" that occupy many pages within his book. However, clear distinctions between hardwired and programmable devices are not as straightforward as Buckwalter or the instruction manuals would lead us to believe.

I already mentioned the full name of the Channel F and should also point out that it was not alone, neither then nor now, in its preference for positioning the game console within a larger system for its product. This chapter does not permit a comprehensive list of US home video game releases from the late 1970s to present. Nonetheless, I must offer a fairly extensive list of official product names (left column) and accompanying descriptions for their "consoles" (right column) as identified by the product's manual, to demonstrate how manufacturers delineated their product lines and the specific device in question.

Atari Video Computer **System** (1977)	Console Unit
Bally Professional Arcade Expandable Computer **System** (1978)	Arcade Console
Mattel Intellivision (1979)	Master Component
Emerson Arcadia 2001 Micro Computer **System** Video Game (1982)	Main Unit
Colecovision Arcade Quality Video Game **System** (1982)	Console Unit
Vectrex Arcade **System** (1982)	Arcade System Console
Atari 5200 Advanced Video Entertainment Super **System** (1982)	Console Unit
Nintendo Entertainment **System** (1985)	Control Deck
Sega Master **System** (1985)	Power Base

Atari 7800 Video Game **System** (1986)	Console
Atari XE Game **System** (1987)	Console
NEC Turbo Grafx 16 Entertainment Super**System** (1989)	Unit
Sega Genesis Video Entertainment **System** (1989)	Console
Neo Geo Advanced Entertainment **System** (1990)	Hardware
Super Nintendo Entertainment **System** (1991)	Control Deck
Panasonic Goldstar 3DO Interactive Multiplayer **System** (1993)	System
Sega Saturn (1994)	Unit
Sony PlayStation Video Game Console (1994)	Console
Nintendo 64 The Fun Machine (1996)	Control Deck
Sega Dreamcast The Ultimate Gaming **System** (1999)	Dreamcast
Sony PlayStation 2 Computer Entertainment **System** (2000)	Console
Microsoft Xbox Video Game **System** (2001)	Console
Nintendo GameCube (2001)	GameCube
Microsoft Xbox 360 Arcade **System** (2005)	Console
Nintendo Wii (2006)	Console
Sony PlayStation 3 (2006)	System
Sony PlayStation 4 (2013)	System
Xbox One (2013)	Console

The term *system* (given in bold text here) predominates in the names of product lines. *Console*, too, is a widely used term, though not unanimous in that *unit*, *deck*, and even in certain cases *system* are used as well (in the product names, the term *console* was used only for the Sony PlayStation Video Game Console in 1994).

In an attempt to pragmatically distinguish the terms *console* and *system*, it is permissible to argue that *system* refers not to a discreet object but to an aggregation of interdependent things: a network of intermingling social practices and technological processes as well as actors necessary for powering, running, and playing the "games console." The "games console," after all, does little without an "RF switch," "UHF and VHF," "Channel 3 or 4," "F-Type Plug," "AC Adaptor Power Supply," "AV Cable," "S-Video," "HD AV Cable," "TV inputs," "controllers," "Memory Sticks and Cards," "hard drives," or "playable storage media," not to mention an "outlet" for electrical power. This list does not even begin to consider today's HDMI-compatible televisions, cable/satellite boxes, USB ports, modems, routers, Ethernet connection, Bluetooth, social media, and assortment of apps. Nor does it begin to consider the human actors working within this chain of nonhuman actors that are both part of a global system of raw materials, manufacturing, distribution, government standards for electrical devices and patent protection, and the inevitable afterlives of hardware and software as obsolete technologies, ewaste, and possible museum/archive acquisitions.

Such diverse yet interactively stabilized elements characterize the scale of the game system, while the "games console" connects to this system as a "centralizing" "control panel" or "control unit"—as in the *OED*'s mid-twentieth-century definitions—not only for issuing instructions to the system related to gameplay but also, in the era of "new generation" games consoles, for online communication, access, and multiple delivery-platform media. The descriptor used for the Xbox One attempts to capture this process: "The All-in-One Entertainment System." "All-in-one" and even "box," "cube," or "station" imply that a "console," in its post-seventeenth-century definition, is imagined less as a "bracket" and more as a "console cabinet." Its specific role (and meaning) is dependent on while enabled by the various technical objects and heterogeneous networks upon which it relies and interfaces/instructs for gameplay. The console, in this context at least, does not so much materialize a system as provide a material means to access that system.

III. Games Console as Control Panel *and* Cabinet

What of the physical object that once rested on a coffee table (or carpet) and now is the centerpiece in one's "entertainment center" or "media center"—that industrial-designed thing taking an intended form, preinscribed with user expectations and functions, expressive of a design concept while embodying social and cultural attitudes about technology? Does the distinction of a "control unit" enabling access to a system render the games console less an object and more an action or processual means? Do we demonstrate a particular bias for researching the games console only as a control panel or control unit to the detriment of its other etymology, console cabinet? That is, are we apt to focus on the game console only as a complex of technical processes rather than also as an intentionally designed thing, an artifact?

Such questions are not necessarily hypothetical. Now that collections of video games—arcade, console, personal computer, handheld device, storage media, and so on—are commonly displayed in museums, we often find ourselves peering at inert electronics encased in plastic. We seldom see what is "under the hood" or experience the game software that a games console was designed to run. We instead observe the persistent presence of wood veneer on games consoles produced in the 1970s with the aim to blend in with existing interiors. We also note that Nintendo released its GameCube in a variety of colors to tempt a collector's culture weaned on *Nintendo Power* and *Pokemon* edition Nintendo 64s. Microsoft would follow suit with its *Halo* edition consoles. In a broader context, we witness how more recent consoles—the Nintendo Wii—have adopted a minimalist, clean, and thin form to match the shrinking scale of consumer home entertainment devices, and our Xbox 360s and PlayStation 3s now stand vertically upright in contrast to their lumbering horizontal predecessors. The disassembly of a games console reveals a complex system of many black boxes within the bigger one—for example, "inside" the console of the Microsoft Xbox 360 Arcade System we find the assembly of its central processing unit, graphics processing unit, DVD-ROM drive, RF unit, hard disk-drive unit, the motherboard, circuitry, X-clamps, fans, and chips, just to name a few elements in Microsoft's Pandora's box. Museums are good at displaying games consoles so that we can examine their external features, but they are not so good at displaying their internal features, let alone the larger (social, technological, material) system to which they connect.

Perhaps if we reconnect with previous definitions of the term *console*, we might begin to offer additional strategies for such curatorial challenges while broadening the histories we bring to the critical study of games. Not long ago, the definition for *console* described both the material container—"a desk, cabinet, or the like, incorporating switches, dials, etc."—as well as the object's intentions "for the control of electrical or other apparatus; a control panel; a switchboard." This dual meaning delineates the game console both as a "cabinet," a designed object, a material container that encloses assembled technical component parts for running game software *and* as a "control unit," a nodal machine, a system itself, for accessing and engaging with another complex system for playing a video game (among other actions today). Perhaps even the earliest recorded definitions of *console* still possesses value: if the term *cabinet* sounds too much like a vessel or container, maybe the sixteenth- and seventeenth-century definition of *bracket* has something to offer our modern understanding of *game console*: a supporting structure (to interact with a system) and an aesthetic feature (with a nonneutral and historically changing physical and technological design). Xbox One: The All-in-One Entertainment Bracket?

Works Cited

Atari, Inc. 1976. Form S-1 Registration Statement under the Securities Act of 1933. for the Securities and Exchange Commission, July 8, Washington, DC.

Boddy, William. 2004. *New Media and Popular Imagination: Launching Radio, Television, and Digital Media in the United States.* Oxford: Oxford University Press.

Bristow. Steve D. 1977. "The History of Video Games." Paper presented at "TV Games—Just the Beginning," Electro 77 Professional Program, New York, April 19–22.

Buchsbaum, Walter H., and Robert Mauro. 1979. *Electronic Games: Design, Programming, Troubleshooting.* New York: McGraw-Hill.

Buckwalter, Len. 1977. *Videogames.* New York: Grossett and Dunlap.

Hawkin, William J. 1976. “TV Games Turn Your Set into a Sports Arena.” *Popular Science* 209 (5): 88–91.

McCarthy, Anna. 2001. *Ambient Television: Visual Culture and Public Space*. Chapel Hill, NC: Duke University Press.

Spigel, Lynn. 1992. *Make Room for TV: Television and the Family Ideal in Postwar America*. Chicago: University of Chicago Press.

“Video Game Console.” n.d. *Wikipedia*. https://en.wikipedia.org/wiki/Video_game_console. Accessed June 30, 2015.

9 CONTROL

Peter Krapp

Who or what, in the setting we describe as gaming, controls whom or what? When game studies seek to describe how technical and other protocols structure gaming actions, a focus on "control" needs to look beyond the input device as such—beyond the history and the affordances of controllers. To control means to surveil something or someone, suggesting a means of domination or power, but it also means to exercise restraint, to regulate, to curb the spread of something. As the etymology of the word *control* suggests, it is also a managerial activity—you control, check, manage, test, and verify processes. All three registers—a hands-on manipulation, a selective process, and managerial oversight—are pivotal in gaming, and this chapter seeks to unfold them successively. To the extent that games are an adaptive response to the omnipresence of computing (teaching us how to interface with our gadgets), it could be shown that the affordances and strictures of our devices control what we see as possible with them. Put another way, a closer look at gaming controls needs to go beyond the twitchy feedback of input devices and include a sense of the historical and technical conditions of possibility of the gaming setup that we are familiar with today.

That is not to say the actual control input devices are not an important aspect of debugging the history of gaming. Whether one sees the graphic user interface (GUI) as a Taylorist discipline teaching ergonomic interaction or as a yielding to reductions of interaction to resemble what users had already learned, the GUI is pivotal for our culture. And the discursive formation of computer games runs parallel to that of the GUI—interaction revolves around perception, hand–eye coordination, and discerning errors. Input devices and operating routines must be tightly controlled, or they appear faulty. If my clicking or pecking can yield unpredictable results or create the (correct or erroneous) impression of a variable and perhaps even uncontrollable situation, then a fundamental principle of interaction is at stake: a peculiar blasphemy against widely held beliefs about designing human–computer interaction. Observers of computing culture and human–computer interface (HCI) design have joked that if HCI experts were to try their hand at game development, the result would be a big red button with the label "press here to win." Inversely, if my user interface were chaotic and irregular (regardless whether it is a browser window or a game, a buggy beta version or an intentionally artsy deconstruction of a browser or a game), then I would be highly unlikely to accord any value to accidental elements. Whereas a typical GUI—from the desktop metaphors of Vannevar Bush's Memex to the Xerox Star or Apple Lisa and beyond—aims for visibility and patient acceptance or even anticipation of user error, games probe the twitchy limits of reaction times and punish user error with loss of symbolic energy. Audiovisual cues pull users into feedback loops that a game might exhibit proudly, and other, perhaps less playful user interfaces will tend to hide them in redundant metaphor.

Few people can claim to remember playing with the original *Tennis for Two* (Higinbotham, 1958) button, the original *Spacewar!* (Russell/MIT, 1962) levers, the knobs of the Coleco Telstar or the Wonder Wizard. And not too many recall their use of the Odyssey Shooting Gallery rifle, the Nintendo Entertainment System (NES) Zapper pistol, the Sega LightPhaser, or the Atari XG-1 and Sega Saturn pistols, though the Namco GunCon3 and the more recent PlayStation Move Sharpshooter are perhaps more readily recognized. Controllers such as the NES PowerGlove and the Sony EyeToy have only recently been rehabilitated as progenitors of what the Xbox Kinect and the PlayStation Move now offer. Music and rhythm game controllers hark back to the plastic maracas for the Sega Dreamcast game *Samba de Amigo* (Sonic Team 1999), not to mention GameCube bongos or the *Guitar Hero* (RedOctane and Harmonix, 2005) and *RockBand* (Harmonix, 2008) instrument simulacra. So of course there are musical instruments, guns, racing wheels, and trackballs, but for the most part control devices that offer alternatives to the generic keyboard have congealed around the joystick and button design. In general, consoles progress, if that is the right term, from a simple stick-and-button mode, as seen in the Tandy or Atari 2600 controllers, to the addition of a number pad in the ColecoVision and Atari 5200 controllers and to the introduction of options on the NES, Genesis, Super NES, Sega CD, Nintendo 64, and Dreamcast controllers, some of which also feature shoulder buttons. The RCA Studio II, which came out before the Atari VCS, used only a number pad; the Emerson Arcadia 2001 had a removable stick housed on its dial located below its membrane numeric pad. Starting with the Nintendo 64 controller, console remotes usually feature sticks, and ever since the NES controller most console input devices are organized around a directional pad—a flat and usually four-directional control with a button on each point, known as a "d-pad" and now found on most if not all console and other game controllers as well as on TV remote controls and mobile phones (which are of course increasingly game platforms). A precursor to the d-pad could be seen on the Intellivision by Mattel in 1980, but Nintendo introduced the now common cross design for the handheld *Donkey Kong* game in 1982. The PlayStation 2, GameCube, Xbox, and WiiMote controllers inherited this assortment of buttons, d-pad, sticks, and shoulders. Many personal computer (PC) gamers continue to live with the strictures of the QWERTY keyboard and the venerable WIMP (windows, icons, menus, pointer) setup handed down from decades ago, while others now use console-style gamepads with their PCs.

But beyond the simple consolations of timelines and industry histories, debugging game history requires us to outline conceptual dimensions that underpin the interface culture of gaming today, so we move from controller to control more broadly. As Claus Pias (2002) documents, one may recognize three clusters of technology, rhetoric, and psychological register in what gaming weaves together into its setup, and, of course, different games foreground these clusters to varying degrees. There are differences between controlling a video game race car, controlling an avatar in role-play mode, and controlling the tax rate in a simulation game. Action games rely most obviously on speedy reactions at the GUI, thus providing immersive metaphors; on the psychological level, they tend to emphasize thrills and transgressions. Adventure and role-play games rely less on speed than on decisions and explorations, thus emphasizing not so much the graphics as the database and pivoting less on metaphor than on metonymy. Strategy games, in turn, foreground planning and constellations, emphasizing rules and consequences in a logic akin to allegory, and fostering not so much the rush of the id or the role-play of the ego but the rule-based play of the superego. Thus to abbreviate Pias (2002), control in games that emphasize graphics, immersion, and speed will hinge on the input device, but control in role play will center on interaction design and choice, while control in strategy or god games and simulations will lean more toward the overview and planning perspective. What all three registers share, however, are deep roots in traditions and technologies that go back before the history of computer games.

Modern information technologies took shape more than a century ago, as James Beniger (1996) details. What in the mechanical era was effected by a crank or a lever becomes effected by a mere trigger in the computer era: the desired effect is already stored in the apparatus, hidden in its complication so as to suggest effortless use (of the camera, the ATM, or the car, for example). If we trace the interface design of computer games back to the formation of the three associative clusters that Pias suggests, the GUI clearly owes a great deal to Frank Bunker Gilbreth and Lillian Moller's (1917) applied-motion studies, which sought to transform the conditions of labor into the foundations of scientific management with the aid of chronocyclography and film. The database, of course, has its roots in library and business card catalogs (Krajewski 2011). Strategy games have a much-documented history in military planning and policy consulting (von Hilgers 2012; Crogan 2011). The maps and calculation tables blackboxed into the complex scenarios of strategy games as well as the decision trees and associative links that go into databases of items, paths, and characters for role-play games are of course rarely encountered in isolation, yoked as they usually are to a GUI that tends to gloss over its roots in the radar screens of World War II (Pias 2011). If having simple controls at the tip of your finger dissolves the close relation between cause and effect, it can also remove the finger from responsibility in direct correlation to the obscuring of the informational process. The more user interfaces trivialize information technology and relegate users to button pushers, the more digital culture threatens to resemble the seductive yet dysfunctional placebo of the close-the-door button in elevators, placed there just to give you the erroneous but flattering impression of contributing to the speed of your trip. Computers are complex enough that we have come to accept that their inner workings are hidden behind interfaces that make them "usable." In gaming, this technical provision sets the scene for the feedback loops of input devices, screen events, and the player's conditioning to repeat or progress. Media history can help us understand both the conditions of possibility and the consequences of the "user-interface" culture.

Following the iterative conditioning of the user by WIMP metaphors, we need to interrogate the surfaces and other control metaphors of the GUI in gaming. Doug Engelbart's groundbreaking demonstration at the Fall Joint Computer Conference in San Francisco on a Monday afternoon in December 1968 was followed by access to remote terminals to the SRI computers in Menlo Park (30 miles away) "in a special room set aside for that purpose." It became enshrined as "the mother of all demos" because it influentially showcased graphics in windows, navigation and voice input, video conferencing, the computer mouse, word processing, and online collaboration. Once it became evident that the WIMP model had taken over almost completely, some experts started deploring the lack of innovation in interfaces, calling for an HCI based on language, richer representation of objects, expert users, and shared control (Gentner and Nielsen 1996). Arguing that the WIMP interface was for naive users, the expert and "anti-Mac" user interface was supposed to supplant metaphors with reality, seeing and pointing with describing and commanding, WYSIWIG with representations of meaning; offer an overdue departure from the stale icons of file, desktop, trash can, and so on; and institute instead an interface for the "post-Nintendo generation" that might expect to play in a richer environment. Playful critiques of the control afforded by tired metaphors inherited from Vannevar Bush's Memex include Dennis Chao's (2001) modification of *DOOM* (id Software, 1993) for Unix process management, and Chris Pruett's (2000) Marathon Process Management using the game engine by Bungie (1994).

Furthermore, game designers and game critics alike have been calling for increased literacy about command-and-control structures that, like learning to swim or to read, enable us to move past fear and isolation into active engagement with the unknown and the incalculable. Key to what Alexander Galloway (2004) calls "countergaming" would be the quest to create alternative algorithms; as he diagnosed, many so-called serious games tend to be quite reactionary at the level of their controls, even if they cloak themselves in progressive political desires on the surface. Critical gaming would tend to move away from the infantilizing attraction of play as obedience to rules and toward a "progressive algorithm"—if there can be such a thing. In a primary mode, it would teach its players the lessons of the game's controls; in a secondary mode, it would build on the discovery and exploration of the game's regularities and irregularities and invite creative play with the game. This difference between playing a game and playing *with* a game is crucial to gaming culture: whereas the former teaches one the game through navigating the game's commands and controls, the latter opens one up to critical and self-aware exploration. Like learning to swim, gaming means learning to learn and initiates us into training systems; yet like learning to read, it does not stop at a literal obedience to the letter. Computer games tend to obey a certain set of design rules, chief among them being efficiency, expediency, and mastery of interfaces and interactions—because we value in such games that they "work."

This technologic embedded in control devices can also be said to install their program in the user. This means that the more gaming withdraws access to the hardware and software behind ever more "superficial" yet "intuitive" interfaces, the less actual control the users have, despite and because of the illusion of being in control of their devices—the Wii that can never be fully turned off, the Xbox 360 that won't accept older Xbox games, the mobile phone that reports your location and consumption to the publishers and developers of the games loaded on your device. Thus, control also has a social dimension. Where Gilles Deleuze (1992), along with Michel Foucault (2009), points to a historical shift from sovereignty to discipline and then on to what he calls control, he sees power located in an increasingly decentralized way, no longer embodied or built into social structures but anonymously and autonomously regulating and manipulating the flow of all social exchanges. In the medieval order of sovereign power, rulers taxed and threatened rather than disciplined and organized production. In Foucault's conception of the disciplinary societies that arose later, individuals used to pass from the family to school and to training barracks and factories, from one institutionalized space of enclosure to another, including of course the hospital and the prison. After World War II, environments of enclosure and interiority entered into crisis mode, marked by endless waves of reform—school reform, military reform, prison reform, health-care reform, and so forth—while new free-floating forces of control emerged: pharmaceutical production, genetic manipulation, information processing and aggregation, as the corporation replaced the factory, continuous education replaced school, ankle bracelets replaced the prison, and perpetual control replaced the rites of passage that used to separate the various spaces of enclosure. As Deleuze comments, "Control is not discipline. In making freeways, for example, you don't enclose people but instead multiply the means of control" (1998, 18). Galloway and Eugene Thacker seek to translate this logic into the realm of technoculture, where protocol is "a totalizing control apparatus that guides both the technical and political formation of computer networks, biological systems, and other media" (2004, 8). Surely game culture today fully participates in the same paradigm shift, including but not limited to the phenomenon known as gamification, whereby the masses are being enticed into quasi-playful digital labor that is often little more than a new kind of crowd control with an ever more invisible hand. Along the same lines, multiplayer games see a great deal of crowd control, channeling what is and is not possible in games by dint of constraints that are less technical than political and economical. "As governments increase their control, they replicate their vices on the Internet" (Goldsmith and Wu 2006, viii)—and surely much the same goes for the multinational providers of gaming software and hardware.

To conclude, a cursory account of control in gaming can profile three strands, each of which brings its own historical depth to bear. One is the heritage of ergonomics and usability testing since Gilbreth, where the image source is the reality of the workplace, the tasks are mechanical in a setup regulating energy, the motivation is physical production, and the peculiarity of the interface is that reality is treated as a surface. Control in this sense is reduction of complexity and of fatigue, harnessing repetition. Second, in the WIMP era, the image source is not reality but a flattened representation, the tasks are textual in a setup regulating information, the motivation is mental production, and the image is symbolic screen action. Control in this sense is choice, testing options. Finally, in the current gaming register, the image source is a stale metaphor (e.g., pretend this pixel is a monster; pretend this is a button; this tree means forest; this tank means army), the task is selection in a setup regulating errors, the motivation is consumption, and the image simulates a surface. Control in this sense is the regulation of scenarios, management. Under these progressively flattened conditions, access to and interaction with gaming interfaces makes human control only one facet of a tightly constrained and, indeed, controlled environment that sometimes threatens to reduce the human in the loop from a true controller to a mere "ebkac"—an error between keyboard and chair.

Works Cited

Beniger, James. 1996. *The Control Revolution: Technological and Economic Origins of the Information Society*. Cambridge, MA: Harvard University Press.

Chao, Dennis. 2001. "*Doom* as an Interface for Process Management." *SIGCHI* 3 (1): 152–157. http://cs.unm.edu/~dlchao/flake/doom/chi/chi.html.

Crogan, Patrick. 2011. *Gameplay Mode: War, Simulation, and Technoculture*. Minneapolis: University of Minnesota Press.

Deleuze, Gilles. 1992. "Postscript on Control Societies." *October* 59 (Winter): 3–7.

Deleuze, Gilles. 1998. "Having an Idea in Cinema." In *Deleuze and Guattari: New Mappings in Politics, Philosophy and Culture*, ed. Eleanor Kaufman and Kevin Jon Heller, 14–19. Minneapolis: University of Minnesota Press.

Foucault, Michel. 2009. "Alternatives to the Prison: Dissemination or Decline of Social Control?" *Theory, Culture, & Society* 26 (6): 12–24.

Galloway, Alexander. 2004. *Protocol: How Control Exists after Decentralization*. Cambridge, MA: MIT Press.

Galloway, Alexander, and Eugene Thacker. 2004. "Protocol, Control, and Networks." *Grey Room* 17 (Fall): 6–29.

Gentner, Don, and Jakob Nielsen. 1996. "The Anti-Mac Interface." *Communications of the ACM* 39 (8): 70–82.

Gilbreth, Frank Bunker, and Lillian Moller. 1917. *Applied Motion Study*. New York: Sturgis & Walton.

Goldsmith, Jack, and Tim Wu. 2006. *Who Controls the Internet? Illusions of a Borderless World*. Oxford: Oxford University Press.

Krajewski, Markus. 2011. *Paper Machines: About Cards and Catalogs 1548–1929*. Cambridge, MA: MIT Press.

Pias, Claus. 2011. "The Game Player's Duty. The User as the Gestalt of the Ports." In *Media Archaeology: Approaches, Applications, and Implications*, ed. Erkki Huhtamo and Jussi Parikka, 164–183. Berkeley: University of California Press.

Pias, Claus. 2002. *Computer Spiel Welten*. Berlin: Diaphanes. Translated as *Computer Game Worlds*. Amsterdam: Amsterdam University Press, 2015.

Pruett, Chris. 2000. Projects web page. http://www.dreamdawn.com/chris/projects.html.

Von Hilgers, Philip. 2012. *War Games: A History of War on Paper*. Cambridge, MA: MIT Press.

10 CONTROLLER

Steven E. Jones

The word *controller*, like *typewriter* and *computer*, once referred to a person doing something and only later came to refer to a device. According to the *Oxford English Dictionary* (*OED Online*, 2015), as early as the Victorian period the term was used for a nautical mechanism for regulating or controlling the speed of a chain cable as it ran through its channel, and shortly thereafter the word showed up by analogy in electrical engineering to refer, for example, to devices with handles and switches for regulating the flow of current in a motor.[1] The term *controller* increasingly came to refer to the device and not to the human operator, although it continued to imply such an operator at the controls (despite the development of various automatic regulation and control systems). The design of controller devices, from factory dials to airplane sticks and steering wheels, became in the twentieth century a common focus of ergonomics and human-factors engineering, focused on the place where human meets machine in the most direct way, where the machinery is literally within the operator's grasp. By midcentury, the new field of cybernetics investigated control and communication systems involving humans and machines, with an emphasis on computers.

The *OED* records the earliest use of the word *controller* to refer specifically to a video game controller in an advertisement for a home console system with "individual game controllers" in 1976, but variants of the term were used to refer to games years before that, even in discussions of the earliest computer games. In a 1972 article for *Rolling Stone* magazine, Stewart Brand described players of Steve Russell's game *Spacewar!* (1962), considered by many to be the first computer game, as "numbing their fingers in frenzied mashing of control buttons" and said that the game "consists of two humans, two sets of control buttons or joysticks, one TV-like display and one computer" ([1972] 1974, 39–40). In the article, Brand also quoted Alan Kay's suggestion that the game controls can be built by going to "a radio control store—like for model airplanes—and get the front end of the controller, which has two sets of joysticks," thus linking earlier kinds of flight controllers to game controllers (83). Brand was writing in 1972 about players at Stanford to tell the story of the creation of *Spacewar!* at MIT a decade earlier, in 1961–1962. Although there's no direct evidence that those first players of *Spacewar!* referred explicitly to controllers, it seems likely, given that the word was already common in computing and engineering. The MIT students played on a Digital Equipment Corporation Programmed Data Processor-1 computer with software loaded from paper tapes, viewed the game on the circular oscilloscope screen, and controlled it at first using the computer's existing vertical console of toggle switches, mapped to effects in the game, like the later mapping of computer keys (such as the WASD keys for navigation) in personal computer (PC) games. Like tiny digital joysticks, the four toggle switches could be flicked to produce clockwise movement, counterclockwise movement, acceleration, and the firing of torpedoes, tiny dots of light representing rockets and torpedoes against a field of stars. But when using the toggle switches became too uncomfortable physically, the student hackers of MIT's Tech Model Railroad Club (TMRC) built two of the first dedicated controllers as input devices for a computer game: "So one day Kotok and Saunders went over to the TMRC clubroom and found parts for what would become the first computer joysticks. Constructed totally with parts lying around the clubroom and thrown together in an hour of inspired construction, the control boxes were made of wood, with Masonite tops. They had switches for rotation and thrust, as well as a button for hyperspace" (Levy 2001, 64). Buttons, switches, joysticks—the basic elements of most future video game controllers were already present in this first cobbled-together example.

In the arcade era that followed, games had built-in controllers, occasionally extended from the main cabinet by a connecting wire, many probably based on the design of earlier pinball cabinets, with their physical plungers (based in turn on billiard cue sticks) and flippers that foreshadowed the joysticks, buttons, and paddles of arcade video games (Kent 2001, 2–3). On the control panel of the arcade version of Atari's game *Pong* (1972) and of its many knock-offs, players twisted knobs to maneuver on-screen "paddles." The same scheme was used in the home version of *Pong* by Atari (1975), with two knobs on the top of the controller, as it had been in the earlier *Table Tennis* and related home sports games for the Magnavox Odyssey (1972), a system with two peripheral "player controls," boxes with two knobs each, set on the sides, as well as a "master control module."[2] Paddle controllers of this kind—a rotating knob beside a firing button—first appeared in another candidate for the first computer game, William Higinbotham's *Tennis for Two* (1958). By the early 1970s, there was a wide array of controllers for arcade games—often based mimetically on the game's theme: light guns for shooting games, steering wheels for driving games, and so on—and many of these controllers were re-created in home console systems. Controller devices are as old as computer games themselves, predating the advent of home video game consoles and their peripherals. But a larger segment of the population probably came to associate the controller with gaming through the widespread adoption of television-based home systems. Devices for these systems took different forms and varied over time: joysticks, single or dual, and knobs and buttons in various combinations (including joysticks with built-in buttons on their ends); circular or rectangular touch pads and, later, directional pads; midway- and arcade-inspired guns, even for home consoles; plastic models of sports equipment; and steering wheels for driving games. There were also numeric keypad controllers (ColecoVision, Intellivision, RCA Studio II) as well as pressure-pad membrane keyboards (Magnavox Odyssey 2).

The diversity of input devices—including gloves with built-in sensors, wired pads or platforms to be controlled with your feet or body weight, as well as devices controlled by fingers and thumbs—continued to develop into the 1980s. But once the joystick controller was introduced in 1977 in the popular Atari Video Computer System (VCS, or Atari 2600), it quickly became iconic, a kind of visual synecdoche representing video games in general in the popular imagination (Bogost and Montfort 2009, 14). Then, with the advent of Nintendo's Famicom system (or Nintendo Entertainment System [NES] in North America, 1983), the small, thumb-controlled, cross-shaped directional pad (d-pad) on the gamepad controller became a standard feature of a number of home consoles and handheld consoles. The Famicom's d-pad was based on an earlier controller built into Nintendo's handheld Game & Watch LCD games (from 1980), and a dual d-pad controller would be used with the Virtual Boy three-dimensional (3D) gaming system designed by the same Nintendo developer, Gunpei Yokoi, in 1995. The ability to rock the four-way digital switch, with the left thumb to navigate, in combination with other buttons on the horizontally aligned rectangular pad—in the Famicom/NES, A and B buttons for the right thumb plus Start and Select buttons in the middle—helped establish button arrangements for a number of controllers, which codeveloped with the 2D platformer games and overhead-view role-playing games of the era. The Famicom/NES controller with its d-pad soon replaced the joystick as the synecdoche for video gaming and with the combinatorial possibilities of its button combinations was arguably the "more universally applicable device" as well as one that many people found "easier to use and more comfortable to hold" than the joystick controller, which at any rate had always required two hands—one to steady the box and one for the stick (Kent 2001, 279).

The digital d-pad persisted in some systems, but its centrality as a controller device for home consoles gave way in the 1990s to the analog thumbstick, versions of which had been used in some arcade games and home systems (e.g., the Atari 5200) during the 1980s. Rather than on/off digital switches, analog sticks rely on potentiometers to measure subtler gradations of electric current so that a range of motions in any direction by the stick (most often a stubby rubber protrusion with a wide top on which the player's thumb rests, but in some cases a depression into which the thumb fits) can be mapped onto in-game actions, especially the navigation of player-characters, whether from a first-person or third-person point of view, through a 3D game world. Once dual analog sticks became widely used in the mid-1990s for home consoles, they also controlled the in-game camera. Twin analog thumbsticks can be seen as descended from the use of twin joysticks, one for movement and one for aiming a weapon, in the arcade game *Robotron 2084* (Vid Kidz) from 1982 (Kent 2001, 221–222). Thumb-size and combined on a single handheld pad, dual analog sticks provided more precise and subtle mapping of controller movements to in-game actions, so they were especially suited to increasingly realistic and immersive 3D games, especially when they required careful targeting and in-game object manipulation. Exceptions and experimental peripherals continued to be produced, such as the notorious Atari Jaguar with a 12-button keypad on the front (1993), which looked like a small personal digital assistant with a keyboard. But by the mid-1990s, the two-handed gamepad, ergonomically curved for a better grip—a change for which earlier rounded-end or "dog-bone" shaped controllers served as transitional forms—and with variant combinations of action buttons (usually signified with letters or numbers, but sometimes with colors or shapes, as in the Sony PlayStation), sticks, and left and right shoulder buttons or triggers, had become standard for all major home consoles, with versions available for PCs as well. The variant forms of these gamepads often served as shorthand symbols of different console systems by Nintendo, Sega, Sony, and Microsoft.

Especially when complex, immersive 3D games are being played, the designed affordances, constraints, signifiers (the signs of those affordances and constraints), and feedback mechanisms of the gamepad controller are central to the player's experience (Norman 2013, 10–13). This may help to explain why, as indicated earlier, the controller has for many become a synecdoche for gaming as a whole. After all, the controller is the device the players actually handle; by definition, it gives them control over gameplay, but it also serves as an interface device more immediate and more somatic than the screen, even without the addition of haptic force-feedback effects, such as "rumble" vibration or peripheral sound effects. Any game controller is an important practical and symbolic object. It is experienced through direct contact, and so it exemplifies HCI in the most literal sense: it is a threshold between player and game, both a prosthetic extension of the player's responsive attention and the manifestation of the game itself (literally) closest to hand. It provides input and affords control, but within its own constraints and within the constraints of the system and the game.

Game development and controller design are closely related, but not in any simple, linear way, and the relationship is often determined by commercial and marketing considerations. The affordances and constraints of controllers, even sometimes before they are released, shape the design of new games. In pursuit of compatibility and because new generations of home consoles can be released years apart, developers make games to fit existing controllers more often than controllers are redesigned to allow for new kinds of gaming. A notable exception was Nintendo's Wii console, codenamed "Revolution," which for the first time made motion control the focus of a major home console (Jones and Thiruvathukal 2012). The rectangular

Wii remote (or "Wiimote") resembles a TV remote control and became the iconic symbol of the new system. The slim controller has a button arrangement that, when held horizontally in two hands, recalls the Famicom/NES controller, and it can be used to play emulated versions of older games. It makes use of a micro electronic-mechanical system (MEMS) device: an accelerometer that measures player movements on an *x-y-z* grid and maps them into the game. When the player swings or thrusts the controller, his avatar swings or thrusts a sword, for example. To increase the "spherical" range of motions, in 2009 a gyroscope sensor was added as an upgrade in an extension that plugged into the base of the controller, a MotionPlus sensor later incorporated in the Wiimote.

Motion control had been used previously in various devices but never as the focus of the primary controller for a home console, and in this way the Wii marked a significant departure in the history of game controllers. And yet, despite the Wiimote's iconic status, during the development process it became clear that other controls would be needed, especially for more complex 3D gameplay, so a curved pistol-grip-shaped device was designed, the Nunchuk controller, with Z and C buttons underneath and one analog thumbstick on top. It attaches to the primary controller with a three-foot cable and rides on the Wiimote's wireless connections, enabling flexible two-handed gameplay (with more freedom of hand and arm movement than a two-handed gamepad allows) and exploiting the affordances of the analog stick on the Nunchuk for navigation, the A and B buttons on the Wiimote, as well as the additional trigger buttons on the undersides of both. The Wii Balance Board controller, a rectangular platform shaped like a bathroom scale and using digital sensors to measure the player's weight, followed, as did a collection of peripheral accessories, including a plastic steering wheel and a gun into which the Wiimote and Nunchuk can be inserted, combining imaginative props with actual controllers.

Nintendo's own games for the Wii exploited motion control in various experimental ways, but only a few third-party games were developed for the system during its first five years. The follow-up console, WiiU (2012), allowed for the use of Wiimotes but reintroduced a more traditional two-handed gamepad with two analog thumbsticks as its main controller, plus motion sensors and a touch screen at its center. The design made possible not only precise control in using the buttons but also a kind of distributed controller system: the controller can serve as a separate handheld console for a player or as a separate window into a shared multiplayer game world or into an otherwise hidden portion of an augmented-reality game (turning sideways in the room shows the side streets the player is passing, for example). Sony introduced the PlayStation Move in 2010, which worked much like the Wii, and the same year Microsoft released the Kinect, to be added to the Xbox 360. The Kinect was marketed as "controller free" because it relies on cameras and sensors in a T-shaped device that sits near the screen and aims back out at the room to capture the player's gestures and the space of the room, mapping gestures in three dimensions into games. As with the Wii, third-party games developed for the Kinect have been limited, but once open-source software for a PC-connected version became available, the device was adapted for various nongame experiments, including augmented-reality art.

The Wii's motion-control mimetic interface ushered in a larger trend, what Jesper Juul (2012) has called the "casual revolution," the expansion of gaming to a wider population in the form of simpler games that can be played for shorter stretches, often designed for social interaction in the physical world as well as online. A year after the Wii was launched, Apple released the iPhone (2007), and Android phones followed within another year. With their multitouch screens, smartphones allowed for the spread of inexpensive application-based mobile games, using a combination of built-in touch and accelerometer-sensing gestures—tapping, swiping, tilting, and so on—and built-in virtual controls in the software interface. By design, the small screen serves as both monitor and controller.

In an interesting counterdevelopment, some controller shells for smartphones have been released based on traditional two-handed gamepads. The phone snaps into place in the center of the shell, connecting buttons on the sides that turn the phone into a nongestural controller, allowing for the more precise use of buttons to control mobile games. Indeed, as gestural control systems have emerged, largely for casual and mobile games, the accuracy and precision of button combinations on a gamepad have remained desirable. Valve's Steam, a digital-distribution platform for PC games, announced in 2013 plans for a dedicated console and a two-handed gamepad controller, but with two circular trackpads in place of analog sticks or d-pad and a centered touchscreen.

As all of these examples suggest, a pattern of interconnected and alternating emergence and convergence in design schemes—buttons and gestural, motion control, gamepad and touchscreen, abstract and mimetic, even virtual representations of controller icons on touch screens—has characterized the history and ongoing development of game controllers. This jostling and competitive diversity belies simpler evolutionary tales and reveals in its particulars the complicated variety that has for decades characterized the design of game controllers. This wider perspective on the history of game controllers has implications for understanding the history of computer interfaces in general. For example, Nick Montfort has cited paper-based systems to refute "screen essentialism," the assumption that the screen is "an essential aspect of all creative and communicative computing—a fixture, perhaps even a basis, for new media" (2004). The use of screens is actually relatively recent in computing, he reminds us, although he concedes that *Spacewar!* is one exception, an example of a very early cathode ray tube (CRT) screen interface in 1962. But, as I have shown here, the powerful symbolic and practical role played by the game controller as a figurative synecdoche for gaming and as the direct interface between the player and the game challenges the exclusive or essential role of the screen even in the example of *Spacewar!* Its players put together their own controller, and that wood and Masonite box with its buttons and switches—the device through which they experienced the game literally in their hands—was clearly as important as the CRT screen on which the gameplay was represented.

Notes

1. As I write, the first references to video game controllers have been added to the "draft additions" of the *OED* for March 2013. It's interesting that in classical game theory, as used in economics and military training, a "game controller" is what might be called a "gamemaster" (in the tradition of live-action role-playing games), the person running and partly designing an exercise.

2. As described in the advertising pamphlet *Odyssey—the Electronic Game of the Future* (1972), at http://www.magnavox-odyssey.com/Advertising.htmandhttp://www.magnavox-odyssey.com/Odyssey/Collector/Pamphlet2_2.jpg. As a reminder of the eclectic diversity of controllers in the 1970s, the Odyssey already had a *Shooting Gallery* game that used a peripheral gun controller, called the "Electronic Rifle."

Works Cited

Bogost, Ian, and Nick Montfort. 2009. *Racing the Beam: The Atari Video Computer System*. Cambridge, MA: MIT Press.

Brand, Stewart. [1972] 1974. "Fanatic Life and Symbolic Death among the Computer Bums." *Rolling Stone*, December 7. Reprinted in *II Cybernetic Frontiers*, 39–95. New York: Random House.

Jones, Steven E., and George K. Thiruvathukal. 2012. *Codename Revolution: The Nintendo Wii Video Game Platform*. Cambridge, MA: MIT Press.

Juul, Jesper. 2012. *A Casual Revolution: Reinventing Video Games and Their Players*. Cambridge, MA: MIT Press.

Kent, Steven L. 2001. *The Ultimate History of Video Games*. New York: Three Rivers Press.

Levy, Steven. 1994. *Hackers: Heroes of the Computer Revolution*. New York: Penguin.

Montfort, Nick. 2004. "Continuous Paper: The Early Materiality and Workings of Electronic Literature." Paper presented to the Modern Language Association, Philadelphia, December 28. http://nickm.com/writing/essays/continuous_paper_mla.html.

Norman, Don. 2013. *The Design of Everyday Things*. New York: Basic Books.

11 COOPERATIVE PLAY

Emma Witkowski

In public forums visited by video game players, the terms *cooperative* or *co-op gameplay*, *collaborative play*, and *multiplayer games* are revealed as befuddling yet conflated descriptors that signify a similar experience—playing together. In describing experience, using this terminology is an acceptable convention, one that I continue in this essay, though each term carries with it an assortment of hardware specifics, social practices, and generic game modes—myriad things orienting just how people might engage with video games. From the start of screen-delivered play, sports have been a familiar gateway to games and have left a notable mark on how people play together. As such, the making of multiplayer sports-themed games offers a snapshot of how and why playing together mattered in these otherwise competitive ("striving against") constructs, where cooperation ("striving with") is also at the heart of play, or, as philosopher James Keating writes, "the goal of sporting activity, being the mutual enjoyment of the participants, cannot even be understood in terms of exclusive possession [of a sole value or objective] by one of the parties. Its simulated competitive atmosphere camouflages what is at bottom a highly co-operative venture" (1964, 30). Team-based play as well as synchronous and asynchronous multiplayer simulations flourished in early game designs. In 1972, Ralph Baer's console Magnavox Odyssey provided multiple sporting options to ordinary consumers of this exciting new leisure technology. Traditional team sports such as football and hockey as well as ball and paddle games translated well into game mechanics and the two-controller system. With playing together in mind, Bill Rusch (a research-and-development engineer at Sanders Associates) achieved this mode of gameplay by adding a modest yet groundbreaking feature. He essentially developed a "third spot," an additional on-screen image implemented alongside the existing player-controlled "paddles" (the two other "spots" on-screen). Recognizing the immense value of his added machine-controlled spot, Rusch declared its purpose: it would be the ball (Baer 2005, 45).

With the ball put into play, the landscape of multiplayer video games blossomed. Two-player game designs and then four-player co-op and team-based options were regularly implemented modes of play and production. As games moved from dedicated circuit boards (such as in *Indy 800* [Kee Games, 1975]) to microprocessors (where every byte mattered), such perseverance in designing for multiplayer involvement could be considered a questionable practice. More hands on the machine called for additional processing power and increased complexity in the overall design. Such production issues would not have been easy sells when pitching a new (and potentially risky-to-develop) game, so why this push toward sports and a bigger roster of players on the playing field? How did multiplayer, sports-themed games manage to make such a mark on a burgeoning industry and new cultural form? And how can we feel that influence today? Through the context of sports, this essay explores such questions surrounding how we play games together. Jon-Paul Dyson's entry on the term *toys* in this collection reminds us of the connection through a small but significant moment: Sears's initial sales contract for Atari was not with the toys division but rather with the sports division. Sports provided players and designers a smooth transition into playing and working with the new medium. Players of all ages could, in most cases, grasp the gameplay when placed in front of a steering wheel on an arcade cabinet or a racetrack on-screen. Designers, too, had some established conventions to riff off of, such as familiar rule sets and established player-interaction formats. In some cases, they even had an embodied familiarity with team sports. Al Alcorn was a self-proclaimed "jock" and a reputable football player, with access to co-players as prominent as O. J. Simpson, prior to becoming an engineer and the creator of *Pong* (Atari, 1972) (Alcorn 2008). And when Alan Miller, Atari game designer and senior programmer and cofounder of Activision, was asked about his sources for game ideas, he revealed his high school basketball days as one of several experiential sources of inspiration (Miller n.d.). Indeed, Miller's final Atari game, *Basketball* (1978), is an indication of the deep sociohistoric links between traditional sports and video game design, a link continued in modern video game development. I return to the case of early sports-themed games, along with other turns to multiplayer designs in video game history, in a moment, but first I want to give some attention to the amalgam of terms used to describe cooperative play.

Despite the everyday player and designer practices that place multiplayer, collaborative, and co-op gameplay in the same silo, the terms alone have significant linguistic arcs. The etymology of the term *cooperation* draws from the positive expression "working together," where notions of sharing and reciprocity saturate its orientation and use. Such meaning is structured into modern game design practice as co-op player-interaction modes, characterized as working toward shared goals or cooperating against a system. These games are highlighted by player-versus-environment modes, such as the four-player, co-op option in *Left 4 Dead* (Turtle Rock Studios/Valve South, 2008), which pitches a team of individually vulnerable but collectively strong ("balanced") player-characters against an unpredictable artificial intelligence. Working together in *Left 4 Dead* is nothing short of necessary for survival because your "life" depends on other players' persistent attention to "watch your back."

The term *collaboration*, however, has a more linguistically devious beginning, in particular connoting "treacherous cooperation" with an enemy during a time of war. Although socio-technical collaborations in video games aren't always treacherous, the notion of engaging as collaborators during play has some hints of this negative interpretation. *Collaboration* in fact can differentiate some of the more recent developments encapsulated under the broader term *multiplayer*. In massively multiplayer online game *EVE Online* (CCP Games, 2003), the existence of co-players who back-stab other players for information, for equipment, or even just for fun is a designed possibility within the open-ended game, where such trickeries are nurtured as a legitimate form and culture of play. *J. S. Joust* (Die Gute Fabrik, 2014) shows another articulation of collaboration. A part of the *Sportsfriends* local multiplayer game bundle (which combines speed and slow motion, music, and pushing people), this screenless game sees alliances formed and broken between players with the tilt or strike of a move controller. During play, the pleasures of mutable player relationships are highlighted, where sudden role adjustments (with you or against you) are left up to the players as they decide just when, if, and how these relationships will be established in a face-to-face setting.

Although cooperative play is explored in this essay, with attention to games that are started, continued, and finished with other players, there are critical distinctions between cooperation and collaboration that warrant deeper inspection. Such recent attunements in playing games together, *EVE* and *Joust* included, need to be taken stock of. In exploring such provocative designs and play cultures, which engage directly with treacherous collaboration in the space of play (as a cultivated element), we may perceive an additional fine distinction in how we produce play together. Turning away from formal distinctions, let's move on to significant designs that support multiple participants in play as well as the cultural turns and technomaterial moments that have bolstered how we play together with games.

In the public spaces of arcades in the 1970s and 1980s, multiplayer games had a persistent place on the floor. The multiplayer mode essentially represented player(s) versus player(s) or head-to-head competition. And sports titles of the time championed this game mode, making it a prevalent way of playing together in the first decade of video games. The standard sporting mode of participating side by side at the screen was certainly not the only option, and alternate genres and formats of play brought with them new directions for shared play. Atari's fantasy maze game *Gauntlet* (1985), designed by Ed Logg, pushed the multiplayer mode into a popular new configuration—cooperative gameplay. In designing toward co-op mode, Logg introduced multiple character roles (Wizard, Warrior, Elf, and Valkyrie). By programming characters with particular abilities, he attended to game balance, which ultimately encouraged friends playing together or pick-up players to work together for the survival of the team and to make the quarters they had spent on the game last longer. Depleted health could be recovered only via shared resources found in the game or a coin drop. This challenging relationship is emphasized in Atari's *Gauntlet Operators Manual*: "As players cooperate to fight off common enemies and try to find their way out of the various mazes, they must also compete with each other for food, treasure, magic potions and other helpful items" (Atari, Inc. 1985). In *Gauntlet*, collaborating and competing exist side by side, a tension offering limitless interpretations and practices of the game that involve continuous reflection on whether, how, and when to engage cooperatively. As such, it might be said that focused cooperation as an intended game mechanic (whether that be via chatting with other players or "buffing" them with beneficial spells or covering your teammate's back) is interdependent on players' orientations to one another in multiplayer games and the values they put into play (Taylor 2006). In this new game mode, *Gauntlet* also exemplified how to make use of the design components, notably with an attention to audio. During play, narrative voice prompts nudged players to cooperate—for example, by announcing "Elf needs food" or "Your life force is running out"—illustrating by design how to put the full-game package to work toward co-op activity. The arcade game cabinet itself extended this complete design attitude. Arcade managers, for example, were keenly aware of the economy of floor space as an imperative calculation to their business' survival. The return on investment from any given machine could be measured by coin drops per square meter. *Gauntlet*, wider than most arcade cabinets at 72 by 29.13 by 37 inches, was an innovative design in terms of the arcade context and economics. Where two-player co-op sports games encouraged double the coins on a standard-size cabinet, *Gauntlet* had the potential to quadruple that amount. And harnessing those extra coins was a highly social affair because on-screen players were afforded the additional opportunity to recruit others midgame. With *Gauntlet*'s unique "drop-in" device, new players could join the game while it was already in progress: add a quarter to an unfilled character's coin slot, and the game continued seamlessly with the freshly configured team (Logg 2012). Such nuanced co-op designs were significant statements about what playing together could mean for players and hosts (arcade owners) alike.

One arcade game stands out as a marker of the reflective design process for co-op games at the time, with particular attention paid to the sociomaterial context of play. *Indy 800*, a racing game for eight simultaneous players, brought with it shrewd implementations that heightened a sense of the broader arena of players, both those in the game and those waiting to play. At 16 square feet, it was a big game, but it was big also in terms of its interior processing electronics, which involved a card rack layout of one dedicated circuit board per car. The grandeur of the cabinet with its eight steering wheels provided a vast setting for playing bodies on all sides. Eyeballing and challenging opponents across the flat-top monitor was an added visceral pleasure of the race. But the octadic, in-your-face competition was not the only significant move toward a collective experience; mounted to the top of the cabinet were periscope-like mirrors, reflecting the tabletop game view to players in waiting. The traditional arcade symbol of "I've got the next game," a coin placed on top of the machine, thus would have been a subtle reminder to players that future competitors were waiting, watching, and learning in the wings. Such moments in design resonate with more recent turns in e-sports (organized, high-performance computer game play), in which event organizers and game developers alike have struggled to align challenging competitions with the ability to spectate at local scenes (through local area networks) and to show the competitions to an online live audience. *Indy 800* is a rubric to bear in mind in terms of thinking holistically about the design of play for multiple bodies, where the context and multiplayer mode are seamlessly attended to in the materiality of the game, as is the elusive involvement of future players.

More conventional factors, including market forces and advances in technology, certainly also had a direct hand in designing toward cooperative play. Some of those key interactions fed into the development of new forms of team sports video games. The push–pull of copycat and innovating companies as well as links between system-circuitry advances and marketing departments articulate two distinct relationships that shaped the way games were made and played. Several striking moments occurred with respect to team-ball sports video games in particular, and the effects of the release of *Atari Football* in 1978 still resonate today with iterations in sports video games as cooperative practices. Dressed in a cocktail cabinet for the original two-player format, the complete *Atari Football* game package showed the fitness of Atari's pre-1979 innovating and replicating work environment. Atari founder Nolan Bushnell recalls a conversation with management around the strange "business" of invention, where a Warner executive declared, "Nolan, why don't you innovate kind of like you did last year, none of this new stuff?" (qtd. in Fulton 2008). With an influx of companies riffing off of established game designs, Atari was pushing to innovate to stay ahead of the pack, but development was pressed to rely on tried and successful designs. *Atari 4-Player Football* (1979), followed by *Atari Soccer* (1979, for two or four players), highlights this relationship in which extending the options toward multiple players onscreen implies the innovation itself. In early sports simulations, cabinet layouts as well as the core of the game design were enhanced from the benchmark set by *Atari Football*, and an increased complexity in the technology was demanded in the bigger, multiobject game innovations. Fitting developments such as the additional four-player game mode set Atari at a distance from copycat companies, who, with fewer resources for development, were engaged mostly in reverse engineering established ball and paddle games, essentially *Pong* rip-offs (Burnham 2001). In creating the 6502 black-and-white raster central processing unit (CPU) and the Motion Object Control 16-objects circuitry (MOC-16), Atari distanced itself from other developers by constructing the mechanisms to handle multiple moving objects on screen. By engineering new play forms and developing the cutting-edge technological innards to support them, Atari transformed the team video game experience at that moment in time and beyond. The 6502 CPU quickly became a staple at Atari for running team games, and the MOC-16 became a backbone in both North American and Japanese companies for handling multiple moving objects in games (Albaugh n.d.).

In striving toward bigger and better video games that sought to include multiple players at the screen, more complex sociotechnical arrangements were experimented with and put into play. Modifications that moved toward team play, including a push toward more physically demanding controllers that would provide a "sporting feel" beyond the screen, and toward a direct association with traditional sports figureheads locate a starting point for this still popular formula of development. An intuitive feel for play was central to the design and reception of *Atari Football.* By deploying an overhead scrolling perspective, players could maneuver up and down the recognizable playing field and have a coherent and equal field of view. Combined with the new "trak ball" technology, a tangible free-flow translation of movement was facilitated. Marketing honed in on this detail in Atari's promotional material, with Atari's corporate newsletter *Coin Connection* declaring in muscular fashion in 1979, "The TRAK BALL™ makes the difference. Player interaction, physical involvement, and direct competition are at a maximum with this ultimate control device and how it interfaces with the play action" (Atari, Inc.1979). In modifying how players physically engaged with the machine, the game extended the experience of computer gameplay between friends and players to include the shared pleasures of perspiration and postgame aches (the product of vigorous arm swings across the trak ball). But it also brought another mode of skill, control, and physical engagement to the experience, which is why Atari designer Mike Albaugh argued vehemently for the recently released controller technology in the face of high cost. Albaugh recollects that in his discussions with Bushnell, who believed that the game would play "perfectly well with joysticks," he "insisted [on] the trackballs [*sic*]. ... I felt, the physicality of the game was an important concept." Albaugh got his way, and the range and experience of physicality (via alternative though "fitting" controller technologies) between players and games were extended (Albaugh 2010, 4). With such intriguing developments toward playing together, *Atari Football* worked its way toward being one of the most successful games of the year in 1978. Its launch date, a vital part of its realization, cannot be divorced from this achievement. The game launch in October corresponded with the start of the American National Football League season. This timing afforded arcade players the opportunity to take the pleasures of spectatorship from the stadium back into their local bar or arcade, augmenting the connection to football writ large by bringing another kind of involvement to the game and for only a handful of quarters. This link was not lost on marketing executives, who explicitly noted the cash potential of the game in the 1978 edition of *Coin Connection*, including the appeal of the familiar sport as a draw card to attract a crowd and a constant flow of coins. Long-running sports games, such as the *Madden* series (Electronic Arts Tiburon), use this same familiarity formula as the epicenter of their designs.

In discussions with the developer Electronic Arts on figureheading a new game, John Madden made the cost of his patronage simple: "If it's not 11-on-11 ... it's not real football" (qtd. in Hruby n.d.). His demand for this basic rule of play pushed the technology and programmer "know-how" to the limit, literally raising the possibility space of how play could be achieved cooperatively in video games. The desire for sports familiarity (or "realism") tested the robustness of systems and programmers, and for players the multiple moving objects brought a relatable aspect of "real football" to the screen. *Atari Football* didn't harness a sports association license in 1978 (Warner courted the license, but the NFL declined), but the potential connection and marketability were established. *Atari Sports*™ was rolled out the following year with newly developed sports titles in numerous two- to four-player incarnations, many of which were intended for the original trak ball cabinet and CPU design developed for *Atari Football*. As such, "innovating like last year" was alive and well with the releases of *Atari Soccer* and *Atari 4-Player Football* in 1979, both of which brought small but significant alterations in the challenge of playing cooperatively but were grounded in the lessons learned from the original two-player football stand-out.

These small but significant structures, designs, and player-oriented moments in the golden years of arcade games clarify some preliminary visions of how games could be played together. Such expressions of cooperative play and sports continue in expressions from games such as the *Sportsfriends* (Die Gute Fabrik) compendium (successfully crowd-funded in 2012), which make a modern statement on sharing in the process of play. With links to both sports (traditional and digital) and the playful aesthetic of the New Games movement in the 1970s, these local multiplayer games sample and celebrate some of the same nurtured, cogent ideas first seen in those early titles and cultures of video game play, from designing for spectatorship engagement to a push toward face-to-face play and other kinds of physical involvement. As one of *Sportsfriends* developers, Ramiro Corbetta, reflects on modern play contexts, "Games have been historically a way to bring people together. ... And then, it somehow became this thing you do at home. There's nothing wrong with that but I want to make games that are a social lubricant" (Khaw 2011). Countless advances in what is essentially cooperative game design showcase sports games and their game cultures as persistent provocateurs in terms of their influence and the crafting of artifacts and code for multiple players. This brief look at the expansive history of cooperative play in video games thus ends with a final reminder—that cooperative and collaborative play is multiplayer, and the social lubricants are a part of sports, "in the game" by design or by way of player cultures, from the very start.

Works Cited

Albaugh, Mike. n.d. "*Atari Football* History." http://www.coinop.org/features/football.aspx.

Albaugh, Mike. 2010. "Mike Albaugh Interview: Interviewed by Dag Spicer." Computer History Museum, X5963.2011, transcript, November.

Alcorn, Al. 2008. "Oral History of Al Alcorn: Interviewed by Henry Lowood." Computer History Museum, X4596.2008, transcript, April.

Atari, Inc. 1979. "Coin Collection." June/July 3 (6). Sunnyvale, CA: Atari, Inc.

Atari, Inc. 1985. *Gauntlet Operators Manual*. Sunnyvale, CA: Atari, Inc.

Baer, Ralph H. 2005. *Videogames: In the Beginning*. Springfield, NJ: Rolenta Press.

Burnham, Van. 2001. *Supercade: A Visual History of the Videogame Age, 1971–1984*. Cambridge, MA: MIT Press.

Fulton, Steve. 2008. "Atari: The Golden Years—a History, 1978–1981." Gamasutra, August 21. http://www.gamasutra.com/view/feature/3766/atari_the_golden_years__a_.php?print=1.

Hruby, Patrick. n.d. "The Franchise: Inside the Story of How *Madden NFL* Became a Video Game Dynasty." *Outside the Lines, ESPN*. http://espn.go.com/espn/eticket/story?page=100805/madden.

Keating, James W. 1964. "Sportsmanship as a Moral Category." *Ethics* 75:25–35.

Khaw, Cassandra. 2011. "Ramiro Corbetta Talks about Hokra, No Quarter and Colors." *Indie Games, the Weblog*, June 2. http://indiegames.com/2011/06/hokra_ramiro_corbettas.html.

Logg, Ed. 2012. "*Gauntlet* Postmortem." Presentation at the Game Developers Conference, March 5–9. http://twvideo01.ubm-us.net/o1/vault/gdc2012/slides/Design%20Track/Logg_Ed_Gauntlet_Postmortem.pdf.

Miller, Alan. n.d. "DP Interviews ... Alan Miller." Interviewed by Al Backiel. *Digital Press*. http://www.digitpress.com/library/interviews/interview_alan_miller.html.

Taylor, T. L. 2006. *Play between Worlds*. Cambridge, MA: MIT Press.

12 CULTURALIZATION

Kate Edwards

Localization is the process of adapting game content for a specific geographical locale. It typically includes translating resources and assets from the original language to the target language, implementing these translated assets into an internationalized build, and then testing, editing, and fixing bugs in the translation. The localized assets can consist of a variety of file types, including translated resource files (art, audio, etc.), localized voice-over files, translated packaging material, and translated metadata, among many other types of content.

Games are most typically localized into languages that represent the primary gaming markets: at least the "FIGS" languages—French, Italian, German and Spanish—as well as Japanese. However, many companies are more frequently launching in additional languages, including Russian, Chinese, Latin American Spanish, Nordic languages (Swedish, Danish, etc.), and others. This continued expansion of launch languages represents the growing popularity and distribution of video game consumption.

The need for game localization is a well-known necessity in the game industry, particularly with the reality that roughly 50 percent of the industry's global revenue is generated from localized versions. For individual companies, the percentage of annual revenue derived from localized versions of their games can range widely, with anecdotal comments indicating from 25 percent to as much as 70 percent. Also, some companies are discovering the value of localization to the point of reviving older game titles and spending a nominal amount on localization for emerging markets. In some cases, the return on investment has been staggering, as much as 400 percent.

So there's no question that localization is critical and necessary. However, the need for *culturalization* is still an often overlooked yet much needed reality. Culturalization goes beyond localization by taking a deeper look into a game's fundamental assumptions and content choices and then assessing their viability both in the broad, multicultural marketplace as well as in specific geographic locales. Localization helps gamers comprehend the game's content (primarily through translation), but culturalization potentially helps gamers to engage with the game's content at a much deeper, more meaningful level. Or, conversely, culturalization ensures that gamers will not be disengaged from the game by a piece of content that is considered incongruent or even offensive.

Culturalization can be divided into two broad types: *reactive* and *proactive*. Reactive culturalization involves identifying and removing content elements that might negatively disrupt a user experience; proactive culturalization entails identifying and adding content elements that may enhance the local experience and relevance. So how does localization fit into this scheme? What we understand currently as "localization" is actually just one aspect of a broader process of content culturalization.

Culturalization can be viewed as comprising the following three high-level phases:

1. Reactive culturalization: avoid disruptive issues.
2. Perform "typical" localization and internationalization.
3. Proactive culturalization: adapt and provide locale-specific options.

Perhaps a simpler and more straightforward way to view these three phases is as follows:

1. Reactive culturalization: make the content viable.
2. Localization: make the content legible.
3. Proactive culturalization: make the content meaningful.

Let's take a closer look at these phases. Making content "viable" essentially means allowing the content (and by implication the product or service that contains it) to remain in the target locale. This means that there is no issue in the content that would cause a local government or gamers or both to react negatively and thus initiate a potentially costly and embarrassing backlash against the game developer. This is exactly what happened in *Age of Empires II* (Ensemble Studios, 1999) when the Saudi Arabian government complained that the "healer" character in the Saracens army waved his hands during the healing process. This action was viewed as necromancy and forbidden by the Qur'an, so the government took action. Culturalization at its most fundamental level is thus simply flagging and removing potentially problematic issues from the risk equation. Most companies that include culturalization in their content development tend to focus on this aspect—that is, risk aversion.

A good example of the viability issue comes from *Fallout* 3, released by Bethesda Game Studios in 2009. In its postapocalyptic landscape around Washington, DC, the game included a creature that was called a "brahmin," a mutated, two-headed Brahman bull. The animal could be used for carrying loads, or it could be killed and consumed (even if radioactive). Brahman cattle, however, are sacred to Hindus, and India maintains laws that protect these animals (the real ones anyway) from harm, so the presence and name of this one creature in the game made the game inaccessible to gaming markets in India. Revising this one aspect would have made the game viable for the Indian market, but that adaptation unfortunately did not occur.

Enabling legibility, or what is known as common localization practices, is more straightforward. Of course, it notably entails the process of doing a text and audio translation and providing proper tools for translators so that the content is better understood by local players. It also includes all the other aspects of "internationalization" that must be implemented for local compatibility, aspects such as considering field size in the user interface, ensuring the use of Unicode to accommodate various scripts, and so on.

Perhaps one of the best negative examples of the need for legibility comes from the game *Zero Wing*, created by Toaplan in Japan and released in 1991, in which an opening-cut scene declares, "All your base are belong to us." The poor translation in the English version has taken a life of its own in popular culture and has been replicated in many contexts as a classic translation error. In a more positive light, some localization efforts are monumental and very successful. In 2008, the Warsaw-based CD Projekt's Polish edition of *Mass Effect*, created by BioWare in Edmonton, Canada, proved to be the largest localization project in the history of Polish media (of any type). It received wide acclaim for its high quality, aided by famous actors from Polish cinema, who contributed to the effort.

Making content more meaningful for local gamers is a step that many developers are only starting to implement and enable via the availability of downloadable content. It is arguably a more time-intensive and research-dependent aspect of content development, yet the return on the investment can potentially be very significant, in addition to gaining local-market share, developing a fan base, and being seen as compliant with local laws. The issue is not just one of translating the language but of cultural tailoring at a more fundamental level so that gamers perceive the game as something "local" in nature or at least very locally relevant. What does that mean exactly? It means creating a gaming environment that can be perceived as being locally developed, accounting not just for the language but for other aspects that can be tailored to local perception—from character design options to time and date formats in the user interface to stories that can create greater empathy based on local concerns or sensitivities, and so on. From a game community aspect, adding more local meaning to a game enhances the game developer's reputation because it becomes known as a studio that truly understands and cares about locally specific expectations.

There are several good examples of adding local meaning. In some car-racing games, such as *Forza Motorsports* developed in 2005 by Turn 10 Studios in Redmond, Washington, care was taken to add locally relevant automobiles for localized versions so that the players would enjoy the types of cars with which they are more familiar. A mobile version of the card game *basra*, which is a very popular game throughout the Middle East, was developed, and then the artwork was culturalized for individual Middle Eastern locales, so the appearance of the Saudi Arabian version was very accurate for that country, as was the version for the United Arab Emirates, and so on. This simple investment in additional artwork resulted in a very positive user experience and affinity for the title, so that the company saw a very positive response and thus increased revenues.

Let's examine these three levels of culturalization a bit further. First, under this framework it should be clear that the common practice of localization as we know it is viewed as a component part of a broader practice of culturalization. As much as localization is emphasized, the process of achieving legibility in other locales through translation is not the only step required in preparing content for consumption in other cultures. This is true for video games as much as it's true for every other type of content.

Second, it can be argued that a game title should be legible before it is viable because the product needs to be understood by local players. However, ample experience demonstrates that the issues preventing a game from remaining viable almost always supersede the need for linguistic understanding. The basic logic is that a government will ban or restrict a game based on a sensitive content issue regardless if it is localized or not; for instance, China banned the World War II title *Hearts of Iron* created by Paradox Development Studio in Sweden in 2002 because the game did not show Tibet and Taiwan as part of Chinese sovereign territory. Before a game is even translated into the local language, it is critical to determine whether the game can even remain within the target market's border. The reality is that many games can be played by people in different countries with little or no localization effort; this fact doesn't downplay the importance of localization but rather emphasizes one of the unique aspects of game content: there is a logic to most games that transcends language and appeals to a broad tool kit of game mechanics with which most gamers are extremely familiar.

And third, even though these levels are described as a type of hierarchy of how deeply the content includes local adaptation, they by no means happen sequentially. As we know, even localization takes place in various stages within the typical game-development cycle, so culturalization is a coordination of various tasks and priorities orchestrated across the entire development process. Although we continue to strive to get game developers to perceive localization as a core and integral aspect of development, culturalization demands that international considerations be taken into account on day one of a project rather than being added on at the end. The most effective culturalization happens at the beginning of the process, with a review and consideration of the game's overall design, direction, theme, and other elements, so that any potential concerns for a local market might be addressed right away. Typically, at least 75 percent of any potential issues can be easily managed in this early stage, which is advantageous considering that any changes late in the process can become very costly.

The core motivation for culturalization is that game developers and publishers adopt "global game development" as their modus operandi for the entire process, so that culturalization is as fundamental and implicit as game design, art, writing, and so on. Content is no longer easily limited to specific locales or constrained by hard-copy media such as DVDs. In today's online environments and digital delivery, everything is everywhere, so we must not only remain conscious of the negative impacts of cultural aspects but also leverage the positive aspects and see how content can be made locally relevant. With English becoming more and more ubiquitous, particularly among the gamer crowds, language alone (and thus "localization") isn't enough to differentiate the gaming experience.

As one may perceive from this discussion, culturalization isn't a straightforward exercise and must incorporate a range of variables, including the ethical and moral sensibilities of the individual game developers and their companies. Also at play is the delicate balance between art and commerce. If pure artistic expression is the ultimate goal, then it will be emphasized at the expense of global reach because the content will inevitably be incompatible with some locales. Conversely, the desire to maximize revenues will require a higher degree of culturalization during the creative process, which might potentially imply greater alterations to the game's original creative intent. Every game developer should feel free to create the game it wants to create, but if it desires to maximize the return on its investment, then it cannot forget the global, multicultural audience that will be participating in its vision and enjoying that vision without any cultural disruption. If well-executed culturalization is pursued as an option for a game's content, it cannot be shortchanged within a development cycle; it takes time and experience to implement successfully. However, the benefits to a company's content quality, government relations, and public image among local gamers will be immediate and obvious.

13 DEMO

Michael Nitsche

Like other terms in the world of video games, *demo* is an evolving one. Through its evolution, it sheds light on the history of games and the emergence of a gaming culture that reaches from commercial sales pitches to hackers and independent artists to the maturing of players into producers.

The term *demo* is used to describe multiple artifacts in video games. First, the demo of a video game is a prototype, a demonstration of technology and design that showcases particular features of a to-be-released game or game engine, such as gameplay or audio-visual achievements. Second, demo recordings can be log files of gameplay activity. These files are traces of play, recorded in-game action, and they can be manipulated and moved independently from the main game engine to be played back on different computers. Third, demos as part of the demoscene are single-standing pieces of real-time animation and consist of often remarkably optimized code to render out largely noninteractive animation sequences.

This chapter touches briefly on all three concepts but ultimately concentrates on the second one to trace divergent evolution as well as overlaps between them.

All three types of demo are defined by the game or render engine's relationship to the data that are fed into that engine. *Tennis for Two* (1958), widely considered the first video game, was a playful exposition for William Higinbotham's expertise in display systems and their circuitry. It was a demonstration of what was possible with the hardware at hand before video games as commercial entities existed. Today, demos are released to attract players to purchase the full version of the showcased game by presenting the specifics of the game engine, design, and overall product quality. They operate as showcases for the full game yet to come. Release mechanisms vary from the shareware releases (e.g., id Software's shareware release policy, which catapulted *DOOM* into consumer awareness in 1992) to releases in publications (such as the *PlayStation Official Magazine*), branded portals (such as Xbox Live), and showcases (for example, game conventions such as E3). As these mechanisms grow more complex, so does the demo's role. The demo's basic function is to present the operational game and support its sales or production. However, emerging formats (such as Steam's Early Access) blur the borderlines between fully functional demos and open beta game releases. The latter are tools not solely for marketing but also for production, and they are testing grounds for developers and financing tools for companies. A traditional marketing demo release aims to attract consumer interest. Thus, possible bugs or glitches should be corrected to avoid any bad first impression. Releasing versions of games that are not fully finished as demos presents games as—at times bug-ridden—technology. These releases are more explorative than commercial. In that way, they are closer to the original presentation of *Tennis for Two*: unfinished, explorative, and speculative instead of finalized presentations of results.

Exceptions to these models include demos that differ significantly from the games they preview. To achieve a favorable experience, fully fledged game demos often include attractive subsections of the to-be-released title, but they can also be single-standing installations. For example, *Half Life: Uplink* (Valve, 1999) is a demo for the first *Half Life* game but contains material that is not available in the actual release. Although elaborate cases like this one might be outliers, the differences between the demo and the final game are often telling and much discussed among gamers, and they highlight technical changes made after the demo is released—one can point, for example, to the controversy regarding the contrast between the demo version of *Aliens: Colonial Marines* (Gearbox Software, 2013) and its original release. In that regard, the demo as a critical artifact in its own right regains importance—not just as a stepping-stone toward the final game release.

Demos as noninteractive, real-time animation sequences are at the heart of the demoscene, a digital art community with its own—often competitive—events, online portals, and growing body of research and publications associated with it. In the 1980s, the demoscene evolved from early hacker communities by means of "crack intros" or "Cracktros": small introductory animation clips showing off the coding and cracking skills of the hacker (or hacker group) that cracked a game's antipiracy mechanisms and lauding his or her technical expertise. Because the hacker often had to get into the game's code to crack the copy protection, he or she could also add special play conditions for the game, such as the option to switch on a cheat. The resulting manipulations were termed "game trainers," which were often distributed with the cracked game.

Hacking was intrinsically linked to commercial game development and its platforms and started with the rise of early personal computers (PCs) in the 1980s. The Commodore 64 and the Amiga were home to countless demos and remain platforms supported by the demoscene to this day. As these crack intros became increasingly complex, they grew into an art form with a focus on optimized code to produce the most extravagant animation sequence possible in real time. From the illegal sharing of cracked software, the demoscene manifested into a community of digital art that applies expert coding, design, sound design, and animation direction. This change was not instantaneous (Reunanen 2010), but different platforms allowed for different development stages along the way. Lassi Tasajärvi defines three main stages: "Oldskool," covering the years 1980–1991 and involving mainly the Commodore 64 and Amiga 500; "Middleskool," covering the years 1992–1996 and involving the Amiga and MS-DOS PCs; and "Nuskool," covering the years from 1997 to the present and including more advanced PCs that included advanced render acceleration and graphic features (2004). Over each stage, the community itself evolved from the first hacker groups (such as JEDI) to large social gatherings (such as Revision) that remain supported by online communities.[1] The community's focus on technical expertise and coding skills has continued, but prolific demoscene members have also developed scaffolding in the form of tool sets that allow easier demo production itself, as seen in Moppi Production's *Moppi Demopaja* (2000) or Farbrausch's *werkzeug* (2004).

The underlying techniques in the demoscene, such as procedural generation and optimized code, have over the years influenced commercial game developers, and various game studios have employed demoscene members. At the same time, the culture developed by the evolving community remains part of gaming. Terms such as *warez* were established during the early days of the demoscene (Polgar 2005) and remain in use today.

A third use of the term *demo* relates to the recording of gameplay activity by players by means of the game engine. Numerous engines allow for the recording of log files from gameplay activity. These files are data from the unfolding play action and can differ in their format depending on the engine on which they are run. For example, the original *DOOM* demo files (.lmp) recorded input commands (tics). In contrast, the *Quake* (id Software, 1996) demo recordings (.dem) were server based and included packages of information about a character's condition at a certain gameplay moment.

Demo recordings do not include much interactive access but instead consist of linear real-time animation sequences triggered by the data from these log files. Unlike the demoscene, a demo recording does not depend on customized code but instead relies on a preexisting game engine. As a consequence, it can be used and played back only within that particular game. Demos use the render logic, sound, animation, and three-dimensional assets as well as any particular features of the underlying game system. A key benefit of these files in the early days of dial-up Internet connections was their size: because only selective data were recorded, the demo file remained small and could be shared easily over the limited bandwidth available at the time.

Demos as play recordings in the form of log files focus on the play as performance and the particular choices conducted by players in their engagement with the game system. The multiplayer game *Quake* depended entirely on this kind of player performance because it did not provide any elaborate narrative and instead relied largely on in-game collaboration and competition. If the demoscene offers technical coding skill, and game demos focus on particular game features, then the recorded demo is a showcase for players keen to present their playing skills as performances.

Demo recordings thus often serve as documentation and verification for certain play styles, matches, or play strategies. For example, such recordings were used in the late 1990s as a key tool by individual players as well as by emerging *Quake* "clans" to improve their playing styles. Next to the actual live games, the demos were the main means to archive, revise, improve, and brag about a player's performance in *Quake*. One tradition in this play-as-performance demo recording is the Quake Done Quick community.

Not only did demo recordings work before screen capture was easily feasible, but the focus on demo recording also served as proof of a particular performance's validity. For play performance, the value of the demo recording is important because cheats can be more easily detected in this format than in a screen capture, which can be postproduced—that is, edited after being recorded. The ability to record and replay demo files has supported the evolution of video games as a spectator sport (Lowood 2007).

But even the .dem files used by game engines such as *Quake* can be tampered with. Uwe Girlich's demo-manipulation tool Little Movie Processing Center and David Wright's Keygrip offered exactly that option: "With such a tool it is very easy to analyze a DEM file but you can change it as well and so create a DEM file of a *Quake* game you never played" (Girlich 1996). If such a file can be tampered with, then the legitimacy of the play performance is put into doubt. One example deals with a seemingly legitimate speed run through all 30 levels of *DOOM 2: Hell on Earth* (id Software, 1994) on the Nightmare setting. The first demo showing this performance was uploaded by Steffen Winterfeldt in 1996 with the notation "30nm6520" ("30" means that all 30 levels were mastered; "nm" indicates the Nightmare setting; and "6520" shows the length of the demo: 65 minutes and 20 seconds), but the performance was later discredited even though it was delivered as a demo recording. Today, Thomas Pilger's notation "30nm4949" in 1998 stands as the first unchallenged demo run for this setting to this day and can still be replayed with the proper game installation.

But games such as *Quake* provided not only new gameplay functions but also introduced groundbreaking visuals. As the *Quake* clans started to explore the possibilities, "a critical shift occurred—*the viewpoint of the player became the viewpoint of a director*" (Marino 2004, 4, emphasis in original). In 1996, the *Quake* clan the Rangers recorded the demo *Diary of a Camper*, representing the turn in the use of demo recordings to a narrativization of the play performance and a decisive moment for what would later become the machinima community. *Diary of a Camper* does not tell about the gamers' playing skills but instead about a simple dramatic set up that resides solely within the game world. It illustrates the shift from a play performance as a showcase of gaming skills to the play performance as an expressive art form that utilizes the game engine as a production tool.

Technically, this shift was further supported by tools such as Keygrip 2, which simplified recamming for *Quake II* (1997) through its own user interface. Recamming allows adjustments of the camera angles and behaviors in an already recorded demo. The play performance itself, including the game avatars' movements and actions, stay fixed, but their visualization can be changed for a more engaging replay. The same feature remains available in modern game engines such as Valve's *Source* engine, which includes not only demo recording but also a demo player with functions to change the camera behavior. Here, the playback device becomes part of the image selection and presentation and feeds off the expressive freedom granted by the demo format.

Where functionality is more foregrounded, as in real-time strategy titles, the recorded game files are used to improve performance through better analytics. The replay-recording feature of *StarCraft II* (Blizzard Entertainment, 2010) allows players to collect detailed data about a play session. These files can be analyzed to provide in-depth analysis of the game session and inform the player's performance. In this case, the demo file not only serves as a documentation of play but is also already designed as an analytical data-collection tool that can be evaluated with other software such as Scelight that allow for data analysis and visualization as players parse the replay files.

All three meanings of the term *demo* reflect changes to gaming culture and have left their traces over time. Their end points might diverge, but their evolution also shows a shared pool. David "crt" Wright, the coder for the Keygrip demo-manipulation tool, stood out as an expert player as well as promoter of his own *Rocket Arena* modification. Mikko "Memon" Mononen of the demoscene group Moppi Production joined the game company CryTek, producer of the *Far Cry* game series. The *Far Cry* game demos are regular showcases of new technical feats—in particular their visual quality. Of course, when *Far Cry* (2004) launched, the game offered its own demo-recording option. All three meanings of the term *demo*—a new approach to game demo releases, an independent animation scene within digital arts, and a player-based machinima community—are only the tip of a huge iceberg in gaming's development toward a complex cultural phenomenon.

Note

1. For one such online community, go to http://www.pouet.net/.

Works Cited

Girlich, Uwe. 1996. *The Unofficial DEM Format Description*. Vol. 1.0.2. http://www.gamers.org/dEngine/quake/Qdem/dem-1.0.2-3.html. Accessed August 21, 2015.

Lowood, Henry. 2007. "High-Performance Play: The Making of Machinima." In *Videogames and Art*, ed. Grethe Mitchell and Andy Clarke, 59–80. Bristol, UK: Intellect Books.

Marino, Paul. 2004. *3D Game-Based Filmmaking: The Art of Machinima*. Scottsdale, AZ: Paraglyph Press.

Polgar, Tamas. 2005. *Freax: The Brief History of the Computer Demoscene*. Vol. 1. Winnenden, Germany: CSW.

Reunanen, Markku. 2010. "Computer Demos—What Makes Them Tick?" Licentiate thesis, Aatlo University, Helsinki.

Tasajärvi, Lassi. 2004. *Demoscene: The Art of Real-Time*. Helsinki: evenlake studios.

14 DIFFICULTY

Bobby Schweizer

There are numerous ways to think about difficulty as a design choice that affects the experience of a game. Difficulty can determine accessibility, be tuned to ensure that players of different skill levels are satisfied by challenges, be based on financial motivations, function as an aesthetic choice, or be an unintentional consequence of the design process. As Jesper Juul argues, the possibility of failure is both part of why we enjoy games and "what gives us something to do in the first place" (2013, 45). Difficulty relates challenge to experience and expectations and typically involves ideas about fairness, failure, and punishment (Juul 2013, 70). Difficulty is an aspect of game design in which the purpose of skill-based challenge in play comes under consideration.

Because the challenge of a game is a matter of context, the history of difficulty concerns the concept as a product of design values and choices. Consider the coin-operated video arcade, whose design considerations are transparently financial. Arcade games in the 1970s and 1980s were designed with play duration and quarters in mind. If a game was too easy, a player could monopolize a cabinet, thus reducing its revenue, whereas a game that was too difficult might put players off, causing the owner to lose money (Loftus and Loftus 1983). Games on home consoles were already sunk costs, which meant that though arcade design sensibilities may have persisted, short play durations were not necessarily a factor in home console game design. With the recent emergence of "free to play" games, difficulty can be used to encourage players to purchase content within the game that might speed their progress through challenges. But genres of difficult games have also emerged to satisfy a growing market of players looking for new experiences.

Difficulty is not merely challenge. It is a way of thinking about game design. Understanding difficulty requires looking at variations in challenge, how interface and mechanics contribute to difficult gameplay, design implications, and what it means for a game to be meaningfully difficult.

Variable Difficulty

Xerox Palo Alto Research Center researcher Thomas W. Malone (1980) outlined three types of "variable difficulty" in computer games: player-selected settings, adaptive adjustments based on performance, and opponent's skill in multiplayer games. Evidence of these types of difficulty are present throughout game history in which the option to vary difficulty is available at both the hardware level and the software level (Therrien 2011). As computational artifacts, games can often be made easier or harder by manipulating variables in the code that determine permissible margins of error. The video game difficulty setting lets the player choose his or her own destiny—balancing the risk of challenge with the reward of high score or personal satisfaction. The Taito arcade driving game *Speed Race* (1974), which provided players with the option to choose between "beginner's race" and "advanced player's race," was the first popular arcade game with selectable difficulty.

Though these kinds of settings were not standard practice, consider the way Atari popularized these variations in both coin-op arcades and the company's home console hardware to satisfy founder Nolan Bushnell's belief that the "best games are easy to learn and difficult to master." The Atari games *Tempest* (1981), *Star Wars* (1983), and *RoadBlasters* (1987) all included options for players to choose their starting difficulty level from a menu; the risk for choosing a higher level was that the player's quarter might not last as long because the player might be quickly destroyed by the enemy, but the reward was that there was a greater score bonus for reaching more difficult stages. From the expert player's perspective, this option was a way to skip the early, less-challenging stages that they were guaranteed to complete to move on to a greater challenge. From the arcade operator's perspective, the bold player, skipping all the early levels, would have a shorter play time, which would net the arcade operator more quarters. Arcade operators were also able to affect difficulty by accessing hardware DIP (dual in-line package) switches inside the game cabinets to toggle variables that would affect the challenge a game posed. For example, in *Star Wars* game cabinets, owners could set how many shields the player started with, how shield bonuses were awarded, and overall "gameplay" difficulty level.

The value that Atari placed on offering different players different skill levels is also evident in its Video Computer System (VCS) catalog from 1978, which claimed that the VCS home console had "difficulty options so the games get better as you get better" (Atari, Inc. 1978). The Atari VCS had physical "Skill/Difficulty" switches for both the left and right players that could be set to A "Expert" or B "Novice." In *Combat* (1977), setting the switch to "A" decreased the range of the tank's missiles. Similarly, the manual for the game stated that while playing the Bi-Plane and Jet-Fighter games, the "player in Position 'A' will fly slower than [the player in] Position 'B'" (Atari, Inc., 1977). These hardware difficulty settings allowed players of different skill levels to play together.

In addition, individual VCS developers could present difficulty options at the software level as game variations accessed with the console's "Game Select" switch. In Activision's game *Freeway* (1981), the different variations determined the pace, composition, and direction of the traffic that the chickens had to avoid as they crossed the road. In Warren Robinett's *Adventure* (Atari, 1980), the player could choose from three modes of different complexity. In the first, a large portion of the game space was removed to make finishing more difficult. The second mode restored absent rooms and enemies, and the third randomized the locations of the objects the player needed to collect.

Hardware-based difficulty settings remained a part of coin-op arcade cabinets well into the 1990s but fell out of favor much sooner with home consoles. By the time the Atari 7800 ProSystem came around in 1986, the difficulty-switch options in the VCS had been replicated in software, and no other major home console manufacturer would incorporate physical toggles into their hardware. Thus, unlike Atari, other video game hardware manufacturers were not necessarily concerned with providing variations to account for different player skill levels, leaving the option to the software developers.

"Easy, normal, hard" emerged as generic descriptions for difficulty variations, though it has become common practice to assign thematic synonyms for these options. The taunting nomenclature for the options in playing *DOOM* (id Software, 1993) included "I'm Too Young to Die," "Hey, Not Too Rough," "Hurt Me Plenty," "Ultra-Violence," and "Nightmare." *Civilization* (MicroProse, 1991) figured the settings as different feudal rulers, from Chieftain level through Emperor level. This type of naming convention has persisted in more recent games, as seen in the Recruit, Regular, Hardened, and Veteran levels of the *Call of Duty: Modern Warfare* series (Infinity Ward, 2007, 2009, 2011). For someone familiar with previous games in the series or other first-person shooters, these descriptors are convenient. However, the limitation of these options is that players are being forced to choose their skill level before knowing how well they will do in the game. Other games offer increased difficulty options as rewards for completing the game: after finishing their first *Mass Effect* (BioWare, 2008) campaign, players can attempt the game on Hardcore difficulty and then on Insanity difficulty following that.

The alternative to self-selected options can be found in the ways games can adapt to player performance. Simple early implementations of this persistent technique include *Breakout* (Atari, 1976), which increased the ball speed at predetermined intervals, and the *Asteroids* (Atari, 1979) clone *Astrosmash* (1981) for Intellivision, which would make the game easier when the player was close to running out of lives. In rare cases, games have also offered the option to change skill level midplay through a menu so that the player may choose a more appropriate challenge. Similarly, games such as *God of War* (SCE Santa Monica Studio, 2005) and *Ninja*

Gaiden 3 (Team Ninja, 2012) will respond to a frequently dying player with a prompt that provides an option to make the game easier. Implicit in both of these scenarios are game design decisions that value players' being able to complete objectives—either to keep them satisfied with the product they have chosen to purchase or to allow the player to experience the entirety of the game that has been designed.

Dynamic difficulty adjustment is another implementation of adaptive difficulty that evaluates and increases or decreases the challenge of given obstacles based on user performance during play. Citing Mihaly Csikszentmihalyi's concept of "flow," Katie Salen and Eric Zimmerman have argued that good game design accounts for challenges that are not too easy and not too hard in order to manage boredom and anxiety (2003, 350), and Noah Falstein (2005) has posed that rising and falling difficulty within this flow channel proves more satisfying. Difficulty can be increased gradually as the player is assumed to become better at the game (Schell 2008, 208). However, it is hard to design for a pleasurable level of challenge that can account for players of different skill levels. Robin Hunicke and Vernell Chapman (2004) argue for "dynamic difficulty adjustment" systems to intelligently account for an "optimal experience" without the player noticing. Naughty Dog's game *Crash Bandicoot* (1996) reacts to player performance at failure points and will reduce the number of enemies or provide improvement items at tough spots, and Remedy's game *Max Payne* (2001) assists with aiming accuracy to make it easier or gives enemies more health to make it harder. One of the major selling points of *Left 4 Dead* (Valve, 2008) was the "A.I. Director" system that actively managed enemies to control the game's pace and to vary replay challenges. Adaptive difficulty is a design decision that is highly conscious of the implications of challenge in games.

Last, when a game is played as a competition between two sides, it is the skill of the opposition that determines the challenge. Thus, designing for difficulty in opponent-based games involves accounting for possible variations that might arise. In *Combat*, for example, Atari encoded software rules into its game such that expert and novice players could complete. But as Bernie DeKoven describes in *The Well Played Game* (1978), social rules are also injected into game rules to make the activity enjoyable for players of different skill levels. Just as kids in schoolyards negotiate handicaps to make the game fair between players of different abilities, players often decide their own rules when there is room for flexibility (Sniderman 2006). In multiplayer games, then, there is value in not hardcoding every rule in order to allow players to collaborate and play together (Wilson 2011).

What Makes Games Difficult?

Jesper Juul and Marleigh Norton (2009) cite a "mythical border" between the two things that are seen as making games difficult: interface and gameplay. In their paper "Easy to Use and

Incredibly Difficult," they explain how gameplay is the execution of interface and identify games such as *WarioWare: Smooth Moves* (2006) for the Nintendo Wii and pick-up-sticks as games in which the challenge lies in the interface. I have expanded the definition of difficult gameplay (Schweizer 2012) to include the experience of difficult *interfaces* and *mechanics* composed of determining factors.

Difficult interfaces can involve obtuse interactions, inaccessible information, and manipulation issues. Interface relates to how we perceive and interact with objects on the screen, textual displays, feedback mechanisms such as score or health bars, instructions, and the physical control input (be it controller, mouse, or finger). The notoriously difficult game *Dark Souls* (FromSoftware, 2011) demonstrates obtuse interactions by obscuring basic information about the meaning of the statistics it displays, the purpose of items, and the function of objects in the environment. Part of the challenge of *Dark Souls* is decoding how to interact with the game. Games that rely heavily on the player's understanding of feedback loops are more difficult when information about the gameplay state is hidden either deep within menus or in plain sight. *Knights in the Nightmare* (Sting Entertainment, 2008) is a complex real-time strategy game that relies heavily on manipulating the user interface during battle, but the screen is often so cluttered with numbers, meters, and visual feedback that the difficulty derives from sorting through all the information being presented. Manipulation issues can be illustrated by *GIRP* (Bennett Foddy, 2011), a rock-climbing game that requires concentration and dexterity as the player's fingers play keyboard Twister trying to get the rock climber's arms to stretch, grab, and pull. In tandem, obtuse interactions, inaccessible information, and manipulation issues can make even controlling a game a difficult proposition.

More often than interface, mechanics—here defined by Miguel Sicart's description of "methods invoked by agents, designed for interaction with the game state" (2008)—are identified as making gameplay difficult. Margins of error, simultaneous actions, rule complexity, interrelated dynamics, and critical thinking skills contribute to this component of gameplay. *Margins of error* arise largely from graphical logics that govern how objects on the screen interact with each other. At the end of a castle stage in *Super Mario Bros.* (Nintendo, 1985), Mario can die when he collides with a flame, the hammers being tossed in the air, or Koopa's body. A screen filled with more deadly objects leaves less room for mistakes. Other margins of errors are implemented in familiar conventions such as health, lives, permadeath (permanent death), continues, and time limits. Because of the number of *simultaneous actions* the player is asked to understand, it has become more common for games to scaffold possible interactions slowly as the game progresses. *Portal* (Valve, 2007) exemplifies this careful pacing, but games can challenge the player by throwing everything at him or her at once or by limiting the amount of time the player has to process new concepts. *Rule complexity* can compound the number of factors the player must consider before being able to act or make decisions in a game.

Typically, tile- or gem-matching games such as *Bejeweled* (PopCap, 2001) ask players to match single colors, but the iOS game *Bauhaus Break* (Michel McBride-Charpentier, 2012) uses the rule complexity of the card game *Set* to expand the possible tile-clearing matches to include all of one color, shape, or pattern or even one from each of those three categories. This large rule set presents the player with an overwhelming number of choices to make the game more difficult. In games with large *interrelated dynamics*, for example *StarCraft* (Blizzard Entertainment, 1998), even basic decisions such as choosing which units to build at any given moment can have consequences as the game progresses. In a game where professional *StarCraft* players' skill is often measured in actions per minute, understanding broader interrelated systems is crucial to success. Last, certain games present *critical thinking* challenges instead of contests of dexterity. *The Secret of Monkey Island* (Lucasfilm Games, 1990) requires both thinking through the puzzles with the information presented and bringing to the table knowledge of wordplay and clever reasoning.

Interfaces and mechanics contribute to a game's difficulty and are thought to be fair when success can be obtained through skill (careful planning) or labor (perseverance) as opposed to chance (lucky dice roll) or capital (spending money to progress) (Juul 2013, 74–79). The question of whether a game is too hard or too easy is, of course, a matter of player perception based not only on the player's skill level but also on his or her personal taste. Therefore, difficulty is a design choice that intersects with many facets of game development.

Design Implications

Valuing challenge (or, alternatively, valuing accessibility) is an aesthetic choice that responds to conventional game design (Asad 2012). "Bullet hell" games push the genre of top-down shoot-em-ups ("shmups") by exponentially increasing the number of projectiles on the screen that the player must dodge. Games described as "masocore" (masochism/hardcore) or "platform hell," such as *I Wanna Be the Guy* (Michael O'Reilly, 2007), *Cloudberry Kingdom* (Pwnee Studios, 2013), and the *Super Mario* parody game *Shobon no Action* (Chiku, 2007; also known as *Syobon Action* or *Cat Mario*) are intentionally inaccessible, brutally difficult, and flagrantly unfair (anthropy 2008). Games with permadeath offer harsh consequences for failure, wiping all of a player's accomplishments and progress. But even extremely difficult games continue to rely on modern game design conventions, which has prompted Douglas Wilson and Miguel Sicart (2010) to call for "abusive" game design as a critical practice. At the opposite end of the spectrum, the absence of difficulty is also an aesthetic choice. Casual games are often (though are not exclusively) intended to be accessible and have low consequences (Juul 2009). Their interfaces are designed to be understandable and their mechanics quickly learnable.

Most games exist somewhere in between these two ends of the spectrum. Though difficulty settings can often be tweaked, creative players have also taken to constructing their own constraints on games to introduce new forms of challenge. Self-imposed permadeath can be thought of as a kind of "expansive gameplay" that uses difficulty as a form of player expression (Parker 2011). Speedrunning, in which players attempt to complete a game in the shortest amount of time possible, adds a new layer of challenge to games that they have already deeply explored. And some players try to combine difficult gameplay with difficult interfaces—for example, attempting games such as *Ninja Gaiden* using the entirely inefficient plastic *Dance Dance Revolution* (Konami, 1998) pad as a controller.

Meaningful Difficulty

What does it mean to design a game such that its difficulty prompts players to think about its design? "Meaningful difficulty," as I define it, occurs when modifications to the challenge of interface or mechanics cause us to reflect on gameplay variation. This type of difficulty makes us aware of the processes that are occurring in the game and induces us to adjust our strategies accordingly. In some games, increased difficulty might prolong a task without causing the player to change strategy.

Halo 3 (Bungie, 2007) exemplifies meaningful difficulty through its Easy, Normal, Heroic, and Legendary play variations. On Easy and Normal, *Halo* is a run-and-gun first-person shooter. Players can play offensively, moving into open spaces and facing enemies head-on. But highlighting the Heroic difficulty on the selection screen displays text that says, "This is the way Halo is meant to be played." Heroic introduces the elements of strategy and defensive play. *Halo*'s combat is designed around encounters; the challenge is not just that there are enemies of varying toughness on the map but that these enemies have been strategically tuned with shields and weapons and placed in intentional groupings to produce different combat scenarios. Switching from Normal to Heroic, the player considers her failure in the face of old strategies, and, in doing so, she will have to realize that the game is asking her to conceive of the gameplay in a new way.

Works Cited

anthropy, anna. 2008. "Masocore Games." *Auntie Pixelante*, April 6. http://auntiepixelante.com/?p=11.

Asad, Mariam. 2012. "Proceduralizing Difficulty: Reflexive Play Practices in Masocore Games." Paper presented at the Society for Cinema and Media Studies Annual Conference, Boston, March 21–25.

Atari, Inc. 1977. *Combat Instruction Manual*. Atari Age database. https://atariage.com/manual_thumbs.php?SoftwareID=935.

Atari, Inc. 1978. *Atari 2600 VCS Catalog—Atari (USA)*. Atari Mania database. http://www.atarimania.com/catalog-atari-atari-usa-co12737_65_2.html.

DeKoven, Bernie. 1978. *The Well-Played Game: A Player's Philosophy*. Garden City, NY: Anchor Press.

Falstein, Noah. 2005. "Understanding Fun—the Theory of Natural Funativity." In *Introduction to Game Development*, ed. Steve Rabin, 71–98. Boston: Charles River Media.

Hunicke, Robin, and Vernell Chapman. 2004. "AI for Dynamic Difficult Adjustment in Games." In *Proceedings of the Challenges in Game AI Workshop, Nineteenth National Conference on Artificial Intelligence (AAAI '04)*, 91–96. San Jose: AAAI Press.

Juul, Jesper. 2013. *The Art of Failure: An Essay on the Pain of Playing Video Games*. Cambridge, MA: MIT Press.

Juul, Jesper. 2009. *A Casual Revolution: Reinventing Video Games and Their Players*. Cambridge, MA: MIT Press. http://www.jesperjuul.net/casualrevolution/.

Juul, Jesper, and Marleigh Norton. 2009. "Easy to Use and Incredibly Difficult: On the Mythical Border between Interface and Gameplay." In *Proceedings of the 4th International Conference on Foundations of Digital Games*, 107–112. New York: ACM. doi:.10.1145/1536513.1536539

Loftus, Geoffrey R., and Elizabeth F. Loftus. 1983. *Mind at Play: The Psychology of Video Games*. New York: Basic Books.

Malone, Thomas W. 1981. "Toward a Theory of Intrinsically Motivating Instruction." *Cognitive Science* 5 (4): 333–369.

Parker, Felan. 2011. "In the Domain of Optional Rules: Foucault's Aesthetic Self-Fashioning and Expansive Gameplay." Paper presented at the 2011 Games and Philosophy Conference, Athens, April 6–9.

Salen, Katie, and Eric Zimmerman. 2003. *Rules of Play: Game Design Fundamentals*. Cambridge, MA: MIT Press.

Schell, Jesse. 2008. *The Art of Game Design: A Book of Lenses*. Amsterdam: Elsevier; Boston: Morgan Kaufmann.

Schweizer, Bobby. 2012. "Easy, Normal, Hard: Meaningful Variable Difficulty in Videogames." Paper presented at the Society for Cinema and Media Studies Annual Conference, Boston, March 21–25.

Sicart, Miguel. 2008. "Defining Game Mechanics." *Game Studies* 8 (2). http://gamestudies.org/0802/articles/sicart.

Sniderman, Stephen. 2006. "Unwritten Rules." In *Game Design Reader: A Rules of Play Anthology*, ed. Katie Salen and Eric Zimmerman, 476–502. Cambridge, MA: MIT Press.

Therrien, Carl. 2011. "'To Get Help, Please Press X': The Rise of the Assistance Paradigm in Video Game Design." In *Think Design Play: The Fifth International Conference of the Digital Research Association (DIGRA)*. DiGRA; Hilversum, Netherlands: Utrecht School of the Arts. http://www.digra.org/wp-content/uploads/digital-library/11312.44329.pdf.

Wilson, Douglas. 2011. "Brutally Unfair Tactics Totally OK Now: On Self-Effacing Games and Unachievements." *Game Studies* 11 (1). http://gamestudies.org/1101/articles/wilson.

Wilson, Douglas, and Miguel Sicart. 2010. "Now It's Personal: On Abusive Game Design." Paper presented at FuturePlay 2010, Vancouver, May 6–7.

15 EDUCATIONAL GAMES

Anastasia Salter

"You have died of dysentery." This was a common ending screen for players of *The Oregon Trail*, a game first developed by a teacher, Don Rawitsch, for his history students. *Oregon Trail* translated the experience of traveling as part of Westward expansion to a series of decisions, mostly in route and supplies, with a minigame for hunting and the constant risk of death from disease or drowning. Students controlled a party of would-be settlers through choices: Buy bullets or spare wagon parts? Ford the river or wait to see if conditions improve? In successive versions of the game, these choices would be accompanied by better and better graphics, but the fundamental mechanics stayed the same as in the classic. The Minnesota Educational Computing Consortium (MECC) hired Rawitsch to complete the game in 1973, making it one of the most memorable parts of this early initiative in educational computing (Hoelscher 1985, 62). The game endured as a cornerstone of MECC's educational gaming enterprise, with new editions released every few years through new publishers to classrooms across the country. *Oregon Trail* defined expectations for personal computer (PC) educational games, particularly for the generation of students encountering it in the classroom when it was first released. It melded clearly defined educational content and learning objectives with familiar game mechanics and graphics. The game made use of the affordances of simulation, with enough left to chance to make it enjoyable. Elements of play outside the learning objectives made it particularly compelling as students could choose and input the names for members of their party and thus watch friends, celebrities, or teachers "die" their gruesome but historically accurate deaths. Despite its tendency to be used not as a learning environment but as one of the few teacher-provided arcade shooter games available in elementary schools, *Oregon Trail* became for many the iconic experience of an educational game experienced on a computer or video game console.

Is this, then, what an educational video game is? The term itself seems like a contradiction or at the very least a redundancy. Do the structures of games make them inherently valuable for some forms of learning, or are games and education more fundamentally opposed? All games might be termed educational by a broad definition: they demand learning as part of mastery of the system. Depending on the genre of game, success might be possible only through solving puzzles and problems, making observations and judgments of an environment or situation, employing reflexes and mastery of a control system, or working in cooperation with a team whose members employ different skill sets. However, educational games are often understood as something separate from video games in general, with a history that predates the introduction of the computer into educational spaces. Educational games marketed specifically for the socialization and skills development of children predate video games, with popular examples ranging from Milton Bradley's *The Game of Life* (1860) to the Indian board game *Snakes and Ladders*. The term *educational games* must likewise encompass games employed for the development of skills and knowledge among adults, including the genre of wargames, particularly as used for military training but also as played more broadly for simulating strategy, tactics, and history (see the chapter on *Kriegsspiel*). The history of wargaming is too vast to relate here and is well chronicled in the archives of *Strategy & Tactics* magazine, but it is particularly essential for the lessons in simulations of complex systems that would be brought to digital games. Many games and simulations used for experiential learning emerged from military routes, such as R. Garry Shirts's *BaFá BaFá* and *Starpower* (Simulation Training Systems Inc.), games for cross-cultural training developed in the 1970s, and their many successors in the field of simulations and gaming. In these games, the explicit use for training and the context of a professional learning environment defined their structures and mechanisms.

Likewise, educational video games can best be distinguished from video games more generally by their imposed external learning objectives, but drawing that distinction can suggest an artificial relationship between games and learning. At worst, it can mean reducing our understanding of educational games to a marketing category and accepting at face value the good intentions behind the genre. Educational games are associated with a number of genres: serious games, persuasive games, games for change, and news games. Although these genres are games with intention, educational games are particularly associated with the intent to achieve learning, perhaps even without the learner's intention. This association recalls the most meaningless experiences of gaming in the classroom, such as test review *Jeopardy!*, which is primarily an opportunity for questions placed in categories and given point values instead of printed on test forms. Such games act as delivery systems without any meaningful integration of content and mechanism. At the other extreme, game systems such as *Reacting to the Past* (developed in the 1990s by historian Mark Carnes for Pearson-Longman) resemble war games in the use of detailed historical simulations as a method to ask students to engage by playing historical roles in pivotal events ranging from the French Revolution to the trial of Galileo. Mark Carnes described how the game transformed the classroom into a "liminal setting," a "threshold region where the normal rules of society are suspended and subverted" (2004, B8). The term *educational game* gets applied to games that fall somewhere on a spectrum between *Jeopardy!* and *Reacting to the Past*. Such games put forth learning objectives combined with everything from the most rudimentary of mechanics to the most complex of simulations or imagined worlds and thus can be hard to pin down as a category or even as a set of design principles. The intent to educate is enough for a game to be categorized as an educational game, regardless of platform: success in achieving this intent, however, is more elusive and difficult to assess, particularly in PC and video games.

Defining educational games by marketing categories gives us a starting point, but many of the games labeled in this way are not notable for their gameplay or for the learning they are meant to prompt. One subgenre of educational games that particularly embodies the philosophy of "chocolate-covered broccoli" or rote learning apparently made palatable by gameplay is "edutainment" (Bruckman 1999, 76). As Mitchel Resnick (2004) points out, the very notion of edutainment suggests the player is a "passive recipient," which undermines any activeness in play. Edutainment is not a category exclusive to games: television programs such as *Sesame Street* and *Mr. Rogers* also fall under the term and often include lessons and drills as part of their narratives. Several companies colonized this genre in gaming in the 1990s, capitalizing on the new popularity of the home computer and parents' desire to buy games that would act as enrichment materials for their children. Chief among these companies were Davidson, known for the *Math Blaster* series; Knowledge Adventure, creator of the *JumpStart* series; Brøderbund Software, creator of the *Carmen Sandiego* series; and The Learning Company, creator of the *Reader Rabbit* and *Super Seekers* series. To an outsider, these games would look similar in that they were packaged in bright, primary-colored boxes with animated characters promising to teach children mathematics, reading skills, geography, or logic. In most cases, the animation and game mechanics provided a facade for familiar drills. In the *Math Blaster* games, players would pilot a ship and shoot the right answer to math problems. The space setting adds no meaning to the problems, and so, as Bruckman points out, the game falls into the "drill and practice" model (1999, 76). This was a common problem for games designed by educators, even those unsatisfied with the traditional modes of teaching, such as Learning Company founder Ann McCormick (Ito 2006, 140). Many of these games used the traditional educational system as scaffolding: the *JumpStart* and *Reader Rabbit* series included recommended ages and grade levels as a primary part of their advertising.

Other examples of early edutainment differed thanks to their affiliation with companies and creators concerned primarily with game design. These games were rarely tied so directly to traditional curricula and grade levels and often included more abstract learning objectives. One series came from Sierra On-Line, a company already known for making adventure games, including the best-selling *King's Quest* series. It brought the same story-based approach to edutainment, with titles in the Sierra Discovery series such as *The Island of Dr. Brain* (1992) and *Pepper's Adventures in Time* (1993). The latter resembled a classic adventure game, with an inventory system and object- and dialog-driven puzzles. However, it placed the player in the role of a girl whose uncle had traveled back in time to 1764 and influenced the would-be American revolutionaries to adopt a laid-back "hippie" lifestyle. The animation and puzzles of *Pepper's Adventures in Time* were similar to those of other Sierra games of the era, including *Gabriel Knight* (1993). Most of the time the player is immersed in the narrative, but each act ends with jarring quizzes that bar progression until the player proves he or she has also met the educational objectives. The connections with learning objectives were better realized in the work by a team from LucasArts, Sierra's main competitor in the adventure game market. Former LucasArts employee Ron Gilbert cofounded Humongous Entertainment and created several series of character-driven children's point-and-click adventures, starring Putt-Putt, Pajama Sam, Freddi Fish, and others. In these games, the story came first, with the educational goals integrated as obstacles for problems relevant to the main character's goals. As Mizuko Ito notes in a survey of edutainment, the list of games and other children's software that can be included in this category is very long, particularly when we include all of the titles where entertainment is coded first but with learning goals for "critical thinking" or "creativity" taking center stage in the marketing (2008, 89). These examples are a reminder that games excel at fostering learning mindsets that "are intrinsic to the game while the students are learning the content. Through game playing, students learn how to collaborate, solve problems, collect and analyze data, test hypotheses, and engage in debate" (Klopfer 2008, 19). This type of learning is not limited to children, but it is only rarely explicitly marketed as a benefit for older audiences, as in *Brain Age* (Nintendo, 2005) and its many sequels and counterparts.

Meanwhile, educators were adopting to the needs of their classrooms titles that had never been marketed as educational. Kurt Squire has analyzed the potential of the *Civilization* games (MicroProse) for classroom use, with attention to both the clear learning outcomes (attentiveness, geography, technology, and civilization) and the consequences of experiencing a biased and unrealistic portrayal of leading a civilization (2011, 113–130). The mechanisms of *Civilization* show roots in the history of wargaming and thus offer a complexity that benefits from more extensive exploration. Because *Civilization* functions as a game first, historical accuracy is often ignored in the favor of mechanics, particularly when it comes to balancing the strengths and resources afforded to different civilizations. However, the game offers direct content for discussions of political influence, war, and even the impact of geography and resources on the development of international boundaries and allegiances. This content correlation is not necessary for the successful use of a game as part of a formal education: other educators picked up titles ranging from *SimCity* (Maxis, 1989), *World of Warcraft* (Blizzard Entertainment, 2004), *Age of Empires* (Ensemble Studios, 1997), and *Rollercoaster Tycoon*

(MicroProse, Chris Sawyer, (1999) to *Neverwinter Nights* (BioWare, 2002) for everything from examining structures and assumptions of businesses to exploring ideas through narrative or working on language and reading skills. In these cases, educators co-opted already successful titles and presented them within a new context and scaffolding, using even the "inaccurate" or fantastical aspects of video games as a starting point for discourse. In some cases, the game's structures serve as motivation for learning outcomes—for example, in the use of *World of Warcraft* as an environment for developing language acquisition and literacy (Peterson 2010). The highly text-driven mechanics of the game's quests and social spaces, combined with the many visual and audio cues, are augmented by the feeling of urgency and the need for cooperation to succeed in the game's more advanced challenges. The presence of so many curriculum materials themed on these types of games suggests the need for a broader understanding of educational games: although games designed for educational environments or intentional learning are part of the picture, games as a whole offer much more valuable opportunities for unintentional learning.

At the beginning of the new millennium, James Paul Gee brought into focus this secondary use of games as an educational force: "Good video games have a powerful way of making players consciously aware of some of their previously assumed cultural models about learning itself" (2003, 162). Gee's work fueled the fire for a broader interest in games as a way to think about learning and particularly as a way to rethink assumptions about what constitutes a "good" educational or classroom experience. This new era of educational games is not about making games push the lessons and structures of the classroom but instead about making education more like a game. As Jane McGonigal points out, "Compared to games, reality is too easy. Games challenge us with voluntary obstacles and help us put our personal strengths to better use" (2011, 3). This way of thinking about educational games encourages an adaptation of game structures. Some educators implement this type of strategy literally, as Lee Sheldon (2012) encourages them to do in his book based on his own experiences redesigning the classroom as a multiplayer game, drawing mechanics and social structures from role-playing games, and asking students to embrace the metaphor of "leveling up" and "questing" in the classroom. This type of experiment might more broadly be labeled part of *gamification*, a term that emerged as a marketing practice but has since been defined as the use of elements and mechanics from gaming within nongaming systems as a way of raising user engagement and investment in the experience (Deterding, Sicart, Nacke, et al. 2011, 2425). Gamification is not without its critics: Ian Bogost has suggested that gamification would be better termed "exploitationware" because of a reliance on easy aspects of games and a focus on "extrinsic incentives" (2011). Such critique recalls the classroom *Jeopardy!* and *Math Blaster* side of the educational game spectrum and serves as a reminder that when the implementation of games for educational purposes is a matter of the easy fusing of game mechanics and learning without any corresponding interrelationship, the use of games is likely to offer little value to the educational context.

Educational games continue to evolve from this broad history, rooted in both educators' and game designers' efforts to make better use of the intrinsic qualities of games as a way to promote learning. Organizations such as Games for Change have drawn attention to the power of games as a way to explore topics ranging from climate defense to colonialism. Although the legacy of edutainment and, more recently, gamification continues to play a role in the movement in the direction of educational games, the visibility of good game design as part of educational experiences is on the rise. The relationship between games and learning is much more complex than the first forays into the space suggested, and although the value of games as an educational paradigm has been established, methods for assessing and promoting that value are still under construction.

Works Cited

Bogost, Ian. 2011. "Persuasive Games: Exploitationware." Gamasutra, May 3. http://www.gamasutra.com/view/feature/134735/persuasive_games_exploitationware.php. Accessed January 9, 2013.

Bruckman, Amy. 1999. "Can Educational Be Fun?" *Game Developers Conference* 99: 75–79.

Carnes, Mark C. 2004. "The Liminal Classroom." *Chronicle of Higher Education* 51 (7): B6–B8.

Deterding, Sebastian, Miguel Sicart, Lennart Nacke, Kenton O'Hara, and Dan Dixon. 2011. "Gamification: Using Game-Design Elements in Non-gaming Contexts." In *PART 2: Proceedings of the 2011 Annual Conference Extended Abstracts on Human Factors in Computing Systems*, 2425–2428. New York: ACM.

Gee, James Paul. 2003. *What Video Games Have to Teach Us about Literacy and Learning*. Hampshire, UK: Palgrave Macmillan.

Hoelscher, Karen. 1985. "The Making of MECC." *Computers in the Schools* 2 (1): 61–64.

Ito, Mizuko. 2008. "Education vs. Entertainment: A Cultural History of Children's Software." In *The Ecology of Games: Connecting Youth, Games, and Learning*, ed. Katie Salen, 89–116. Cambridge, MA: MIT Press.

Ito, Mizuko. 2006. "Engineering Play: Children's Software and the Cultural Politics of Edutainment." *Discourse* (Abingdon) 27 (2): 139–160.

Klopfer, Eric. 2008. *Augmented Learning: Research and Design of Mobile Educational Games*. Cambridge, MA: MIT Press.

McGonigal, Jane. 2011. "Playing Games Is Hard Work: An Excerpt from *Reality Is Broken*." Gamasutra, January 26. http://www.gamasutra.com/view/feature/134640/playing_games_is_hard_work_an_.php. Accessed August 1, 2014.

Peterson, Mark. 2010. "Massively Multiplayer Online Role-Playing Games as Arenas for Second Language Learning." *Computer Assisted Language Learning* 23 (5): 429–439.

Resnick, Mitchel. 2004. "Edutainment? No Thanks. I Prefer Playful Learning." *Associazione Civita Report on Edutainment* no. 14. https://llk.media.mit.edu/papers/edutainment.pdf. Accessed August 11, 2015.

Sheldon, Lee. 2012. *The Multiplayer Classroom: Designing Coursework as a Game*. Boston, MA: Cengage Learning.

Squire, Kurt. 2011. *Video Games and Learning: Teaching and Participatory Culture in the Digital Age*. New York: Teachers College Press.

16 EMBODIMENT

Don Ihde

This will be a postphenomenological analysis of variations on embodying screen games—video, computer, arcade, and Nintendo Wii-style bodily play skills. Gaming with screens, in contrast to pinball and other earlier mechanical arcade games, began in the mid-twentieth century under the dark designs of artificial intelligence combined with wargaming, but the approach taken here concentrates on entertainment games and looks at typical variants in mid-twentieth- to twenty-first-century screen or projected-image games. In each variant, the player must develop bodily skills, some of which have far-reaching implications with respect to contemporary life practices.

The analysis here is interrelational and parallel to the many styles of such analyses found in science and technology studies but focuses on the postphenomenological interest in bodily perception and action so that the player remains one focal concern. But a game presents a "world" in the sense that some scene of action is presented or imaged—for my purposes here, usually on a screen or visual display device. The interaction involves bodily perception, movement, and skill development, which varies with the multiple "screen worlds." Embodiment, in relation to technologies, however, takes different shapes. Indeed, each type of technology calls for different actional skills. Degrees of bodily engagement, ranging from amateur to virtuoso skills, also come into play. In my own case, I have done studies of such engagement in relation to musical instruments (Ihde forthcoming) as well as in relation to a range of technologies in my books *Listening and Voice: Phenomenologies of Sound* (2007, chaps. 12–13) and *Experimental Phenomenology: Multistabilities* (2012, chaps. 12–14).

Embodiment is a dynamic and complex phenomenon. It entails learning bodily skills and thus is developmental, and different technologies call for different degrees of engagement. Take two keyboard examples. Typing or word processing calls for speed skills and eye-hand coordination, but it does not ordinarily call for more complete bodily engagement. In contrast, a virtuoso piano player engages much more with whole body motion, as any prime performance shows. The same spectrum applies to game skills, although the simpler and earlier games remained closer to the word-processing example, except with respect to speed.

Two-Dimensional Games

The simplest early games were often two-dimensional (2D): *Pong* (Atari, 1972), *Tic-tac-toe*, *Pac-Man* (Namco, 1980), and *Raster Blaster* (Apple, 1981) were paradigmatic. Here the image worlds included abstract figures that moved on a flat background screen: figure *on* a screen. The figure—an abstract ping-pong ball, an X and an 0, or an abstract gobbler—moved on the screen. The player, using keys or a joystick, controlled the motions according to the game's scheme. Note that this game world is an analog of a *writing technology.* Cuneiform was an early hard-technology writing process that employed a stylus, clay or pottery as the proto-screen for an inscription, and the skilled scribe who made the inscriptions. These elements had to follow the writing game to be intelligible and thus were learned forms of embodiment. Almost as old is soft-technology writing, which utilizes brushes or quills and paint or ink to produce images on usually soft-screen analogues—papyrus, parchment, canvas, and so on. Then by industrial times a two-handed approach, a typewriter with keyboard, was used to make images on paper appear. But note that the writer takes action through a device—to produce an image world on a screen-type tablet. The 2D game worlds remain analogous to this history; their textual counterpart is, of course, word processing, which produces its text world on the screen and is later transferred to printed form. *On* screen (or on tablet or inscription surface), figures typically are located on stable, opaque backgrounds. Here the bodily motions are minimal. Eye–hand motion is focal. The player remains seated, in a stationary relation to the game screen, and thus whole-body motion is minimized. Indeed, the earliest forms of games often involved monochromatic figures against a contrasting monochromatic and opaque background.

Before leaving this style of player–technology–game world, I want to make note of two features. First, through game capacities—for example, the speeds at which image balls, gobblers, and the like could be accelerated to beyond human reaction times—skilled players could improve their relative eye–hand coordination times to a personal maximum. Second, by the reduced but specific eye–hand movements, new skills could emerge among expert players who were quite different from a general populace. The unpredicted outcome, not foreseen, was a preskilling of rapid eye–hand motility, which eventually became useful in nongame contexts, such as laparoscopic surgery. The unintended effects of gameplay can be applied to virtually any technology, and they may be destructive, simply unexpected, or, as in this case, surprisingly useful. Eye–hand coordination skills are, from a perspective of partial bodily movement skills, seemingly reduced bodily actions, but they are also more useful for the meticulous movements called for in microsurgery. My auditory surgeon tells me he gave up actual surgery some years ago because younger, game-skilled surgeons now perform better.

Two-dimensional games such as those previously described remained abstract in image and highly reduced from more full human–game interaction. Hybrid attempts to "jazz up" such games sometimes took the shape of adding colors to the earlier monochromatic game worlds and 3D backgrounds (such as clouds or flat buildings). But a larger shift occurred, with games taking on a new dimensionality through a different player–screen variant: *through the screen imaging spatiality.*

Shooters and Simulations

A marked shift in game worlds began once the figures and characters became more than cartoonlike and took on 3D image characteristics. The games selected here are "shooter" and "simulation" games, of which there are hundreds. Again the focus is on the player–game world relationship, with special attention to embodiment skills. The first and dramatic difference was that the game world began to appear *through* the screen rather than on it. The screen became mostly transparent (back glare or smudging could occur as an irritant, of course), and the images moved through the screen in a spatiality that is today usually called "cyberspace." The two variations of spatiality now noted are screen opacity (2D) and screen transparency (3D), which are multistable spatialities. Multistability occurs with most human–technology interrelation in that there are multiple possibilities of uses and outcomes in corresponding shifts in spatialities. Such multistabilities are quite dramatic in that what counts as figure and ground change. In *on-screen* spatiality, the opaque screen is stable and flat ground, but with the introduction of even partial three dimensionality the ground can become dynamic with the figures, which clearly are mobile. With Wii-type games, both player and field become dynamic.

In shooter games—for example, *Duke Nukem* (3D Realms, 1991), one of the longest-running of such games—the player "shoots" whatever counts as an enemy. These games also introduced a multistable set of shooter points of view (POVs). The player simply sees a weapon immediately in front of him or her, close up or embodied. Different weapons may be used, but all are "as if" being held in the player's hand. Or an avatar variant may be introduced ("piggy back" is the medium-close variant, by which I mean that the avatar is immediately in front of the implied player, not "out there" in the field), in which some figure stands in for the player, holds the weapon, and shoots the enemies while either walking or riding a vehicle. Or the avatar may be farther away, in the field itself, but again the controller-player fires the weapon of choice. Thus, the screen worlds introduced many more multistabilities than the simpler games. The player's bodily actions are more complicated, although in these games the player still remains relatively stationary, seated in front of the screen, and bodily action is confined mostly to eye–hand motility. (Note that here I am still attending to embodiment and the development of player-world skills, not to plots or narratives. In the case of *Duke Nukem*, both fairy-tale and science-fiction plots prevail. These games also frequently entail multiple deaths, from which the player may restart the game, and the previously noted speed-up so that no player can keep up with the attackers. One wonders—no answer attempted here—how this redying and loss in the game process affects the player's psyche?)

Flight Simulator (Microsoft Flight Simulator II, 1980) is the game of choice for a second related category: simulations. In this simulation game, a variety of planes are offered—a Spad, a Cessna, an F-14, and so on—each with a program determining its flight capacities. Note that this set of choices parallels that of weapons in the shooter games. Multiple POVs are also variants: one may be seated in the cockpit, the instrument panel and controls immediately before one and the scene outside viewed through the virtual cockpit window, or one may simply have the airplane fly at a distance with an avatar pilot projected inside the plane. In playing the game, it is possible to get lost, to exceed the airplane's capacity, and thus to crash into trees, buildings, and the like.

These games, like the games that can lead to "Nintendo surgery" (the name given to game-skilled eye–hand surgery), also have had impacts on new technology uses. For example, highly complex and "realistic" simulators are used today to test and train airline pilots' skills. Programs are introduced for emergency situations that are unlikely to occur in actual flights but that can teach pilots how to react and deal with such emergencies. Actual pilots report sweating and feeling stress during such training. In contemporary drone warfare, the pilots are often skilled gamesters who, even though a continent away from the actual battlefield, control flights over the battlefield (in space exploration, pilots control space vehicles on Mars from even longer distances away!). In short, the spatiality transformations possible in games and remote sensing relocate the capacities of human–technology relations to megaspace.

Such virtual and remote control again shows phenomenologically how geometrical spatiality differs from experienced spatiality. The near-distance of remote control is a type of bodily or bodily extended, experienced embodiment. It is distinctively an embodied experience.

Learning Lab Denmark

LEGO has long been a major toy manufacturer in Denmark. One of its spin-offs was a gaming think-tank called Learning Lab Denmark. Participants in the lab came from many disciplines and were frequently assigned experimental tasks that involved imagining possible games.

During the 1990s and early 2000s, a major concern was that children were spending very large stretches of time seated stationary before screens. Indeed, recent surveys show that US college students average some 37 hours per week in front of screens (Central Connecticut State College survey, July 13, 2014). Danes became concerned that the resultant lack of whole-body activity would lead to obesity, and so LEGO launched a study to imagine modes of engaging more complete bodily motility in electronic games. (I was a frequent participant in this study for more than a decade.) Workshops included analyses by neurologists, gym teachers, physiologists, and philosophers. Simple ideas such as stationary bicycle races tied to simulation games, 3D ball hit-back games, and the like were imagined. Wii games were invented independently.

In later lab meetings, some of these games were demonstrated, and, indeed, much greater degrees of bodily motion were called for. For example, players had to stand inside a designated space while various imaged 3D balls were fired at them, and they had to hit them back. A certain irony emerged among the players: the adults seemed to enter quite fully into the action and used their arms and legs, soccerlike, to return the balls. But the younger children—probably more accustomed to the minimovements of screen games—quickly realized that a small and bare-handed swat was sufficient to return the ball, and so they tended to use much more minimal motions. Embodiment in games like this does increase from eye-hand movement, but only incrementally. Also, if a player steps out of the designated space, he or she effectively leaves the game. What we are seeing here is both a change in relation to limited, partial-body motions and greater "whole"-body motions.

During this same period, virtual-reality caves were also popular. Here, either with goggles or multibeamed location sensors carefully tracking the player and again in delimited space, bodily action was called for in the goal to discover keys for hidden chests or other in-the-game actions. These games were sort of in-game variants of *Dungeons & Dragons* (Gygax and Arneson/Tactical Studies Rules, 1974) and yet were also another way of enticing whole body motion.

Pterosaurs

In 2014, the New York Museum of Natural History had an elaborate exhibit on pterosaurs, the many species of extinct flying reptiles of the age of dinosaurs. Of pterosaurs, there are roughly 186 genera, from those with a 30-foot wingspan to those with a very small wingspan. The exhibit included two pterosaur games through which one could "fly" two different kinds of pterosaurs. The first was a fairly large animal. As in Wii games, the player had to stand in a specified area and then mimic with his body the pterosaur's flight action. Flapping his arms made the pterosaur takes flight, tilting made it turn or glide down to catch a fish and splash into the sea. (Playing this game was the closest experience of embodied flight I have ever had!)

The second pterosaur was a forest dweller, much smaller, and the task was to catch flying insects. Again, standing in the prescribed position and mimicking flight motions, the player had to avoid smashing into trees and the like. This game was much harder to embody, and failures were much more likely than in the first game. It became obvious to the player that considerable learning time would be needed before success could be attained. It should be noted that the projected 3D image was not screen bound but was much closer to a virtual-reality projection. Such game worlds remain limited, and if the player either tries to escape the game world or steps out of the designated space, the game crashes. These constraints relate to embodiment because game embodiment *is distinctly different* from ordinary embodiment. Similarly, game embodiment entails what all aesthetic experience calls for: a suspension of disbelief.

Embodiment Trajectories

What this progression shows—from the simplest 2D games to the most complex, more fully embodied Wii-style games—is that each game world is interrelated to player embodiment skills. And, as noted, the player can emerge from each skill context into the ordinary lifeworld with a newly practicable skill, from eye–hand coordination or "Nintendo surgery" to the ability to manipulate distant and remote-sensing robotics. This pattern is one of game to new modes of life. Games in the contemporary world have tended to take up more and more "real" concerns. For example, the Information Technology University in Denmark has some twenty faculty whose academic role relates to e-gaming. Of course, this focus is a reflection of what is today a multi-billion-dollar industry. Similarly, a robotics company in Nara, Japan, hired two hundred "roboticist PhDs" in 2008, another multi-billion-dollar endeavor. Gamely embodiment, however, clearly retains a human–technology interrelation. From this interrelation can flow new lifeworlds that are limited only by the constraints of the imagination.

Works Cited

Ihde, Don. Forthcoming. "Postphenomenology: Sounds beyond Sound." In *The Routledge Companion to Sounding Arts*, ed. Marcel Cobussen, Barry Truax, and Vincent Meelberg. London: Routledge.

Ihde, Don. 2012. *Experimental Phenomenology: Multistabilities*. 2nd ed. Albany: State University of New York Press.

Ihde, Don. 2007. *Listening and Voice: Phenomenologies of Sound*. Albany: State University of New York Press.

17 EMULATION

Jon Ippolito

For institutions tasked with ensuring the survival of digital culture, emulation is the new kid on the block. In a blog entry summarizing the conclusions of the "Preserving.exe" conference in 2013, former Library of Congress staffer Leslie Johnston declared, "I was convinced this week that emulation may serve our needs better than hardware." This brand of digital mimicry, easily the most promising new technology shared at "Preserving.exe," suddenly seemed poised to find its way into every preservationist's tool kit. What made this pronouncement anachronistic, if welcome, was that the technique of emulation had already been around for decades without most curators or librarians paying it any mind.

To emulate is to translate the code for one hardware or software environment into semantically equivalent instructions for a new environment. The emulator Virtual PC, for example, might reinterpret a command that originally added a number to the storage register of a Windows 95 central processing unit to mean that it should add that number to a variable running inside Virtual PC in OS X; an instruction to read a game off a floppy disk might instead be interpreted to read the game from a virtual-disk image that exists only in the emulator's random-access memory. As computer programs that fool original code into assuming that it is still running on its original equipment, emulators enable software from an out-of-date device to run on a contemporary one.

In the early 2000s, recommending that collecting institutions preserve their collections using emulation was like recommending they power their heating and air-conditioning units with nuclear fusion. When they heard the technique explained by some gearhead at a conference or by their own information technology department, many archivists and conservators just scratched their heads.[1] Emulators seemed far too complicated and nerdy compared to the apparently simpler task of mothballing the computer guts necessary to run Apple II programs or Voyager CD-ROMs. It was not until 2013 that nationally prominent archivists such as Johnston realized that "we cannot all become museums of computer hardware"—that is, maintain a suite of working floppy drives, cathode ray tube (CRT) monitors, and other hardware has-beens. Internet Archive staffer and emulation enthusiast Jason Scott posits the official turning point for institutional adoption of emulation in November 2014, when 2 million users accessed emulators at the Internet Archive within a mere three days.[2]

So what happened in those intervening years that turned this bastard child into the favorite son of the cultural heritage community? In short, video games happened. Or, rather, they were born, then died, then were born again.

From the mid-1970s to the 2000s, video games increasingly occupied a major place in living rooms from Tokyo to Tacoma. (Perhaps not in the living rooms of librarians and curators, but those professionals were increasingly on the losing side of that demographic shift. Though many of those 2 million visitors to the Internet Archive ended up exploring other software collections, they really came there to play games.) Much of Generation X had grown up shooting aliens and stomping mushrooms and after adolescence pined to relive that entertaining childhood; the younger Millennials, meanwhile, yearned to play the games they saw their older siblings enjoy. Thanks to the advent of the Internet, information spread quickly about some little programs hacked together by game fans that would allow a modern Windows user to play the arcade version of *Space Invaders* (Taito, 1978) or the Nintendo Entertainment System (NES) cartridge version of *Super Mario Bros* (Nintendo, 1985). One of the most popular programs, the Multiple Arcade Machine Emulator (MAME), started as a side project by Italian programmer Nicola Salmoria in documenting hardware. When Salmoria was called for national service in 1997, he passed MAME on to fellow programmers, whose ranks swelled to the hundreds over the succeeding decades and whose efforts have reached the point where MAME currently supports more than 7,000 games. The emulator community makes the source code free to use for everyone from aficionados who embed computers running MAME inside vintage arcade cabinets to casual gamers who just want to play Atari games on their personal computers. So while museum conservators and librarians were busily archiving digital files on CDs for the long term—with "long term" meaning merely half a decade in the case of self-pressed CDs—their eight-year-old cousins were wielding a powerful technology that in theory could extend the life of bits into the indefinite future.

So it should not be a surprise that video games were the catalyst for museums to accept this new technique, which was both more geeky and more folksy than anything institutions of high culture were accustomed to putting in their white exhibition halls. Carl Goodman's exhibition *Computer Space: A Digital Game Arcade* at the American Museum of the Moving Image in 1995 would have been difficult to mount without emulators powering the games on view, and others have followed suit. Chris Melissinos's blockbuster exhibition *Art of Video Games* in 2012 proudly displayed games on their original software (if not hardware), but that exhibit was organized by the Smithsonian American Art Museum, the world's largest museum. Even "the nation's attic," as the Smithsonian has been dubbed, will be hard-pressed to maintain life support for aging Atari and Xbox consoles, which for the long-term preservation of this software makes emulation the only game in town.

What will be lost when emulation is the only solution for reliving out-of-date software? The Guggenheim exhibition *Seeing Double: Emulation in Theory and Practice* in 2004 set out to find the answer to this question.[3] Unlike previous shows involving emulators, in this exhibition video games represented only two of the eight works on view, and those two were artists' hacks of popular games. Rather than focus on games per se, the exhibition was designed to test the viability of emulation as a preservation strategy for creative works of all kinds, from interactive video to performance art.

To say there were eight works on view is misleading, however, because most appeared in pairs (and sometimes triples)—the first in each set running in vintage hardware and the others running in a migrated or emulated version of that original. The show's premise was that viewers would have to decide whether the pairs exhibited side by side constituted the same work or different works based on the re-created version's perceived fidelity to the original. An audience survey conducted at the exhibition concluded that in most cases emulation produced copies that were spiritually, if not physically, equivalent to the originals.

Yet age once again played a role in the relative acceptance of emulation. Respondents who were the least digital savvy—who tended to be the older viewers—preferred re-creations that aped not just the behavior but also the exact look and feel of the original. For example, older viewers gave high marks to the re-creation of *The Erl King*, a 1985 video installation by Grahame Weinbren and Roberta Friedman, for which the Guggenheim emulated not just the source code but also the appearance of its fiberboard kiosk and bulky CRT monitor (in this case by embedding a flat screen inside a wooden box). The re-creation looked so similar to the original, in fact, that the curators had to outfit both versions with Plexiglas windows so museumgoers could see that the second had different guts—the ancient Sony SMC-70 computer and its rack of analog laserdisc players replaced with a single Linux box.

Older viewers were less approving of the re-creation of John F. Simon Jr.'s wall piece *Color Panel* (1999), which ran the same code but was displayed on a somewhat larger and brighter laptop from 2004 in place of the original PowerBook 280c from 1994. By contrast, younger viewers who had grown up playing 8-bit games on 24-bit screens accepted the changes in look and feel as long as the code behaved the same.[4] Neither *The Erl King* nor *Color Panel* is a game, yet it is tempting to ascribe younger viewers' tolerance of changes introduced by emulation to their years of experience drinking old digital vintages from new bottles. This difference matters because we are not used to looking to youth for expertise or for that matter to popular pastimes as a model for preserving elite culture.

To neglect the opinions of older viewers, however, would be to spurn those who have actually felt an Atari joystick or Nintendo Game Boy in their hands rather than wearing out their fingertips pressing the WASD keys. True to its mission, *Seeing Double* surveyed creators familiar with the original hardware about what was lost in translation. For his work *I Shot Andy Warhol*, a modification of the original Nintendo game *Hogan's Alley* (1984), artist Cory Arcangel released a downloadable read-only memory online but nevertheless felt it important at the Guggenheim exhibition that museumgoers experience the heft of a plastic light gun. Joan Heemskerk of jodi.org authorized the curators to show her modded game on both the original ZX Spectrum computer from the 1980s and a Spectrum emulator running in Windows XP but lamented how the emulated version lost many of the particularities that she had admired about the original hardware. For example, seeing the game on a Windows PC gives the viewer no clue that Spectrum games once ran off audiotape rather than on a disk drive. In her presentation for the associated symposium "Echoes of Art: Emulation as Preservation Strategy," Heemskerk also described how the crisp look and antiseptic feel of a flat screen differs from the warm buzz of a CRT: "The ZX works with a TV signal, so the screen is fed by antenna cable. A line is not a line. A piece of red on an LCD display is just straight, one color, but on a TV it's totally lively. Even if you put a white against a black, the TV tube cannot hold the line, and it bleeds or bows" (2004).[5]

Now, some game fans have created software deliberately designed to mimic the look of CRT screens or even earlier vector-based displays. To feign the look of a game such as *Asteroids* (Atari, 1979) or *Tempest* (Atari, 1981), the popular MAME emulator can simulate the way vector scans would slow down while drawing multiple objects on the screen, causing the phosphors that lit up on the tube to fade before the scanning beam could refresh the line on the next pass. Emulating dying phosphors to re-create a defective display may seem like adding hiss to make an MP3 sound like a scratched LP. Even so, if the persistent enthusiasm for vinyl records is any indication, fans from the future who never experienced the evanescent glow of ancient monitors may nonetheless harbor a sort of inherited nostalgia for such imitated imperfections.

Of course, it is the express mission of many museums and libraries to capture such hallmarks of history. Whether emulators are up to the task was one of the questions raised by Preserving Virtual Worlds, a consortium project begun in 2008 by the University of Illinois at Urbana–Champaign, Maryland Institute for Technology in the Humanities, the Rochester Institute of Technology, and Stanford University in association with the Library of Congress. After examining eight case studies—including *Spacewar!* (Russell/MIT, 1962), *Colossal Cave Adventure* (Will Crowther, 1976, *Star Raiders* (Atari, 1979), *Mystery House* (On-Line Systems, 1980), *Mindwheel* (Brøderbund, 1984), *DOOM* (id Software, 1993), *Warcraft III: Reign of Chaos* (Blizzard Entertainment, 2002), and *Second Life* (Linden Lab, 2003)—the consortium's final report lamented "the disconnect between active collectors and programmers building software such as emulators" and encouraged "museums and other collectors to start the exploration process into emulation," with a goal "to unite various grassroots development teams into a larger community dedicated to the preservation mission" (McDonough, Olendorf, Kirschenbaum, et al., 2010).

As of the mid-2010s, too few official custodians of culture have invested in emulation.[6] This may not be surprising given the limited resources and broad mandate of today's collecting institutions. Emulating *The Erl King* was possible only thanks to a "perfect storm" that brought together a dedicated museum, a team of experts, an eager sponsor, and a willing artist with access to the original source material. Renewing *The Erl King* was a heroic effort, but heroism, as preservation expert Richard Rinehart (2001) reminds us, can't rescue everything because by definition it is unrepeatable.

Fortunately, for common platforms, emulation can be repeatable and, even better, extensible. Some emulators emulate only an operating system—masquerading as, say, the NES cartridge system while in reality running as an application in Mac OS. More sophisticated emulators, however, which simulate an actual computer chip such as the Pentium or PowerPC, can be nested together to transition from one platform to another. Computer scientist and emulation champion Jeff Rothenberg is fond of demonstrating this principle by running a Windows-based emulator for the EDSAC (Electronic Display Storage Automatic Calculator), a forerunner of modern computers built in 1949, inside the Windows 2005 emulator Virtual PC, which runs on the Macintosh.[7] This daisy chain of emulators, in which a Mac impersonates a Windows machine, which in turn impersonates the EDSAC, makes it possible to view a computing history of 50 years on a single device.

Fifty years from now, when the Macintosh as we know it will be long dead, a preservationist might write an emulator to simulate a Macintosh 2005 on the prevailing platform of 2065. At that point, she will be able to embed Rothenberg's daisy chain inside this new emulator and so on into the future. By overcoming the dependence of an emulator on its parent platform, this stratagem resolves the paradox whereby tools designed to liberate artifacts from obsolescence eventually fall prey to obsolescence themselves.[8]

Another avenue that might lead to platform independence is "emulation as a service," an emerging paradigm whereby interacting with vintage software requires nothing more than a web browser. There are at least three different ways forward. Based on the emscripten framework, JSMESS (JavaScript Multi Emulator Super System) is a client-side emulator that requires no plug-in but is limited by the performance of JavaScript in the browser. The emulator bwFLA (Baden-Wuerttemberg Functional Longterm Archiving and Access) is less dependent on the client because it operates on the server, though this configuration can delay real-time streaming updates. The server-side Olive project, meanwhile, requires a special plug-in but is robust enough for remote users to fight off demons in *DOOM* in real time.

Gamers have not just helped develop emulators; they have helped test them, too, by pushing emulators to their limits while exploring dimensions of games that would be inaccessible without emulation. For example, an advanced emulator lets users slow down the speed of the game to a single frame at a time. This superpower has made possible tool-assisted speedruns—optimized gameplays recorded as screen videos and shared via a set of timed key presses that can be played back on the actual console. Game historians may be interested to learn about the "walk through a wall" glitch by which emulator-equipped speedrunners can shave an hour off *Super Mario 64*, but digital preservationists will be more interested to learn the defects of particular emulators that might be exposed by speedrunning. The keystrokes saved by speedrunners should re-create their perfect game every time they are executed, but if the emulator adds any randomness that wasn't present in the original console game, then this infidelity to the original system will be exposed when the emulator fails to reproduce the perfect game. Speedrunners also use emulators to save states of a game so they can re-record an optimal performance at every stage of the game. When such "save states" do not capture the state of the system correctly, again the perfect game will fail, revealing another defect that can be fixed in future releases of the emulator.

Even if most emulators to date have been developed by game-loving geeks, a side benefit of chip-level emulation is that nongaming software can be run as well. At the JSMESS emulation portal, you can collect Reese's Pieces in Atari's game *E.T. the Extra-Terrestrial* (1982), but you can also list your expenses in Texas Instruments' Household Budget Management software or rearrange folders in the Macintosh SE operating system. For its part, the bwFLA website lists use cases that include reproducing scientific experiments, accessing historic digital documents, reenacting business processes to understand past decisions, and enabling "crowd curation."

Jason Scott captures the opportunity represented by emulation's coming-out party:

> For the contingency of people who were hoping that one day society would understand what Emulation was, and would be able to articulate, even in the lightest fashion, its underlying theories, well, that day has arrived. ... Through it all, one thing is obvious: software is more than code. Programs are more than executables. This medium we have formed forms us, and returning to it brings back joy and the agony, no different than a book, a poem, or a face. ... Soon, a stunned populace, an alerted establishment, is going to come forward and have a lot of questions and want to debate. (2014)

That debate may end up tipping over some of high culture's sacred cows, from its privileging of an artifact's appearance over its behavior to its faith in intellectual property as a boon for future generations.[9] For example, Nintendo's official position is that emulators are "the greatest threat to date to the intellectual property rights of video game developers" (Nintendo n.d.). Ironically, in 2014 Nintendo filed a US patent application for a "Hand-held Video Game Platform Emulation"; however, many observers saw the application as a preventive measure to shut down third-party Nintendo emulators rather than as a possible embrace of this innovative technology (Rösti 2014).[10]

Curators at major museums may still argue whether *Sonic the Hedgehog* (Sega, 1991) or *Grand Theft Auto* (Rockstar Games, 1997) deserve a place in their white cubes alongside *The Erl King* and *Color Panel*. But without games and the emulators that resuscitate them, museums of high or low art would be missing the most promising solution for preserving the *materia prima* of digital culture.

Notes

1. For example, Jeff Rothenberg has championed emulation as a preservation strategy since the late 1990s, as evidenced in his report to the Council on Library and Information Resources in 1999, *Avoiding Technological Quicksand: Finding a Viable Technical Foundation for Digital Preservation*. Emulation was also one of four recommended strategies promoted by the Variable Media Initiative, a Guggenheim effort begun in 1998 and expanded in 2002 into the Variable Media Network, a consortium of museums and archives dedicated to new models for media preservation.

2. Jason Scott describes what he observed at this turning point: "While I've had bursts of somewhat public facing events and wonders over time, none of them have ever have crashed into the lives of millions in just a couple days. And certainly not in the continually ongoing way I am finding people are returning, over and over, to look at the games. Additionally, it hasn't taken people long to find other software collections, and other places where the emulator has been utilized for consoles and general computers. ... I am hesitant to use a term like 'turning point' in regards to the archive in general [but] [n]ot so with emulation, or the state of software preservation. The last three days have been, unquestionably, a turning point in emulation. Emulation made the local TV news" (2014). In a comment on this blog entry, Scott revised his original estimate of 2 million, suggesting the actual visitorship might have been twice as large.

3. The analysis of this exhibition and its attendant symposium is drawn from Rinehart and Ippolito 2014. See also the exhibition website, http://www.variablemedia.net/e/seeingdouble, and Margaret L. Hedstrom, Christopher A. Lee, Judith S. Olson, and Clifford A. Lampe (2006). I'm grateful to Raiford Guins and Henry Lowood for pointing out the latter article.

4. In a related observation, Hedstrom and her colleagues found that test subjects asked to play an emulated game valued usability over other factors more closely tied to historical authenticity. The age of their subjects ranged from 18 to 44, with a mean of 24 years old.

5. For a more general analysis of how monitor variants affected the evolution of video games, see Bogost and Montfort 2009.

6. The lag for cultural institutions to adopt emulation is clear to anyone familiar with its longstanding use in the computer or gaming industry. Every copy of Apple's OS 10.4 came bundled with its Rosetta emulator to allow Intel-based Macs to run PowerPC applications. Emulation is arguably even more of a mainstay for game companies. In 1994, the company Digital Eclipse helped inspire the genre of retrogaming with the emulation-based suite of games from the 1980s entitled *Williams Arcade Classics*; since merging with ImagEngine to form Code Mystics, the developers of Digital Eclipse have released emulation software that can run games for venerable platforms such as the Atari on PlayStations and iPhones and in web browsers (Code Mystics n.d.).

7. Private conversation with Jeff Rothenberg, March 15, 2004. I have also run this daisy chain on my own Macintosh to demonstrate the technique.

8. An emulator is technically as precarious as the platform it runs on. However, assuming that platform is common enough to warrant a future emulator, then in principle the current emulator and all its associated nested emulators and platforms will come along for the ride. To be sure, this claim presumes that future hardware will run fast enough to encompass all the nested emulators; bwFLA collaborator and Rhizome digital conservator Dragan Espenschied contests this view, arguing that in the long run Moore's Law will expire, making such daisy chains unworkable (private conversation, October 28, 2014).

9. For more digital-preservation sacred-cow tipping, see Rinehart and Ippolito 2014, passim.

10. I am grateful to John Bell for pointing me to this ironic development.

Works Cited

Bogost, Ian, and Nick Montfort. 2009. "Random and Raster: Display Technologies and the Development of Videogames." *IEEE Annals of the History of Computing* 31 (3): 34–43.

Code Mystics. n.d. "Technology." http://www.codemystics.com/technology.shtml. Accessed November 29, 2014.

Hedstrom, Margaret L., Christopher A. Lee, Judith S. Olson, and Clifford A. Lampe. 2006. "The Old Version Flickers More: Digital Preservation from the User's Perspective." *American Archivist* 69 (1): 159–187.

Heemskerk, Joan. 2004. "Echoes of Art: Emulation as a Preservation Strategy." Presentation at the Solomon R. Guggenheim Museum, May 8. Transcript available at http://variablemedia.net/e/seeingdouble/index.html. Accessed November 29, 2014.

Johnston, Leslie. 2013. "What Are We Going to Do about Hardware?" *The Signal* (Washington, DC), May 24. http://blogs.loc.gov/digitalpreservation/2013/05/what-are-we-going-to-do-about-hardware/. Accessed November 7, 2013.

McDonough, Jerome, Robert Olendorf, Matthew Kirschenbaum, Kari Kraus, Doug Reside, Rachel Donahue, Andrew Phelps, Christopher Egert, Henry Lowood, and Susan Rojo. 2010. *Preserving Virtual Worlds: Final Report.* https://www.ideals.illinois.edu/handle/2142/17097. Accessed February 20, 2012.

Nintendo. n.d. "Corporate Information / Legal Information (Copyrights, Emulators, ROMs, etc.)." http://www.nintendo.com/corp/legal.jsp#emergence. Accessed November 29, 2014.

Rinehart, Richard. 2001. "Preserving the Immaterial: A Conference on Variable Media." Solomon R. Guggenheim Museum, New York, March 31.

Rinehart, Richard, and Jon Ippolito. 2014. *Re-collection: Art, New Media, and Social Memory.* Cambridge, MA: MIT Press.

Rösti. 2014. "Nintendo Files Patent for Game Boy Emulation on Mobile Phones, PDAs, PC, and More." *NeoGAF*, November 27. http://www.neogaf.com/forum/showthread.php?t=940813. Accessed November 29, 2014.

Rothenberg, Jeff. 1999. *Avoiding Technological Quicksand: Finding a Viable Technical Foundation for Digital Preservation.* Report to the Council on Library and Information Resources. Washington, DC: Council on Library and Information Resources. http://www.clir.org/pubs/reports/rothenberg/pub77.pdf. Accessed November 28, 2014.

Scott, Jason. 2014. "Before It All Arrives." *ASCII Blog*, November 6. http://ascii.textfiles.com/archives/4433. Accessed November 28, 2014.

18 FUN

David Thomas

Just Play. Have Fun. Enjoy The Game.
—Attributed to basketball star Michael Jordan

Between 1861 and 1901, the British magazine *Fun* published a blend of humor and commentary in a style made famous by its archrival *Punch*. For 40 years, the periodical entertained Victorian readers and provided a weekly dose of the title's putative pleasure. What *Fun* never did was bother to seriously interrogate or analyze its subject. *Fun* was fun for reasons that were either too apparent or too elusive to warrant discussion.

Much as the literary arts have had their fun without bothering to delve into its deeper meanings, so have games managed a peculiar distance from a notion that seems central to their pleasure. We play games to pass the time, to win, to connect with other people, to learn, to demonstrate skill, but most of all to have fun. At the center of play and games sits the pleasure of fun. But despite designating the most obvious feature of games, the term *fun* floats problematically inside the vocabularies of game studies, design, development, criticism, and history. Even though games are fun, we don't spend much time on the topic of fun. If you want to take your games seriously, it seems, you avoid talking about fun. This creates a serious problem with fun.

Johan Huizinga made the importance of fun clear in his seminal discourse on the nature of play, *Homo Ludens*. As he attempted to move beyond the more instrumental understandings of play, he wrote, "So far so good, but what actually is the fun of play?" ([1938] 1949, 2). As Huizinga realized, we play games because they are fun. So even as the field of game studies was forming through some of its most important thinkers, the concept of fun sat waiting for further inquiry, clarity, and theory.

Long after *Fun* ran its course, Brian Sutton-Smith (1997) explored the "ambiguity of play." Referencing Gregory Bateson and others as his starting point, Sutton-Smith organized his masterful review of the literature of play around the idea that play itself (and presumably its pleasures) defies easy classification and as a result of its inherent ambiguity remains open to a variety of epistemological positions. Play will never be certain because play is necessarily uncertain. At the same time, Sutton-Smith also clearly admits that play is fun: "A feature that is almost unanimously acknowledged to be the hallmark of play is that it is intrinsically motivated (that is 'fun') (188)."

Others have attempted more reductive definitions. Game designer Raph Koster (2005) famously describes fun as a form of mastery and comprehension of patterns. This perspective mirrors at least one formal theory from computer science by machine intelligence researcher Jürgen Schmidhuber, who describes fun as "the discovery or creation of novel patterns" (2010, 230). Serious games, gamification, and funology movements have offered *fun* as a motivating factor, mixing games and play with traditionally serious situations and admitting the motivational pleasure of fun without necessarily defining the term. Dwight Bolinger (1963), for his part, locates some of the confusion around the term in its use as both a mass noun, indicating a quantity, and an adjective. In English, you can "have fun" as easily as you can "have a fun time." It seems that Sutton-Smith's ambiguity remains deeply rooted in the structure of the linguistic term itself.

This summary tour of the very brief literature of fun shows that the pleasure of play, of *fun*, does have a theoretical basis. But the relative dearth of conversation on the topic suggests that the problem of fun lies closer to a willful ignorance, a practiced avoidance, of a difficult but critical piece of our understanding of games and play. And this absence creates a historical dilemma regarding how to talk about the concept when its conceptualization remains marginalized.

Nowhere does this disconnect between fun and games glare more apparently than in the world of professional game journalism. As a critical actor in the discourse surrounding games—in particular video games—game journalism sits between fans and professionals, academic researchers and average consumers. Looking at game journalism, then, provides a useful lens for discussing the seemingly missing concept of fun in the world of games and its evolution from the margins to exclusion in game writing.

Although it would be too much to say that game journalists control the vocabulary of game description and criticism, they clearly hold a specific and integral role in the popular consumption and comprehension of contemporary video games (Zagal, Ladd, and Johnson 2009). Game writers' discourse, whether in magazines or in newspapers or online, reflects conventional attitudes about games, and it directs the conversation by focusing on key modes of understanding. Thus, looking at how game journalists have thought about the use of the term *fun* reveals a specific rhetorical substratum of how we culturally think and talk about games in general. Even more so, looking at the concept of fun in terms of game journalism reviews—the qualitative assessment of the pleasures of games—sheds light on the shifting and marginal nature of the term.

The Videogame Journalism Style Guide and Reference Manual (Thomas, Orland, and Steinberg 2007) provided the first comprehensive effort to catalog and define the language of writing about video games. Among the nuts-and-bolts definitions and usage recommendations for terms such as *first-person shooter* and *console*, the guide briefly touches on the notion of fun in two ways. First, it derides the use of the label *fun factor* as jargon (31). Second, it suggests that "serious games" need not have fun as a part of their focus (58). Beyond these two mentions, however, the guide remains indifferent to the idea of fun and what fun might have to do with games.

This indifference presents a critical absence when you consider the intimate relationship between games and fun. Although people certainly engage in games for a wide range of motivations, fun remains, self-evidently, a consistent feature of games. To say people play games because games are fun is the sort of obvious and perhaps circular logic that must have left the editors of the *Style Guide* focused on more nuanced concerns, such as teasing out the differences between first-, second-, and third-party game developers or accurately defining what qualifies as a "platform game." This lack of attention to the notion of fun in a reference guide designed to support game journalists points to a perhaps unintentional but systematic avoidance of the discussion of fun in this discursive look at game journalism. This systematic exclusion becomes clear through a series of interviews I conducted in 2013 with video game journalists who have game-writing experience spanning a 30-year history of the medium. These interviews revealed a range of attitudes about the use of the term *fun*. Game historian, designer, and journalist Rusel DeMaria has worked in and around the game business since the 1980s. From his formative years as a freelance writer working for early video game publications such as *PC Games* and *GamePro* and his ongoing involvement as a game historian and developer, DeMaria speaks to the broad sweep of game vocabulary when he addresses the idea of fun. "In some ways, the word *fun* is too generic. Why is it fun? What specifically makes it fun, and for whom? To say a game was fun to play is almost meaningless without details. On the other hand, to say, 'I had more fun playing this game than I've had in years,' could be a setup, as long as the reviewer provides the details to describe why it was so fun" (email, February 23, 2013).

Dan "Shoe" Hsu, the editor in chief of the influential game enthusiast magazine *Electronic Gaming Monthly* from 2001 to 2008 and (in 2013) the editor of the industry website GamesBeat, similarly offers about the term *fun*:

> I strongly discourage its use in our writing because it's just too vague. It's a useful term, sure. When consumers read reviews, bottom-line, they want to know if a game is fun or not. But that's something that's easier to hear from a friend or through word-of-mouth versus from

> a professional reviewer. If your buddy says a game is fun, you might give that a lot of weight. But when you read a review from a professional media outlet, you expect much more detail. What does that mean, "fun"? That can mean so many different things to different people. (email, February 13, 2013)

Hsu simultaneously recognizes the importance of fun while making it clear that the word is too loose, too open, and too undefined to mean much in the professional writer's toolbox. Fun, in this view, is less the problem than a lack of a good definition or general theory of fun to anchor the term. In other words, a lack of understanding of fun seems to underpin a lack of interest in interrogating the term further.

Interviews with current writers and freelancers reveal a similar point of view, as freelance game critic Kate Cox explained.

> I haven't received comment from editors specifically one way or another, but it's a word I tend to try to avoid using in my own writing because ultimately it's meaningless as a way of communicating what I need to. (If I'm tempted to write it, I kind of hear my ninth-grade English teacher in my head mentally beating the use of the word *nice* out of us.) With a gallery of more specific words to hand, I'm unlikely to write that something is "fun" specifically unless I'm very deliberately echoing (or rebutting) marketing copy to make a point, which is rare. I strongly suspect the editors for whom I have worked (at *Kotaku* and at the Gameological Society) would generally prefer I avoid [the term] *fun* and support the instinct. (email, February 18, 2013)

Although these examples show a relative consistency of understanding and use of the term *fun*, it would be unfair to suggest that all game writers and editors share a consistent attitude about the term. Whereas the editor of Gamecritics.com, Brad Galloway, makes clear, "[The term *fun*] is not allowed at our site in any review. It's too lazy and subjective" (web forum, August 23, 2013), others are less trenchant. Freelancer Scott Nicols, who writes for the *Official Xbox Magazine* and other game outlets, notes, "I have never been given editorial policy that specifically addressed the term *fun*, but several policies have encouraged me to explain why I think a game is a certain way rather than simply stating that a game is fun, interesting, boring, good, bad, etc" (web forum August 26, 2013). In other words, the term *fun* is considered conceptually empty but not without utility. Finally, Best Buy in-house game magazine *@Gamer* editor Andy Eddy suggests, "While there was no formal or informal policy on the use of the term *fun*, I personally have no issue with using it. ... I came from the *GamePro* alumnus that used the term *fun factor* as one of its rating criteria. ... I'm fine with using the term. It's a 'game' ... it's meant to be fun, right?" (web forum, August 25, 2013).

Despite this diversity of opinion about the utility of the term *fun*, game journalism as a whole has avoided its use or at least demoted the concept to an empty philosophical category. This apparently causal assumption drives a behavior in journalists, who have been taught to write with detail and specificity. And this attitude would carry little consequence except for two points. First, avoidance of the term *fun* by game journalists, as nominal gatekeepers of the discourse around games, echoes and amplifies an indifference to the term as well as reflects and reinforces this attitude in other areas of game discourse. Second and more important, if *fun* is more than a generic valuation—like *cool*, *awesome*, *sweet*, *amazing*, and more—then ignoring the concept simply socializes ignorance. Game journalists ended up agreeing that *fun* is an empty term without bothering to dig for possible meaning, thus echoing and socially reinforcing an unexamined bias.

In their defense, game journalists may simply be reacting to the discursive flow of game conversations outside of the press context. For instance, although there is a diversity of opinion about the value of fun in the realm of game design, a common attitude about the term itself is summed up by one developer writing on the topic: "Fun is probably one of the most (over)used words in game design discourse. It's also a broad, non-specific, subjective term that actually doesn't actually tell us anything meaningful about a game experience" (Garcin 2011). If game developers don't use the term in a substantive way, the people who report on games, often by talking to game developers, may find themselves steering away from it. It does not come up as often as other equally difficult but more commonly used terms such as *interaction*, *engagement*, and even *addictive*. Game researchers can likewise share a similar indifference to the term *fun*. In an article about serious games published on the game developer–focused site Gamasutra, game studies scholar Ian Bogost declared, "For 30 years now we've focused on making games produce fun. Isn't it about time we started working toward other kinds of emotional responses?" (qtd. in Ochalla 2007).

Much as game journalists find the term benign at best and pointless at worst, so do game developers and scholars. So because of this informal yet mutual agreement to act as if this word is not a mechanism for making interesting statements and claims about games, it sits uninterrogated and underused. Games may be fun at their core, but the greater community of gamers, developers, scholars, and critics use the term with a measured hesitancy. In this way, marginalizing it in the professional game journalism vocabulary unnecessarily marginalizes a concept that could help us understand games and other forms of play. Although professional game journalism is nominally a serious business, this exclusion does not mean that the term *fun* disappears.

Escaping the inertia of indifference that seems to swirl around the notion of fun may not be the term's only complication. Not only does the term *fun* carry certain ambiguities and lexical complexities, but it also appears to be a concept largely rooted in contemporary Anglo-Saxon culture.

Huizinga acknowledged more than 75 years ago that finding fun might present a challenge. "Now this last-named element, the fun of playing, resists all analysis, all logical interpretation.

As a concept, it cannot be reduced to any other mental category. No other modern language known to me has the exact equivalent of the English 'fun'" ([1938] 1949, 3). Tackling this essence of play and by proxy the pleasure of play that Huizinga found so perplexing starts from a challenging position. Lacking a formal study of fun and working from a common acknowledgment that *fun* presents a difficult problem at the philosophical, sociological, and linguistic levels, the search for fun has to start off broadly. But once the scope has opened up, a theory of fun seems possible and in many cases has already been explored.

The way to fun may come ironically not through the study of fun but through the much better understood notions of play. If fun is the operational affect of play, then the study of play, whether analytical, sociological, or historical, provides a map of the present shape and structure of fun.

Gregory Bateson's classic essay "A Theory of Play and Fantasy" provides a startlingly clear point of entry and illustrative example. In this essay, he introduced the concept of play as something paradoxical or self-contradictory. When watching monkeys in the zoo and searching for evidence of metacommunication in animals or of communication about communication, he wrote, "'This is play' looks something like this: 'These actions in which we now engage do not denote what those actions for which they stand would denote'" (1972, 180). Bateson showed that in practice animals engage in metacommuncation. Monkeys play at fighting, signaling denoted bites with bites that are not bites. Although not arguing for a specific definition of the term *fun,* Bateson's "is/is not" construction offers a novel entry point for thinking about the term *fun* as having a rich, complex meaning. If play is the "bite that is not a bite," as Bateson argues, then *fun* is the result of this contradictory condition. Pointing to fun then becomes a way of talking about the meaning of play as much as play is a means of talking about fun.

Ultimately, the historical problem of where to find fun may simply be a game of semantics. The term *fun* may act as a generic noun covering myriad features in play. Or, more tantalizing and problematic, following Bateson, perhaps we only lack a better-developed theory and philosophy of fun. Developing these tools provides the means of understanding how this central aspect of games and play has developed and changed over time. And from this point of view of the term *fun,* a historical aesthetics of play emerges. Much as a conceptualization of beauty allows for a blossoming of aesthetic perspectives and theories about art over time, so can a rich discussion of fun facilitate rather than reduce the diversity of understanding of games. Bringing fun into the conversation provides for a multiplicity of solutions to the serious problem of fun.

Works Cited

Bateson, Gregory. 1972. *Steps to an Ecology of Mind: Collected Essays in Anthropology, Psychiatry, Evolution, and Epistemology*. San Francisco: Chandler.

Bolinger, Dwight L. 1963. "It's so Fun." *American Speech* 38 (3): 236–240.

Garcin, Pete. 2011. "Is 'Fun' What We Really Mean?" http://www.gamasutra.com/view/news/127020/Opinion_Is_Fun_Really_What_We_Mean.php. Accessed August 23, 2014.

Huizinga, Johan. [1938] 1949. *Homo Ludens: A Study of the Play-Element in Culture*. London: Routledge & Keegan Paul.

Koster, Raph. 2005. *A Theory of Fun for Game Design*. Scottsdale, AZ: Paraglyph Press.

Ochalla, Bryan. 2007. "Who Says Video Games Have to Be Fun? The Rise of Serious Games." Gamasutra, June 29. http://www.gamasutra.com/view/feature/129891/who_says_video_games_have_to_be_.php. Accessed January 6, 2014.

Schmidhuber, Jürgen. 2010. "Formal Theory of Creativity, Fun, and Intrinsic Motivation (1990–2010)." *IEEE Transactions on Autonomous Mental Development* 2 (3): 230–247.

Sutton-Smith, Brian. 1997. *The Ambiguity of Play*. Cambridge, Mass.: Harvard University Press.

Thomas, David, Kyle Orland, and Scott Steinberg. 2007. *The Videogame Style Guide and Reference Manual*. 1st ed. Lulu.com.

Zagal, José P., Amanda Ladd, and Terris Johnson. 2009. "Characterizing and Understanding Game Reviews." In *Proceedings of the 4th International Conference on Foundations of Digital Games, Orlando, Florida*, 215–222. New York, NY: ACM.

19 GAME ART

Mary Flanagan

For thousands of years, games have helped players pass the time, settle arguments, and have fun. For the past 50 years, playful creative work has proliferated: the entertainment games industry has exploded in popularity. Contemporary computer games often make the news—in part due to debatable associations between video games and violence—but journalists often forget that games have always had an important role to play across cultures. Games function to "abstract" situations—conflicts, relationships, challenges—and transform them into something that can be won or lost. In other words, games model complex systems and imply or provide a form of resolution. Games, through their fundamental element *rules*, are microcosms of the human condition, and they help us reflect on strategies, incentives, and rewards. They also put values such as fairness and cooperation into play. They allow us to fail safely and to try again. Games are especially interesting because they are systems that in themselves have "play" and are set in motion by players. These myriad aspects make games highly complex. One thing is certain, however: games are a form of art. In just what ways they are art, however, needs to be explored.

This short chapter outlines key ideas behind the term *game art* and explores the nuanced way games are discussed as art. Computer-based games have come into their own as a unique form of creative activity, like cinema and photography. They are at the vanguard of media culture, and creators across the globe are making more innovative games than ever before (Lenhart, Kahne, Middaugh, et al. 2008). Does game art represent a break from long-held traditions about beauty and art? Or can we read games alongside a comprehensive history of art? Are all games art under any circumstances?

The term *game art* can confusingly mean several things. A basic map of the terminology must be sorted out. *Game art* can refer to one of five ways of looking at games:

1. The art in a game.
2. The elevation of games themselves as a culturally recognized form in creative spheres.

3. A type of creative work made by people identifying as artists in the art historical sense, working in an art studio practice, and using the game as a conceptual form to reflect on culture.
4. Something that has artistic intentions, also called "art games," typically by those who wish to make complete games with end states that are playable.
5. An exemplar of the form of a game itself; in other words, a game that as a game transcends most other games to reach an ideal, "pinnacle" state with high-quality interactions, aesthetics, and more.

Understanding the term *game art* is made difficult by these sometimes wildly varying definitions, yet it is necessary to combine analysis and discussion of works across all types of game art to hone in on their characteristics. Here, I unpack these diverse definitions of *game art* and offer examples of each. It is essential to note immediately, however, that in emerging domains, taxonomies are made to be broken by exceptions. The spirit of experimentation represented by game art will certainly mean that these categories will shift and combine over time. This is the exciting part of working in and among an emerging form. Let's start from the easiest definition and move out.

Game Art: The Art in a Game

A "game artist" in the commercial game industry is someone who is involved primarily in the creation of characters and environments. The phrase "great game art" might refer both to the aesthetic qualities in a given game and to the creators. If we look to *Shadow of the Colossus* (Team Ico, 2005), for example, or *Okami* (Clover Studio, 2006), we might admire the look and feel of the game in a visual or sonic sense. One might play the game *Limbo* (Playdead, 2010) and marvel at the crispness of the shadow world. Game artists conceptualize the look and feel of a game and render it sonically and visually.

Well-known game artists in this category include Rich Werner, the artist behind *Plants vs. Zombies* (PopCap, 2009); Chris Robinson, senior art director for *World of Warcraft* (Blizzard Entertainment, 2004), lead character designer Kenichiro Yoshimura for *Okami*; and Andrea Wicklund, the artist of "the hand" of *Left 4 Dead* (Turtle Rock Studios/Valve South, 2000).

Game Art: Games as a Creative Form

Following the history of other media forms such as photography and cinema, which at their inception were considered "novelties" at best, games too are finally becoming recognized as a genre of creative expression. In this sense, games are finally called "an art form" in their own right and should be given serious attention. University programs are being formed to study video games, and academic journals are proliferating in the new field of game studies. These cultural shifts recognize that games, like other art forms, are indeed creative forms in and of themselves that involve aesthetics and create new human experiences. Games may involve narrative, or they may not, and as a phenomenon that can trace its way back to before written language, games are unique.

In the chapter on games as a medium in this volume, I argue that a game is not a medium per se, for games themselves can be created through nearly *any* art medium: for example, clay (ancient checkers or go), wood (chess), computer (*Tetris* [ELORG, 1984]), performance (embodied play, such as charades). They are therefore rather unlike a "medium" and much more a high-level creative category of phenomena, like music, narrative, image making, and so on. Their inherent properties include that they can model systems, offer players agency, and lead to dynamic outcomes. Although games are influenced by their material representations and properties, both the game and the medium carry with them their own social debates and practices. A game is thus a larger art phenomenon expressed across various media.

It is easy to find evidence in recent museum exhibitions, such as *The Art of Video Games*, held at the Smithsonian in 2012, that games have indeed been elevated to the status of other creative media forms. In this exhibit, commercial games were grouped by platform and genre, and a few of the games were playable in segments (*Super Mario Bros.* [Nintendo, 1985], *The Secret of Monkey Island* [Lucasfilm Games, 1990], *Flower* [thatgamecompany, 2009], and *Myst* [Cyan Interactive, 1993]). In 2012, the Department of Architecture and Design at the Museum of Modern Art announced its purchase of 14 video games for its collection, which were shown in the *Applied Design* exhibition of 2013–2014. In this curatorial example, popular computer games are categorized as designed products akin to Eames chairs and iPods, but not akin to the game art produced by surrealist artists.

Game Art: Artists Make Games

For more than 100 years, some artists who also craft paintings, sculpture, and performance have used games as part of their approach to their work. Game art is a type of creative work made by people identifying as artists in the art historical sense, working in an art studio practice and using the game as a conceptual form to reflect on culture. Such artists might experiment beyond standard stories, representations, or game mechanics and might include subverted gaming tropes, broken aspects of games, and the use of games in a conceptual way to reflect on culture. In my chapter on games as a medium in this volume, I note a few examples of artists using games in their work. From Alberto Giacometti's game sculptures to Alison Knowles's nonsensical bean games to Marcel Duchamp's play of chess as a form of expression, the art world and artists have labeled certain types of games as art forms (on this topic, see Naumann, Bailey, and Shahade 2009; Higgins 2002; Hickey 1997). Surrealist artists used games as generative creative procedures, and games became part of the vocabulary of Fluxus artists as they made provocative commentary on contemporary life, using properties of the medium to reference and reinforce the absurd.

What is important to note is that not all of the output by any given artist need be a game, nor is it very important to stick within the form of a game from start to finish. This category of artists engaged in creating game art are typically not building small game design studios to make game after game. Rather, games tend to be one of several modes, references, and techniques that these artists might use in the creation of a work. These artists also refer to their work as art and circulate their ideas inside at least some traditional art venues and spaces. Artists making game art in this sense are typically interested in linking their work not just to earlier games but to art and art dialogue from current and past eras. Here, game artists use the ludic language of games and invoke their elements, but they are not so dedicated to making exemplar games "for games' sake." These artists instead make their own forms of games that change rules, question authority, or are literally impossible to win.

Artists have been using play in subversive and disturbing ways, making impossible and grotesque objects or nonsensical game kits whose rules are enticingly unresolvable in the conventional sense of traditional games, where winners, losers, player roles, and game goals are clearly articulated. Swedish artist Magnus Jonsson created the very slow board game *Brainball* (1999), which can be won only by relaxing. Dutch artist Aram Bartholl's game *WoW* (2006) brings a massively multiplayer game over to real-life people on the streets and by doing so calls attention to the ways in which aspects of virtual culture are normalized and accepted in the physical world.

In the game artwork *Waco Resurrection* (2004) by the artist cooperative C-Level, players engage in playing a three-dimensional multiuser game as the Seventh-Day Adventist "Branch Davidian" cult leader David Koresh, who has been resurrected in the game and must continue to gather and keep followers (figure 19.1). During the experience, each player literally becomes Koresh by wearing a large polygon-chiseled Koresh head that has a voice-activated control mechanism and built-in speakers that blast messages from government agents, religious readings, and battle noise. The various Koreshes in the compound say biblical phrases to gain special powers and intensify their aura in order to garner converts. The artists meant to bring a critical eye to the mechanics of religious cults.

Artists from the French art collective One Life Remains created the game art *Generations* (2012) to push the boundaries about thinking about play and digital sociality through time.

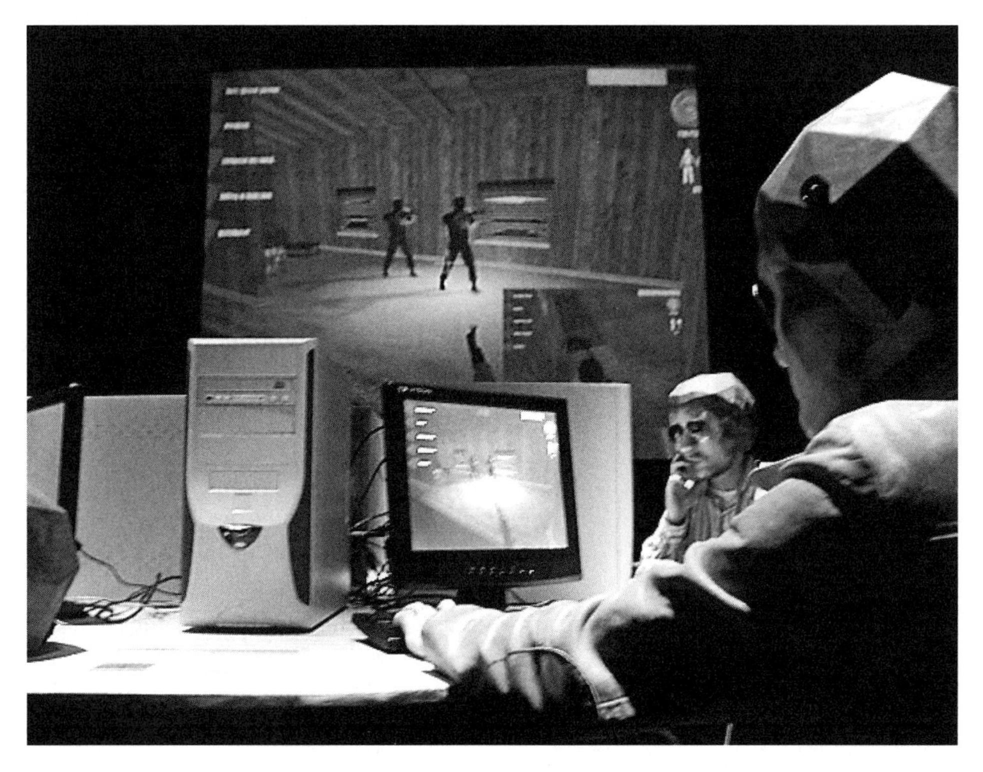

Figure 19.1
Eddo Stern, Peter Brinson, Brody Condon, Michael Wilson, Mark Allen, and Jessica Hutchins (C-Level), *Waco Resurrection* (2004). Courtesy of Eddo Stern.

Generations is meant to be played across entire lifetimes, with one player passing the game's "save file" down to heirs. The game must never end because it is designed to be a digital family heritage object, with future relations engaged with play from the past. Autobiographical games (such as my own game *[domestic]*, 2006) would also fall into this notion of game art, in particular if they are also subverting significant tropes in gaming expectations, such as winning, losing, and competing. Some games help us consider the world and big-picture conceptual concerns. Some reflect on the act of playing games. This "metalevel" of commentary reveals core notions of what makes a game a game in the first place. American artist Cory Arcangel's work *Super Mario Clouds* (2002) removes everything from *Super Mario Bros.* (Nintendo) but the clouds, suggesting the concept of a game that offers no agency, a game reduced to one of its most basic components. The Whitney Museum of American Art has acquired *Clouds* as an artwork.

In this category, one would also find artists making games that have physical presence or physical consequences: with *Painstation* (2002), German artist duo Volker Morawe and Tilman

Reiff, known otherwise as //////////fur////, created a console platform that shocks players while they try to win their *Pong*-like computer game. A final example would be the performance work of Anne Marie Schleiner in *OUT* (2004), for which the artist went into the online networked space of the game *America's Army* (US Army, 2002) and staged in-game interventions while using projections to bring the game out onto the street in New York City. This piece, executed during the days of the Republican National Convention in 2004, offered a critique that spanned the city's in-game and street space and blended virtual and real into one shared space of political conversation.

Game Art: Art Games

Some creative practitioners are intent on creating games—often computer games—that fit traditional criteria for games first and then express ideas within that form. The term *game art* in this sense refers to games that have artistic intentions; as noted earlier, this type of game art is also called "art games," typically by those who wish to make complete games with end states that are playable. Many "indie" games operate in this realm, and award winners at the IndieCade Festival of Independent Games often fall into this strand. One criticism of this use of the term *game art* is that indie game designers, in particular those coming out of more scientific university programs, want to call their work "art" without necessarily knowing the history of the terminology (Warren 2011).

Games that strive for expression while staying within the traditional format of a game include *Papers Please* (Lucas Pope, 2013), a game that challenges players' own sense of moral justice while they act as border guards and allow only certain people to pass into the country. The minimal interaction required in Jason Rohrer's game *Passage* (2007) leads the player to reflect on the brevity of life. Games in this category can be politically motivated as well. To comment on war, the artist-created game *Unmanned* (2012) by the Italian collective Molleindustria puts players in the seat of a military drone pilot. *Dominique Pamplemousse* (2012) by Deirdra Kiai (figure 19.2) subverts the point-and-click genre by staging a game as an opera that involves black-and-white claymation figures who explore the politics of identity while engaged in solving a mystery. These games, however, are solvable or winnable and operate primarily within the game-focused community for their reception.

Game Art: Exemplars of the Form

Sometimes games emerge, independently created or crafted by commercial companies, that embody a shift in the genres of gameplay itself. These games have typically been created by

Figure 19.2
Deirdra Kiai, *Dominique Pamplemousse* (2012).

small innovative teams. Games that win awards at the Game Developer's Conference are representative of this category. *Time* magazine announced in 2012 that the immersive, explorable game *Myst* was one of the best games of all time (Peckham 2012). Yet so was *Zork* (Infocom, 1977), a very different type of game with entirely different aesthetics and player interactions. More recent examples continue to show that the game industry and fans relish the opportunity to point out exemplar games that excel in new graphical styles, novel interactions, or in-depth, rich player choices.

For example, the commercial game industry, through the Game Developer Choice Awards, gave the Game of the Year award to Naughty Dog's open exploration marvel *Journey* in 2013 and then to that same company's game *The Last of Us* in 2014. Choices for best handheld/mobile games were *The Legend of Zelda: A Link between Worlds* (Nintendo, 2014) and *The Room* (Fireproof Games, 2013). In 2012, *The Elder Scrolls V: Skyrim* (Bethesda Game Studios) was designated Game of the Year. This collection of games shows the diversity of recognized genres and styles that push the craft to new forms of excellence but are not necessarily engaged in inventing hybrid forms of play or in engaging with traditional social critique, conceptual challenges, or other notions of art.

A Nuanced Term

This chapter exposes how the term *game art* is used to span the international game industry's creation of exemplar commercial works as well as contemporary artists use of games and play in their work. As artists and games use new techniques and create new playful models, what does the future of gameplay promise for a global society? In the recent past and likely in the future, we will have many examples of games and play spaces that mix digital realities and the physical world. Now that screens are embedded in smartphones and will soon be embedded in fabrics and objects—or, indeed, may not even be needed—what will game art be in the coming years? Digital games will increasingly blend in with their precursors, board games and performative games, out in the world around us as pervasive games, role playing, and more. Games are getting bigger, more flexible, and more personal, and game art will indeed be entering everyday life in profound ways.

Games embody subtle yet powerful methods of enculturation by which social values, interaction styles, and everyday activities are explored, practiced, questioned. The games discussed in this chapter reveal key nuances of understanding games as an art form, how play has permeated the social fabric of contemporary life, and how games as aesthetic and experiential creative forms help us to make meaning in the world.

Works Cited

Hickey, Dave. 1997. "The Heresy of Zone Defense." In *Air Guitar: Essays on Art & Democracy*, 155–163. Los Angeles: Art Issues Press.

Higgins, Hannah. 2002. *Fluxus Experience*. Berkeley: University of California Press.

Lenhart, Amanda, Joseph Kahne, Ellen Middaugh, Alexandra Macgill, Chris Evans, and Jessica Vitak. 2008. *Teens, Video Games, and Civics*. Washington, DC: Pew Internet and American Life Project.

Naumann, Francis M., Bradley Bailey, and Jennifer Shahade. 2009. *Marcel Duchamp: The Art of Chess*. New Haven, CT: Readymade Press.

Peckham, Matt. 2012. "*Myst:* All-TIME 100 Video Games." *Time,* November 15. http://techland.time.com/2012/11/15/all-time-100-video-games/slide/myst-1993/. Accessed September 1, 2014.

Warren, Jamin. 2011. "Game Designers Want to Be Called Artists without Knowing What That Means." *Kill Screen*, March 21. http://killscreendaily.com/articles/game-designers-want-be-artists-without-knowing-what-means/. Accessed September 1, 2014.

20 GAME AUDIO

William Gibbons

Although the term *video games* suggests a predominantly visual medium, sonic elements have made up a central part of the gameplay experience since the early 1970s. The terms *game audio* and *game sound* (used interchangeably) refer to the combination of music, sound effects, and voice acting that make up a given game's soundscape. As in traditional media such as film and television, audio in games establishes both mood and setting and provides players with crucial emotional and narrative cues. But whereas in films music and sound can be carefully timed to correspond precisely with on-screen actions, in games the variability between one player and another makes this type of synchronization difficult, if not impossible. Instead, the aesthetics and technology of game audio have generally evolved in ways that favor adaptability and flexibility—the ability to directly or indirectly respond to player input or changes in game states. Initially a major challenge to game audio creators, this dynamic nature has nonetheless spurred a number of musical and technological innovations and rewards over the course of game history.

Dynamic audio plays a significant role in player immersion and the gameplay experience as a whole in games of a wide variety of genres (Collins 2013; Paul 2013; van Elferen 2011; Jørgensen 2008; Grimshaw 2008). In a distinction that has become relatively standard, Karen Collins (2008) breaks down the "dynamic audio" aspect of games into "interactive audio" and "adaptive audio." The former includes music and sound effects that result *directly* from player actions—such as the sound of a gun firing in response to a button press or the musical pitches produced when players correctly play notes in a music game such as *Guitar Hero* (RedOctane, Harmonix, 2005) or *Rock Band* (Harmonix, 2008). "Adaptive audio," by contrast, changes based on criteria immanent in the conditions of gameplay—for instance, when the music speeds up as players approach a time limit in games such as in *Super Mario Bros.* (Nintendo, 1985) or when blocks reach a "danger" point in *Tetris* (ELORG, 1984).

Because game audio's push toward adaptability is deeply tied to how the music and sound are created and produced, what follows here is a brief overview of the interactions between major technological developments and aesthetic trends in game audio from the late 1970s to the present. For the sake of brevity, this chapter focuses on personal computer (PC) and console gaming rather than on arcade and mobile games. Although arcade machines were initially both influential and technologically advanced in terms of game audio, advancements in more recent years have clearly favored console and PC games. Furthermore, the audio hardware involved in arcade machines lacked the standardization found in consoles and PCs, making it problematic to draw generalizations because sound varied significantly from game to game. Mobile games are equally diverse and differ from other media in that they are often designed for ease of play without sound (for play in public places, for example, or for multitasking with other media).

Although the earliest commercially available video games were free of music and sound effects (aside from the noise naturally created by the act of playing, such as button presses), game audio quickly became an integral design element. Since *Computer Space* (Nutting Associates, 1971), in fact, few games have entirely lacked in-game sound. Early arcade machines featured sound effects both to attract players and to add to their sense of excitement and immersion—effects were often credited with adding an element of "realism" that could supplement the limited graphical capabilities of the time (Collins 2008). Creating these effects initially posed difficulties for programmers because they had to be achieved through the manipulation of existing hardware, meaning that the process could vary dramatically from machine to machine. For the distinctive sounds incorporated in *Pong* (Atari, 1972), for example, the sound designer noted that he "poked around the sync generator to find an appropriate frequency or tone. ... They were the sounds that were already in the machine" (qtd. in Kent 2001, 34).

Music followed a few years after sound effects, but quickly became equally indispensible. Early experiments with musical replication include games such as *Circus* (Exidy, 1977), which featured brief fragments of well-known tunes to indicate progress or failure (Lerner 2014). The practice of using preexisting music (mostly folk and classical but occasionally film, television, and popular music) remained a common strategy in early games to avoid the time and expense of finding and paying composers who were also capable of complex programming or manipulation of hardware. Neil Lerner (2014) has noted connections between early games and the heavy use of borrowed music for silent films, pointing to titles such as *Crazy Climber* (Nichibutsu, 1980), *Crystal Castles* (Atari, 1983), and *Donkey Kong* (Nintendo, 1981). Preexisting music remained prominent in games through the 1980s (and beyond); the Nintendo Game Boy version of *Tetris* (1989), for example, included music by J. S. Bach and a Russian folk song, and the Nintendo Entertainment System (NES) version of the same title (also 1989) included music by the composers Tchaikovsky and Bizet (Gibbons 2009).

Early efforts at continuous music (that is, music repeated constantly during gameplay) included *Space Invaders* (Taito, 1978), which featured a four-note "melody" repeated at increasing speed until the player's inevitable loss. After 1980, continuous music quickly took hold, and by the middle of the decade its presence in new games was an almost foregone conclusion. To balance the demands for more music with concerns over memory limitations, most game composers created continuous soundtracks by writing music that could seamlessly "loop" or repeat endlessly. An early (possibly the first) example of this looping occurred in the racing title *Rally-X* (Namco, 1980) (Fritsch 2013), but the practice quickly spread to other titles. Aside from brief introductions and breaks between levels, the musical score of the popular arcade title *Moon Patrol* (Irem, 1982) consisted entirely of an endlessly looped 12-bar blues bass line (about 25 seconds of music). As home consoles such as the Commodore 64 (1982), ColecoVision (1982), and the NES (1985) became common, the number of games featuring continuous music increased dramatically. Because the home environment was (generally) much quieter and less distracting than arcades, the lack of game audio was much more noticeable. The perceived awkwardness of playing in near silence encouraged game designers to maximize the amount of music to keep players engaged (Gibbons 2014; Bridgett 2008). The nature of game loops (which could be quite complex in structure) conveniently covered silences while still avoiding any need for composers to anticipate the length of time players would take to complete a particular task or level. As a consequence, the technique has far outlasted the technological limitations that spawned it and remains a prominent aspect of game audio.

In quality as well as quantity, the last years of the 1970s and first years of the 1980s brought dramatic improvements to both music and sound effects. Arcade machines began including dedicated sound technology such as programmable sound generators (PSGs), which provided some standardization of audio-production technology between games and for the production of more complicated combinations of sounds. Most typical was what Collins (2008) refers to as a "3 + 1 generator," which enabled the simultaneous production of three musical pitches and white noise (often used for effects). As a result of PSGs and other enhanced technologies, sound effects increased in both quality and quantity—multiple simultaneous effects were possible, for example—and music became increasingly complex texturally (i.e., with more layers of sound). Despite these technological advances, through the 1980s game music remained grounded in the *bleeps* and *bloops* that had developed at the end of the previous decade. Gaming hardware remained incapable of replicating the sounds of traditional analog instruments, but it is unclear whether composers were even interested in doing so. Early game composers likely drew inspiration from the sounds of groups such as Kraftwerk or CyboTron, whose music had little in common with traditional acoustic instruments. Even more directly, the popular Japanese electronic ensemble Yellow Magic Orchestra both influenced and emulated game audio, as in the single "Computer Games" (1979), which extensively sampled snippets of game music and sound effects.

Aesthetic concerns aside, strict technological limitations on both the number of possible simultaneous sounds (music and sound effects) and on the quantity of sound that memory resources could accommodate shaped game audio throughout the 8-bit era. Many home consoles employed 3 + 1 generator chips, including the ColecoVision and the Sega Master System (1986). The Commodore 64 computer system demonstrated some of the most innovative music and sound of the early 1980s (Collins 2006), yet the PSG chip found in the NES was arguably the most influential audio technology of the 8-bit era. It provided five available audio channels: typically two devoted to upper melodic lines, one bass channel, a static channel that was often manipulated to create a percussion effect, and a delta modulation channel reserved for sampled sound effects.

The late 1980s and early 1990s brought numerous innovations in game audio that corresponded loosely with the era of 16-bit gaming. Frequency Modulation (FM) chips replaced PSG chips, allowing for more realistic emulation of instruments, and improvements in voice synthesis technology allowed for brief samples of human voices. The incorporation of voice samples came at a prohibitive memory cost, however; the instantly recognizable "SEGA" chant (later included in many titles on Sega's Genesis console) reportedly took one-eighth of the total memory space available for *Sonic the Hedgehog* (Sega, 1991) but offered an opportunity to showcase the console's advanced audio capabilities. This technology similarly enhanced the quality of available sound sampling, again enhancing the realism.

The implementation of Musical Instrument Digital Interface (MIDI) technology, first in PCs and a few years later in consoles (the latter beginning with the Super Nintendo Entertainment System, or SNES) allowed for additional simultaneous sounds, increased amounts of music due to small file sizes, and a library of digitized instrument samples. MIDI files could also be easily input into music notation software and then read by sound-processing hardware, thus removing the requirement that game music composers be proficient at programming. The library of sound samples also meant that composers could emulate traditional instruments much more effectively than before, with the end result being a move away from *bleeps* and *bloops* toward more orchestral or popular styles. The use of MIDI technology on computers was initially hampered, however, by a lack of standardization. It was impossible for composers to ensure that the correct instrument sample would be triggered, so, for instance, a melody intended for piano might instead be played by an electric guitar. The introduction of General MIDI (GM) in 1991 corrected this issue, standardizing the technology with a library of 128 instruments and up to 24 simultaneous sounds (16 melodic voices and 8 percussion voices).

As a result of the flexibility these new technologies provided, many composers moved beyond looping to creating new game-specific approaches to dynamic music. This shift led to techniques such as "layering" and "branching." These techniques illustrate the general move toward a more adaptive type of game score, in which players' choices could have subtle or obvious ramifications on the musical content, with the intention of enhancing immersion and emotional affect. Layering is a type of dynamic audio in which individual or multiple channels can be turned on or off (instantly or by fading from one to another) in response to in-game states, such as the possession of a certain item, an approaching time limit, or a "danger" state. At its simplest, layering might involve the addition of a single channel, as in *Super Mario World* (Nintendo, 1990), when a layer of percussion is added anytime Mario rides his dinosaur companion Yoshi (Paul 2013; Collins 2008). More complex examples include game audio from LucasArts' innovative iMUSE engine, as in *Monkey Island 2: LeChuck's Revenge* (1991) (Summers 2012; Collins 2008). The iMUSE engine enabled the game audio to add and subtract layers based on a number of possible in-game situations.

The iMUSE engine and other similar programs also focused on musical branching, a technique in which a musical track could seamlessly diverge (in its entirety or only in certain channels) at a smooth transition point to another track based on in-game criteria rather than make an abrupt cut to that other track. Players were thus presented with an illusion of inevitability, as though they were watching their choices unfold in a film. If a player opted to head left at a fork in the road, the music would follow one path; if the player went right instead, the music would reflect that decision. Branching thus eliminates the risk of music unintentionally "running counter to the action," which might create a jarring disconnect with players' on-screen actions. However, branching scores are arduous and time-consuming to compose, requiring a considerable amount of music that a player might never hear. In *Monkey Island 2*, for example, the iMUSE engine allowed the music to branch based on the player's current location, creating seamless transitions even between multiple areas in close proximity. Layering and branching could of course be directly combined, as in *Gears of War* (Epic Games, 2005), in which the music branches based on the player's location, with each branch containing multiple layers to reflect the situation (danger, success, intensity) (Paul 2013).

Running in parallel (and sometimes in opposition) to these approaches to adaptive music was the use of pre-recorded CD-quality audio beginning in the 1990s. Introduced on PCs and then moving to consoles such as the Sega CD (1991, an add-on to the Sega Genesis), followed by the Sega Saturn (1994) and Sony PlayStation (1994), CD audio offered a number of advantages. The most obvious benefit was the ability to use a substantial amount of high-quality music that did not differ from computer to computer. CD audio also provided for true-to-life sound effects as well as for large amounts of voice acting for the first time. The fixed nature of pre-recorded CD audio, however, meant that music could not be adaptive in the same ways MIDI could be: layers could not be effectively added and subtracted, and creating branching scores would have required prohibitive amounts of memory. Even looping was made difficult by the fact that CDs required a pause before restarting a track (Fritsch 2013). For these reasons, few games include only pre-recorded audio; most incorporate at least some music generated by onboard sound hardware.

As many games transitioned to three-dimensional graphics and first-person perspectives, what Mark Grimshaw (2008) refers to as the "acoustic ecology" became increasingly important for helping players to position themselves in the game space, which required the adaptation of cinematic surround sound technology to the game. In genres such as first-person shooters, music and sound function as narrative elements, alerting players to important events happening outside their field of vision; for example, a musical cue or sound effect might alert players to the approach of an unseen enemy or the presence of an important item nearby. In these (and other) circumstances, game sound is essential to gameplay, and its absence results in severe detriment to both the player's immersion and ability to succeed (van Elferen 2011; Jørgensen 2008).

Since 2000, game composers have continued to develop and refine methods of creating dynamic music. Game scores have experimented with *aleatoric* (or chance) elements, for example, both to create aesthetic effects and to further reduce players' aural fatigue. In an aleatoric type of layering, in-game criteria might trigger short excerpts of music, combining either at random or in response to a player's inputs. This indeterminacy results in a soundscape that is essentially unique to each playthrough, a technique that appears in games such as *The Legend of Zelda: Skyward Sword* (Nintendo, 2011) and *Flower* (thatgamecompany, 2009) (Medina-Gray 2014). Similarly, some games have embraced a sound design with close ties between level design and musical score, in which the music (in whole or part) seems to derive directly from the game itself, as in *Rez* (Sega, 2001) and *Dyad* (Right Square Bracket Left Square Bracket, 2011), in which the player's actions effectively create the games' aleatoric, techno-influenced soundtracks (Reale 2014).

Though less common, composers have also experimented with generative soundtracks as a way of avoiding musical oversaturation, particularly in games that can produce hundreds (if not thousands) of gameplay hours. In generative scores, music is "composed" in-game according to preprogrammed probabilities, resulting in a unique experience with each playthrough. The downsides, however, are that the generative process can be demanding on game hardware (as opposed to precomposed music, even of a dynamic nature) and that, despite some promising developments, the technology remains unable to effectively mirror the complex logic of human compositional styles (van Geelen 2008). A more common technique in games with extremely long playtimes is the elimination of continuous music and a general shortening of musical cues, which results in longer stretches of silence—as in, for example, *Shadow of the Colossus* (Sony Computer Entertainment, 2005) and, more recently, *The Elder Scrolls: Skyrim* (Bethesda Game Studios, 2011) (Gibbons 2014).

One final notable trend in game audio in recent years has been the dramatic increase in the amount of popular music in games. Available memory initially limited the amount of prerecorded music and sound in CD-based games, but the shift to DVD storage on computers and consoles beginning with the PlayStation 2 (2000), the advent of compressed file formats such as MP3s, and increased hard-drive space essentially removed these restrictions. The PlayStation 2 game *Grand Theft Auto III* (DMA Design/Rockstar Games, 2001), for example, included extensive in-game radio stations that played mostly popular music, a strategy featured prominently in subsequent games in the series and in similar sandbox-type games such as *Fallout 3* (Bethesda Game Studios, 2008) (Cheng 2014; Miller 2012). This type of "jukebox" game provides a realistic explanation of why the music might not fit the on-screen action and allows players some control over the soundscape even though the music is pre-recorded. Certain game genres (sports, racing, etc.) rely extensively on licensed popular music, and the music industry synergistically employs these games to market music—leading software giant Electronic Arts, for example, developed an in-house recording label, Next Level Music (Tessler 2008). In a related development, games' ability to include massive amounts of CD- or MP-quality music paved the way for games in which the music took center stage as the primary gameplay mechanic, such as the *Guitar Hero*, *Rock Band*, and *SingStar* (London Studio) series of games, in which players participate in what might be described as "schizophonic" music making (Collins 2013; Miller 2012), a hybrid authorship divided between player input and preprogrammed musical elements.

Though speculation is always hazardous, the future of game audio looks highly pluralistic. Technology has advanced to the point where few limits remain to composers and sound designers. The large budgets, impressive production values, and often cinematic aspirations of major retail releases push game audio toward ever more realistic sound effects, voice acting, and musical scores that rival (and frequently surpass) Hollywood films in scope and complexity. At the same time, riding a wave of nostalgia for the 1980s and 1990s, games such as *Fez* (Polytron, 2012), *Megaman 9* (Capcom/Inti Creates, 2008), and *Superbrothers: Sword & Sworcery* (Superbrothers/Capybara, 2011) have returned to the historical *bleeps* and *bloops* of earlier audio (in both their music and their sound effects). In both cases, games continue to demonstrate unique approaches to dynamic audio, embracing the unique possibilities inherent to the medium.

Works Cited

Bridgett, Rob. 2008. "Dynamic Range: Subtlety and Silence in Video Game Sound." In *From Pac-Man to Pop Music: Interactive Audio in Games and New Media*, ed. Karen Collins, 127–134. Aldershot, UK and Burlington, VT: Ashgate.

Cheng, William. 2014. *Sound Play: Video Games and the Musical Imagination*. Oxford: Oxford University Press.

Collins, Karen. 2013. *Playing with Sound: A Theory of Interacting with Sound and Music in Video Games*. Cambridge, MA: MIT Press.

Collins, Karen. 2008. *Game Sound: An Introduction to the History, Theory, and Practice of Video Game Music and Sound Design*. Cambridge, MA: MIT Press.

Collins, Karen. 2006. "Loops and Bloops: Music on the Commodore 64." *Soundscapes: Journal of Media Culture* 5. http://www.icce.rug.nl/~soundscapes/VOLUME08/Loops_and_bloops.shtml.

Fritsch, Melanie. 2013. "History of Video Game Music." In *Music and Game: Perspectives on a Popular Alliance*, ed. Peter Moormann, 11–40. Wiesbaden, Germany: Springer.

Gibbons, William. 2014. "Wandering Tonalities: Silence, Sound, and Morality in *Shadow of the Colossus*." In *Music in Video Games: Studying Play*, ed. K. J. Donnelly, William Gibbons, and Neil Lerner, 122–137. New York: Routledge.

Gibbons, William. 2009. "Blip, Bloop, Bach? Some Uses of Classical Music on the Nintendo Entertainment System." *Music and the Moving Image* 2 (1): 40–52.

Grimshaw, Mark. 2008. *The Acoustic Ecology of the First-Person Shooter: The Player Experience of Sound in the First-Person Shooter Computer Game*. Saarbrücken, Germany: VDM.

Jørgensen, Kristine. 2008. "Left in the Dark: Playing Computer Games with the Sound Turned Off." In *From Pac-Man to Pop Music: Interactive Audio in Games and New Media*, ed. Karen Collins, 163–176. Aldershot, UK: Ashgate.

Kent, Steven L. 2001. *The First Quarter: A 25-Year History of Video Games*. Bothell, WA: BWD Press.

Lerner, Neil. 2014. "The Origins of Musical Style in Video Games, 1977–1983." In *The Oxford Handbook of Film Music Studies*, ed. David Neumeyer, 319–347. Oxford: Oxford University Press.

Medina-Gray, Elizabeth. 2014. "Meaningful Modular Combinations: Simultaneous Harp and Environmental Music in Two *Legend of Zelda* Games." In *Music in Video Games: Studying Play*, ed. K. J. Donnelly, William Gibbons, and Neil Lerner, 104–121. New York: Routledge.

Miller, Kiri. 2012. *Playing Along: Digital Games, YouTube, and Virtual Performance*. Oxford: Oxford University Press.

Paul, Leonard J. 2013. "Droppin' Science: Video Game Audio Breakdown." In *Music and Game: Perspectives on a Popular Alliance*, ed. Peter Moormann, 63–80. Berlin: Springer VS.

Reale, Steven Beverburg. 2014. "Transcribing Musical Worlds, or, Is *L.A. Noire* a Music Game?" In *Music in Video Games: Studying Play*, ed. K. J. Donnelly, William Gibbons, and Neil Lerner, 77–103. New York: Routledge.

Summers, Tim. 2012. "Video Game Music: History, Form, and Genre." PhD diss., University of Bristol.

Tessler, Holly. 2008. "The New MTV: Electronic Arts and 'Playing' Music." In *From* Pac-Man *to Pop Music: Interactive Audio in Games and New Media*, ed. Karen Collins, 13–25. Aldershot, UK: Ashgate.

Van Elferen, Isabella. 2011. "¡Un Forastero! Issues of Virtuality and Diegesis in Videogame Music." *Music and the Moving Image* 4 (2): 30–39.

Van Geelen, Tim. 2008. "Realizing Groundbreaking Adaptive Music." In *From* Pac-Man *to Pop Music: Interactive Audio in Games and New Media*, ed. Karen Collins, 93–102. Aldershot, UK: Ashgate.

21 GAME BALANCE

David Sirlin

Game balance has different meanings, but the common theme is that balance refers to some aspect of the design that would not work properly if it were any stronger or any weaker. In a single-player game, that aspect might be difficulty. Players might call a game imbalanced if it were exceedingly difficult or too easy. Game balance can also refer to system design, as in *Diablo II* (Blizzard North, 2000), where the lack of hotkeys to easily access different abilities means the player wants to put all their talent points into just one ability. This seems imbalanced, as if too little of the system is used—less than the designers intended and the players expect. Game balance can also refer to the objects in a system, such as a particular ability in *Diablo* or a particular card in *Magic: The Gathering* (Wizards of the Coast, 1993) or a particular character in *Street Fighter II* (Capcom, 1991) being too powerful or two weak.

Proper game balance is most critical in multiplayer competitive games because even small balance problems can ruin the experience for a whole community of players. In these types of games, the term *game balance* refers to two different but related concepts: *viable options* during gameplay and the *fairness of starting options*. The "viable options" meaning of game balance refers to options presented to the player during a game all having some use and not being dominated or obsoleted by other options they could choose instead. The "fairness of starting options" meaning refers to anything the player locks in at the beginning of a game, such as choice of character in *Street Fighter* or choice of race in *StarCraft* (Blizzard Entertainment, 1998), being at a similar power level to the choices other players lock in at the start of the game.

These two types of game balance may seem similar, and both do stem from the idea that things can't be so strong that they make other things useless, but there are also substantive differences. To understand why, let's first look at the difference between symmetric games and asymmetric games.

Symmetric versus Asymmetric Games

A symmetric game is one where players start with the same options, whereas an asymmetric game allows players to start with different options. Some examples of symmetric games are *Quake* (id Software, 1997) (players initially spawn with the same weapon), chess (same pieces on both sides, though some slight asymmetry in that white moves first), and poker or *Pandante* (players have an equal chance of being dealt any hand of cards). Some examples of asymmetric games are *Street Fighter* (many different characters), *StarCraft* (three different races), *League of Legends* (Riot Games, 2009; many different champions), *Yomi* (Sirlin Games, 2011; twenty different character decks), and *Chess 2* (Sirlin Games, 2010; six different armies).

Symmetric Game Balance

In a symmetric game, game balance requires having viable options during gameplay. That means each element or choice has at least some use. For example, all the different pieces in chess are used by expert players. Even though some pieces are stronger than others, all pieces have a purpose. If instead some particular piece were invulnerable, that piece would be so powerful that other pieces become extraneous and useless. The game might boil down to just using that piece and ignoring everything else. Whenever there's a serious imbalance relating to viable options in gameplay, the result is a *degenerate* game.

Game balance can also apply to a game's macro level. Chess has a mix of both offensive and defensive opportunities, but we might imagine some alternate version of the game that was so defensive that players were forced to attack as little as possible. This (im)balance would lead to overly long games and shallower gameplay than actual chess because so many of the usual attacking options would be gone. The problem is not that a particular piece goes unused; it's that entire strategies go unused. It's another way that a deficiency in viable options can lead to a degenerate game.

Local versus Global Imbalance

Global imbalance refers to an imbalance that permeates an entire game, whereas *local* imbalance comes up only in certain situations in the middle of a game. Global imbalances are bad for game balance, but local imbalances are actually expected and often good to have. The two examples in the previous section about imaginary chess variants involve global imbalance—those problems would affect the entire game from start to finish. Any chesslike game you play with the invulnerable superpiece or with the overly defensive rules variant would probably have problems. Let's look at some local imbalances instead.

In chess, if you have three pieces left but your opponent has nine pieces left, that is "imbalanced." Your opponent's forces are stronger than yours and you might be left with no viable options other than losing the game. That type of imbalance is a good one rather than a bad one, though. Multiplayer competitive games are supposed to have situations that are better for one player than the other, and the whole point is for players to try to maneuver into those situations.

For another example of local imbalance that isn't a problem, imagine a first-person shooter with eight weapons that spawn in various locations around the map. Two of these weapons are the most powerful overall, three are OK but not as good as the best weapons, and the remaining three are generally weak but happen to be extremely powerful against one or the other of the two best weapons.

Even though the weapons are not identical in power level, that does not necessarily mean the game is lacking in balance. There is no actual requirement in game balance that every object you encounter during gameplay must have equal strength to every other object. In the previous example, each weapon is still a viable choice in the right situation. It might be fine to have two powerful weapons that players compete over, a few medium power weapons that are still OK, and some weak weapons that allow players to specifically counter the strong weapons. There could be a great deal of strategy in deciding which parts of the map to try to control (to access specific weapons) and when to switch weapons depending on what your opponents are doing. Even though the weapons are intentionally different power levels, it is entirely possible that none is useless and that many different strategies are still viable. If set up well, this kind of local imbalance can add more nuance to gameplay and create varied dynamics.

Asymmetric Game Balance

Asymmetric games have a much tougher challenge when it comes to game balance. These games need to meet the same standards of viable options that a symmetric game needs; in addition, an asymmetric game also needs all its starting options to be as close to equal in power level as possible. In other words, to be balanced all of the characters in *Street Fighter* need to be as close as possible to each other in power level, as do the three races in *StarCraft*. That is in addition to making sure that during gameplay there are many viable strategies.

If a game fails at balancing the starting options, it might be *degenerate*, but more specifically it is also *unfair*. If one player chooses Protoss in *StarCraft*, and their opponent chooses Zerg, it is unfair if Protoss is weaker than Zerg. In effect, any imbalance in power level between Protoss and Zerg is a global imbalance because it would affect any *StarCraft* game played with Protoss versus Zerg. Notice how stringent this balance requirement is. In a symmetric game, it is fine to have weapons of very different power levels; even if one weapon is a bit stronger than the others, all players have an equal chance of getting that weapon in a symmetric game. It is also fine for some units of the Zerg race to be weaker than some units of the Protoss race. But for starting options, such as picking the Protoss race at all, even the tiniest imbalance in power level creates unfairness with respect to other starting options and hurts the ability of the game to test player skill.

Historical wargames are notable in that they intentionally have an imbalance in the power level of each side. This intentional imbalance reflects the historical reality of war that one side is usually more powerful than the other. But even in these games, players usually expect to have a fair match with their opponents, so the side of the historical loser might only need to do better than its historical general did to win the game rather than needing to win the entire war to win the game.

Matchups

Asymmetric games have multiple different matchups. In *StarCraft*, Protoss versus Zerg is a different matchup from Protoss versus Terran, and each matchup is as balanced as possible overall. That means we hope that when experts play an asymmetric game at a high skill level, no particular character or race or deck or whatever has an advantage over any other. It's impossible to achieve perfect game balance across so many matchups (or even in a single matchup), but the challenge to designers of asymmetric games is to minimize the unfairness while still offering the wealth of variety inherent in having multiple different sides to choose from.

StarCraft has only three races to choose from, leading to six possible matchups. The *Yomi* card game has 20 decks to choose from, leading to a whopping 210 possible matchups. The fighting game *Guilty Gear* (Arc System Works) has 24 characters in the Accent Core+R version (2013), leading to 300 different matchups.

It is difficult for game designers to achieve fairness in even one single matchup. Remember that even one matchup, such as Protoss versus Zerg, can be enormously complex and it needs the same kind of balanced, viable options during gameplay that a symmetric game needs. On top of that, the two races (or characters or decks or whatever) must also be very close in power level to each other. That power level is difficult to know because designers don't know the best possible ways to play the game; if they did, then so would the players, and there wouldn't be any depth left. Because achieving good balance is so hard and involves so many unknowns about how good players might someday play, designers sometimes use self-balancing forces and fail-safes to make the task easier.

Player Perception: Matchup Charts and Tier Lists

When discussing game balance in an asymmetric game, players often use a tier list. Rather than rank the perceived power level in a linear list, they group them into tiers to better show which characters are similar in power level to each other. A standard format for a tier list goes as follows:

0: God Tier (no character should be in this tier; if one character is, players are forced to play only that character to be competitive)
1: Top Tier (characters who are not necessarily too good but are the most powerful overall)
2: Middle Tier (characters who are pretty good in power but not quite as good as the characters in the top tier)
3: Bottom Tier (it's still possible to win with these characters, but it's more difficult)
4: Garbage Tier (not reasonable to play these characters because even experts can't win with them)

If a game has any God Tier characters, that is a serious imbalance because it makes all other characters in the game obsolete. If a game has any Garbage Tier characters, that's unfortunate, but it's not disruptive. It doesn't negate the viability of any other characters, so it's less of a problem.

When asymmetric games improve their balance, these tiers don't really go away. Even if all the characters were exactly perfectly balanced, human players wouldn't perceive it that way, and they would still claim that there are tiers. What can happen, though, is that the "distance" between tiers gets smaller with better balance. Conversely, if a game has worse balance over time, then tier lists might show more tiers over time, which means that the power-level difference between characters is so large that it is easier to notice even more gradations of power level.

Matchup Notation

Even if an asymmetric game had no God Tier, no Garbage Tier, and relatively compressed tiers that point to overall good balance, it could still have a balance problem in a specific matchup. Players usually rate the balance of matchups with a notation such as "7–3" or "5–5," which means if two experts played ten games of character 1 versus character 2 against each other, the player of character 1 would win seven times. A 5–5 matchup means it's perfectly fair. Sometimes half numbers such as 5.5–4.5 are used, but that is as much precision as is possible with this method because matchup numbers are usually opinions from players, a way to express their perceptions of a matchup's balance.

Although a game with only three sides such as *StarCraft* can get all its matchups inside the 5.5–4.5 and 4.5–5.5 ranges, games such as *Street Fighter* with more characters often have numerous matchups of 7–3, 8–2, and even 9–1. Customizable card games such as *Magic: The Gathering* tend to have even a larger number of unfair matchups because some cards and decks are such direct and strong counters to others. The *Yomi* card game has 20 different decks, with almost all matchups at 6–4 and 4–6, which is a level of balance that no fighting game has yet achieved. In other words, asymmetric games cannot reasonably achieve "perfect" balance, and they make the trade-off of having some imbalance in starting options to offer more overall variety.

Trends in Balance

The rate of change in games is much higher now than it used to be. Chess, for example, is about 2,500 years old, though the ability for pawns to move two spaces on their first move is about 500 years old. This new ability caused a balance problem that required the en passant rule to be added, and although that rule was used by some soon after the pawn change, almost 300 years passed before it was universally accepted.

In the 1980s, video games were expensive arcade machines. Game consoles later offered games on cartridges and read-only discs. In all these cases, even if balance problems were found, it was often not feasible to do anything about them other than release a sequel to the game. Board games are in this same situation right now. Modern video games are not, though: making changes via patches in computer games such as *StarCraft* or *World of Warcraft* (Blizzard Entertainment, 2004) is common. There is also an in-between zone, as in most Xbox 360 games, where some things can be changed via a small patch, but such changes are limited in scope.

As the ability to change games has increased over time, it has affected both players and designers. In times when balance problems in video games could not feasibly be fixed, players accepted that such problems were just part of a game, and they were even against the idea of any changes being made. Now players expect balance problems to be addressed after a game's release. Designers can now create games that might have been too difficult to attempt earlier. For example, *StarCraft* might have been impossible to make in an earlier era when updating games was infeasible because it requires tight balance to work at all as a competitive game.

We can actually see both the "difficult to change a game" and "easy to change a game" cultures alive at the same time right now: board gamers live in one world, and video gamers live in another. Board games might someday include digital components that are more updatable than the cardboard of today is, and there might be a cultural shift in players' attitudes about balance when that happens.

Conclusion

The term *game balance* can mean several different things, though the common theme is that the power level of something in a game is tuned just right. It can be as specific as when all the pieces or moves in a game have a purpose or as general as when a game has several different strategies that are equally effective to use. Not every situation or item that comes up during gameplay need be equally powerful, though; only the overall, global experience should be balanced. In asymmetric games, the various characters, races, decks, and so forth really do need to be of very similar power level to each other, though, and achieving this balance is an extremely difficult task for game designers. As games have become easier to update over time, players perceive balance as more mutable and designers are more able to make games that are very difficult to balance.

22 GAME CAMERA

Jennifer deWinter

In this entry, I define the term *game camera* in two distinct sections. The first section focuses largely on representation in on-screen camera usage. I discuss the history of virtual camera systems, the use of lensing to create compelling stories, and the challenges that three-dimensional (3D) graphics technologies introduce to the in-game cinematic camera. I also consider the camera's relationship to photorealism and representation. In the second section, I consider the game camera as an input device for gameplay. This history considers early uses of cameras to create virtual-reality games, the camera as a controller for home consoles, and cameras in pervasive play, such as geocaching and alternative-reality games. These two sections highlight two different historical trajectories: (1) the computer game's long shared history with photography and cinema and (2) an emergent practice in which the camera as input articulates onto bodies in motion, changing the player's body into a game input and mechanic.

Camera Metaphors and Visual Interfaces

Camera design affects the game experience because it provides visual cues for gameplay. One of the main differences between cinematic camera work and game camera work is the manipulation of lensing, or how people use different lenses to present a view of a picture or world. In the case of traditional camera work, lensing is quite literal; camera people change lenses to affect the representation, focal lengths, and field of view, helping audiences to focus on specific characters or scenes that further a story's purpose. In games, however, lensing is a metaphor that attends to how a world is presented to a player, which affects player motivation and option. The difference here, of course, is that the camera is not focused on the character in the same way that it would be in a movie; the character's emotions or actions are not the purpose of a scene. Rather, the camera is on the character's environment, affecting what actions the character can likely do.

In "Camera Comes First," Tadhg Kelly (2011) discusses design aspects of "gameatography" (photography techniques for games), defining the different cameras that provide players with a viewing field: fixed point, rotating, scrolling, movable, floating, tracking, pushable, and first person. Camera design, according to Kelly, dictates the types of games that people are able to build, and poor camera work creates "a user experience nightmare." For early games, the camera techniques were limited by the hardware's storage and processing power. Thus, when Shigeru Miyamoto was asked to redesign the Radar Scope cabinet into a new game, he had to abandon his idea of a side-scrolling game with a tracking camera in favor of the fixed-point camera of *Donkey Kong* (Nintendo, 1981). Only in his later games, such as *Excitebike* (Nintendo, 1984), was Miyamoto able to develop games that required a dynamic and moving camera. Game cameras can also intimately affect a game's feel or ambiance. For example, horror games often limit where people can see and sometimes shake the camera to mimic popular techniques in cinema (for example, *Silent Hill* [Konami, 1999]), and stealth-based games require players to hide and then look, linking visual cues to skill in gameplay (for example, *Metal Gear Solid* [Konami, 1998]). Traditional game camera design required the designer to set the game camera and define the visual field to best support the game's experience and story. This approach changed with the introduction of 3D games, which challenged game designers to reconfigure representation, animation, and gameplay.

In 1996, Nintendo released the Nintendo 64, and for its release much-admired game designer Miyamoto created *Super Mario 64* (Nintendo, 1996) and *The Legend of Zelda: Ocarina of Time* (Nintendo, 1998). These games designed almost twenty years ago defined the conventions for designing in 3D environments. Although these games are still touchstones, the move to 3D game cameras necessitated redesigning both hardware and software to support these new forms of representation. Indeed, it is often said that Miyamoto designed the Nintendo 64 controller for *Super Mario 64* instead of the other way around, and for players the game feels like this could have been the case. The yellow buttons on the right side of figure 22.1 allow the player to change the camera angle. The analog stick is pressure sensitive, which means that Mario would tiptoe, walk, and run depending on the pressure applied.

Further, users could configure the entire controller: they could switch what the buttons do (customizing hardware), or they could hold the controller in the traditional way or sideways or upside down (customizing use). However, as Jim Merrick explains, Miyamoto did not design the controller for only one game: "Mr. Miyamoto wanted analog control because he had a vision of how he wanted that game to work, but the controller wasn't designed specifically for one game" (2000). Rather, Miyamoto designed the controller because he had a vision of how the analog stick and the 3D controls would affect different possibilities in gameplay. And these analog sticks became the industry standard for controlling 3D cameras in game space, allowing players to look around and direct the cinematography.

Figure 22.1
Nintendo 64 controller. Photograph by Judd Ruggill. Courtesy of the Learning Games Initiative Research Archive.

The challenge facing Miyamoto and his team was translating the 2D side-scrolling *Super Mario* games into a 3D environment. The enemies still occupied the space, and Mario still had to run, jump, and collect, only now he could do so on a *z* axis. "The most difficult part was to realize a virtual 3D world," Miyamoto explained. "It's difficult to give the best 3D angle so that the players don't experience frustration. The game must remain fun to play" (qtd. in Mielke 1998). To address this problem, Miyamoto turned to cinema for the metaphors of vision. The idea of a game camera was already in the design lexicon, yet 3D animation required a major shift in camera manipulation and its consideration in gameplay. Miyamoto was careful to explain that he didn't create cinema on the game console. He saw cinema as passive, whereas games are active. However, he explained that game designers can borrow good ideas from movies, such as camera angles and real-time voices ("Mr. Miyamoto on *Star Fox*" 1997, 118). He

Figure 22.2

Two screenshots showing a Lakitu cameraman in *Super Mario 64* (one close-up and one as a small icon at the bottom-right corner of screen).

noted in 1997, "Since I have been working on 3D games, I have begun to specify camera angles, locations and movement" (qtd. in "Mr. Miyamoto on *Star Fox*" 1997, 118), and he deftly managed these elements in *Super Mario 64* by offering two types of camera control: automatic control to provide "a recommended view" (sometimes the game takes control when enemies or challenges are being approached) and player control, enabling everyone to be "cinematographers."

In addition, Miyamoto designed a film crew into the game in the form of the Lakitu Bros. Prior to the development of *Super Mario 64*, these cloud-riding characters threw spiked turtles at Mario and Luigi, but in this game they follow Mario around, recording his progress (see figure 22.2).

The Lakitu Bros. did not have a long life as cameramen in the Mario-verse, and they did not necessarily influence the trajectory of video games. What this characterization of a film crew did, however, was provide the player with a useful metaphor to understand camera views in the video game environment, to float above a world, as Miyamoto described it, and do whatever he or she wants (Grajqevci 2000). With the launch of the Nintendo 64 and the introduction of a complex controller, Miyamoto and his team at Nintendo had to accomplish two things: (1) teach the player how to navigate this space in order to have a fun and immersive play experience and (2) teach new developers what was possible with the Nintendo 64 platform. Although no other game succeeded in utilizing the Nintendo 64 controller to the same capacity as *Super Mario 64*, this launch title did shape how an emerging 3D video game market would evolve.

Almost simultaneous with 3D graphics was a drive to start representing images with photorealistic qualities. This progression is interesting because it borrows a sensibility developed with photography and traditional aperture-based cameras. The idea is simple: cameras capture and represent the real; however, although this supposition is critiqued and challenged in photography—the real is never "real" in a photograph but is rather staged and framed to manipulate meaning—the idea of real and accurate representation in computer graphics aims for unrecognizable distinction between material models and pixelated representation. To accomplish photorealism is an artistic and technological feat worthy of attention. Textured and random organic matter tends to present seemingly insurmountable obstacles. As photorealistic technologies progress, their visual attributes are being incorporated into gameplay. For example, *L.A. Noire* (Rockstar Games, 2011) used live actors, recorded them with digital camera technologies, and then imported these files into graphics programs to translate the full range of facial expressions into the game. The object of the game: to "read" truth and lies on the different characters' faces.

Photorealism pushes the game industry to develop new graphics capabilities and techniques, but camera-quality representation presents a number of challenges as well. Perhaps the most acknowledged of these challenges is the danger of the "uncanny valley," a theory first proposed by Masahiro Mori ([1970] 2012)) to explain the feeling of revulsion toward humanoid representation that is close to yet not exactly human. In this model, humans feel empathy and sympathy for humanoids' simulated actions and emotion until a certain point where the representation becomes creepy. In addition to the uncanny valley, photorealism coupled with 3D space challenges what animators know about movement and emotion. Animation style focuses on simplified faces and large eyes because it is easier to convey emotions this way in a 2D abstract form. In addition, animators are trained in the twelve principles of animation, first articulated by Disney animators. A number of these principles attend to exaggerated form to simulate, not represent, movement and character development. Thus, what might be understandable stillness in film becomes static and dull in animation. Animators therefore exaggerate movement, starting and stopping, staging, and even building anticipation for movement by holding a pre-pose (knees bent to jump, for example) (see Johnston and Thomas 1981; Lasseter 1987). Photorealism and 3D in games changes these principles because the animator cannot control the scene, cannot anticipate player's movements, and cannot even ensure that the camera angle will translate movement well. Even if the character representation is photorealistic, then, the movement is not film realistic as of yet. Characters running must enter a transition animation to proceed into their next movement, such as standing or changing directions. Regardless of the potentials and challenges, the fact remains that cameras are integral to understanding and defining video game interfaces and representations, and the intersection of these two technologies continues to influence the direction of the gaming industry.

Camera Inputs and the Body in Play

The year 2014 was abuzz with the potential of virtual-reality headsets, the game news focusing on both Oculus Rift and Valve's proclaimed investment in developing this emergent technology. However, a careful look at the game camera as an input device unveils a complicated history in which people have long tried to use the camera to integrate the whole body into the act of play. With controllers, people manipulate the actions of the game with their hands. With the game camera, the body in motion is the body at play: ontological immersion realized.

Early in computer research, individuals were already developing virtual-reality systems and camera-input systems. In 1968, for example, Ivan Sutherland discussed what was arguably the first head-mounted virtual-reality display, explaining that from the beginning the objective was "to surround the user with displayed three-dimensional information" (757). In 1995, almost thirty years after these early experiments, Nintendo attempted to bring virtual reality to the home console in the form of the Virtual Boy, a small console that required the player to put her head into the visor and control the games with control pads. The Virtual Boy was a commercial failure, yet it's important to consider in the history of the game camera because early camera inputs clearly articulated to dreams of virtual reality.

For example, the children's television show *Knightmare* (1987–1994) is ostensibly an adventure game in virtual reality. The game is simple: teams of four participants (live human beings) must navigate a series of three virtual-reality dungeons. Equipped with a knapsack of food to replenish life force, the dungeoneer (one member of the team) walks into a blue-screen room wearing the helmet of justice, which blocks his or her vision except for the *immediate* surroundings to allow him or her to walk safely. Meanwhile, the other three members of the team watch a television set that shows their teammate surrounded by computer-generated environments, objects, and enemies, and their job is to yell out directions, such as "Pick up that object." If the team is successful in navigating through all of the dungeons and collecting all of the items, they win. Few teams won in the tenure of this show. The show is a solid example of a virtual-reality computer game that relies on camera inputs. Although the technology was complicated enough that it could not be scaled to the home, the show itself had a long run on television (seven years). Yet if we consider *Knightmare* an early example of a game camera controller, it laid the foundation for many of the tropes made possible via home console game cameras—providing a visual representation of a body in motion transported into a virtual environment.

As *Knightmare* was concluding its run, research-and-development departments were exploring the possibilities of game cameras both large—the Computer Assisted Virtual Environment (CAVE) virtual-reality environment (1992)—and small—Nintendo's Game Boy Camera (1998). The CAVE, originally developed at the University of Illinois, Chicago, depends on cameras for both inputs and visual displays. Cameras are trained on the body, and camera technologies are carefully considered to develop visual interfaces and increase the sense of immersion. Indeed, in an early conference paper concerning the CAVE, Carolina Cruz-Neira, Daniel J. Sandin, Thomas A. DeFanti, and their colleagues emphasized the connection between view-centered perspective and camera technologies: "The perspective simulation of common visualization systems dates back to the Renaissance, and is based in a mythical camera positioned along an axis extended perpendicular from the center of the screen" (1992, 65). The technologies of the camera inform the developing technologies of stereoscopic computer representations. Add to that the cameras trained on the body in space, and the CAVE provides an environment that exists only because of cameras through which people can play *Quake* (id Software, 1996) and *DOOM* (id Software, 1996).

CAVE was commandeered for game use, but the first game camera designed for a game system was arguably the Game Boy Camera. Developed for the handheld Game Boy systems, the Game Boy Camera was an attachment that slid into the cartridge slot with a lens located in a large sphere. People could take photos, view the photos, and play games that came packaged with the camera. The games were not integrated into the camera—it was not an input device. The Game Boy Camera, like other Nintendo experiments, highlighted a tension in Nintendo's strategy; Nintendo has long conflated digital technologies in ways that fail in the market. For example, it tried to sell its gaming consoles as computers, complete with floppy drives and a mouse, and in the case of the Game Boy Camera, it was selling a camera for its handheld game console. This is not to say that the camera doesn't have a place in game camera history; it has developed quite the connoisseur following among certain photographers. In his article about photographer David Friedman, for example, Alan Devenish notes of the images captured on the Game Boy Camera that "the low-res images made New York resemble a series of Nintendo game title screens" (2014). So even though the Game Boy Camera was capturing data and storing them digitally, it would be some time before digital cameras became an affordable input option for home gaming consoles.

In 2003, Sony released the EyeToy for the PlayStation 2. This small add-on is similar to a web cam, providing players and users with a color digital camera. According to VGChartz (n.d.), only fifteen games were released for the EyeToy between 2003 and 2008, providing players with a mixed-reality experience (a blend of material and virtual worlds). It was succeeded by the PlayStation Eye (2007), which provided similar game inputs and was later augmented with the PlayStation Move (2010). The Move is similar to the Wii sensor and remotes, using handheld controllers to track bodies in motion. When Microsoft entered the game camera market with the Kinect (2010), its aim was to provide players with a hands-free controller, which helped to build a casual gamer market. In addition to an infrared projector and camera, the Kinect used a Light Coding System to 3D scan and reconstruct objects into games. Dancing in front of the camera ensured that a projected image of an avatar dancing mirrored the player's movements. Both PlayStation and Microsoft have continued to develop camera inputs and have been joined by third-party developers such as Ubisoft with its Motion Tracking Camera, and these developments seem to answer the accusation leveled against gaming inputs that game cameras are underutilized and underdeveloped (Dionisio, Burns, and Gilbert 2013). Development in game cameras is moving quickly, and with these quick advancements arise very real concerns about privacy and surveillance.

With the 2013 release of the Xbox One and its Kinect motion-sensing peripheral, a general cry of outrage greeted Microsoft's always-on Kinect camera system. The Kinect allows people to voice-activate their Xbox One, using the game console as the center of their multimedia experiences. This announcement came at the same time as the National Security Agency spying scandals broke in the United States, and the perceived surveillance by the Kinect was immediately linked to heightened concerns about privacy and government data collection. Microsoft published a new privacy statement (Microsoft 2014), and gaming magazines and websites had to assure the gaming public that they could just turn off the Xbox or pause it if they were concerned (Fahey 2013). Although some sites dismissed voiced privacy concerns, the fact is that ethicists have long considered the close relationship between the camera and surveillance (see, e.g., Dubbeld 2003; Lyon 2001; Haraway 1988). Thus, as game cameras evolve to allow for different gaming experiences and inputs, it's important to remember that both games and cameras come with ideological practices and meanings that might have been forgotten or effaced in their almost fifty-year trajectory.

Finally, although much contemporary hype surrounding game cameras focuses on cameras, consoles, and other developing virtual-reality inputs such as the Oculus Rift, the fact is that traditional cameras and cell phone cameras are also being employed in games. For example, both geocaching and alternative-reality games often ask players to take pictures of items, log photographic evidence of completing tasks or quests, or share information with a dispersed group of players so that the players can collectively solve problems using photographs as clues. Alternative-reality games and geocaching have been around in some form for quite some time. However, as digital cameras have become more ubiquitous through dedicated cameras and cell phone cameras, the visual technologies become important components in games that interface with material worlds. In these cases, the camera provisionally becomes a game camera.

The history of the game camera, then, is much like the history of photography; it raises both aesthetic and technological questions, and those questions are embedded in the cultures and histories of practice and play. As research into game platforms continues, it is important to look to these "peripherals" such as the game camera because such peripherals compel people to use systems differently. Game cameras for input devices, for example, can of course be used to play games, but they can also be used to take pictures, Skype with friends, or record short videos. Because they do so through game consoles, however, questions arise: How do games affect photography and film? How do photography and film shape games? Do new practices and aesthetics arise within the articulation of the various forms? How do the technologies create affordances and possibilities, and, contradictorily, how do they limit possibilities? Such historically bound questions have important implications for design and criticism and should be interrogated with the same rigor as game studies considers such terms as *play* and *games*.

Works Cited

Cruz-Neira, Carolina, Daniel J. Sandin, Thomas A. DeFanti, Robert V. Kenyon, and John C. Hart. 1992. "The Cave: Audio Visual Experience Automatic Virtual Environment." *Communications of the ACM* 35 (6): 64–72.

Devenish, Alan. 2014. "Awesome Portraits and Landscapes, Shot with a Game Boy Camera." *Wired Online*, June 20. http://www.wired.com/2014/06/game-boy-camera-color/.

Dionisio, John David N., William G. Burns III, and Richard Gilbert. 2013. "3D Virtual Worlds and the Metaverse: Current Status and Future Possibilities." *ACM Computing Surveys* 45 (3): 34.1–34.38.

Dubbeld, Lynsey. 2003. "Observing Bodies: Camera Surveillance and the Significance of the Body." *Ethics and Information Technology* 5 (3): 151–162.

Fahey, Mike. 2013. "Xbox One Kinect Privacy Concerns? Turn It Off. Pause It While Gaming." *Kotaku*, June 6. http://kotaku.com/xbox-one-kinect-privacy-concerns-turn-it-off-511759241.

Grajqevci, Jeton. 2000. "Profile: Shigeru Miyamoto: Chronicles of a Visionary." *N-Sider*, October 9. http://www.n-sider.com/contentview.php?contentid=223.

Haraway, Donna. 1988. "Situated Knowledges: The Science Question in Feminism and the Privilege of Partial Perspective." *Feminist Studies* 14 (3): 575–599.

Johnston, Ollie, and Frank Thomas. 1981. *The Illusion of Life: Disney Animation*. New York: Hyperion.

Kelly, Tadhg. 2011. “Camera Comes First [Game Design].” *What Games Are*, October 25. http://www.whatgamesare.com/2011/10/camera-comes-first-game-design.html.

Lasseter, John. 1987. “Principles of Traditional Animation Applied to 3D Computer Animation.” *ACM Computer Graphics* 21 (4): 35–44.

Lyon, David. 2001. “Facing the Future: Seeking Ethics for Everyday Surveillance.” *Ethics and Information Technology* 3 (3): 171–180.

Merrick, Jim. 2000. “The Big Gamecube Interview: Part 1.” *IGN*, October 20. http://www.ign.com/articles/2000/10/20/the-big-gamecube-interview-part-1.

Microsoft. 2014. “Xbox Privacy Statement.” http://www.microsoft.com/en-us/privacystatement/default.aspx.

Mielke, James. 1998. “The Miyamoto Tapes.” *GameSpot*, November 11. http://www.gamespot.com/articles/the-miyamoto-tapes/1100-2465458/.

Mori, Masahiro. 2012. “The Uncanny Valley.” Translated by Karl F. MacDorman and Norri Kageki (1970). *IEEE Robotics & Automation Magazine* 19 (2): 98–100.

“Mr. Miyamoto on *Star Fox*.” 1997. In *Star Fox 64 Official Nintendo Player’s Guide*, 116–119. Redmond, WA: Nintendo of America Inc.

Sutherland, Ivan E. 1968. “A Head-Mounted Three Dimensional Display.” In *Proceedings of the December 9-11, 1968, Fall Joint Computer Conference, Part I*, 757–764. American Federation of Information Processing Societies. New York, NY. doi:10.1145/1476589.1476686.

VGChartz. n.d. “*Super Mario 64*.” Games database. http://www.vgchartz.com/game/2278/super-mario-64/. Accessed August 1, 2014.

23 GAME CULTURE

Reneé H. Reynolds, Ken S. McAllister, and Judd Ethan Ruggill

As with other terms in this book, the term *game culture* is permeated by conundra, not the least of which is its strange temporal resonance. On the one hand, the video game medium is barely a lifetime old, with games first appearing as computational and video-based experiments in the decade following World War II and as commercial products a few decades later. On the other hand, scholars have been studying play and games in general since the days of the ancient Greeks and Chinese, if not before. There is thus a concomitant and countervailing antiquity to game culture's potential newness.

Complementing this anachronistic quality of being both recent and ancient is the peripateticism of the term *game culture*. Not only has the term had a number of different activities and associations attached to it over the years, but scholars also remain challenged to authoritatively define *game* and *culture* individually—never mind their conjunction—despite centuries of inquiry and analysis.[1] Making matters more difficult still is that today's most conventional usages of the term—usages that seem semantically disconnected from their historical, cultural, and political implications—are manifold and context dependent, capable of referring to wildly divergent artifacts and phenomena depending on the invoker. For example, the term *game culture* as used in newspaper articles, popular magazines, and various hobbyist and trade periodicals might alternately mean (but is not limited to) (1) the social structure and connections of in-game guilds or clans; (2) the goings-on at game-oriented conventions, trade shows, midnight release promotions, tournaments, and other public events; (3) the social and material products of handicraft, maker, and artist groups; (4) the professional, interpersonal, and technologically mediated relations of game designers; and (5) scholarly books, articles, conferences, and other research devoted to games.

Moreover, the proliferation of possible meanings is constantly expanding. Consider such relatively recent phenomena as the development and monetization of community-management practices associated with massively multiplayer online games; the pullulation of downloadable and user-generated content; the expansion of multiplatform gaming, development, and tool manufacture; and the mainstreaming of laboring in game worlds to produce artifacts with extragamic economic value and other techniques for commodifying play. All are part of game culture in one way or another—and are identified as such—and some are actually robust game subcultures all their own.

As a result of these various conundra and the limited space we have to explore them in this chapter, we can trace only some of the historical and signifying moments of "game culture." Therefore, we limit what we mean by the term *game culture* to the set of shared practices surrounding video games (including their hardware, software, and paratexts), their makers (including developers and players as makers through play), regulators (including public and game-based policy makers), and players. More importantly, we advance the idea that *game culture* (as a term and semiotic domain) is important to study not only because of what it describes—the shared practices noted earlier—but also for what it does: nominates, instantiates, and reifies practice into structure. In short, we suggest that the term *game culture* ultimately functions as a kind of speech act.

Genealogy

The earliest uses of *game culture* as a distinct term—which well precede the emergence of the video game medium—refer to the management of natural resources.[2] In contrast to farming, which generally involves the deliberate cultivation of plants and domesticated livestock by humans, the cultivation of game—that is, nondomesticated animals—was until fairly recently both a philosophy of land management and an ideology guiding human connection. In stewarding (rather than simply using up) local resources, a community shapes not only its habitat but also the relationships among individuals within the group who collectively but differently need to attend to the many details of resource management. In the first half of the nineteenth century, for example, numerous American and British pamphlets, almanacs, magazines, and trade publications began to emphasize the importance of nurturing plants—including vines, olives, melons, grains—and "wild game," which often simply meant fish. Indeed, aquaculture was a rising concern at the time, yielding dozens of book-length illustrated guides on how to raise trout, salmon, and other (primarily freshwater) species. It was not until the late nineteenth century that the term *game culture* began to include other wild animals (e.g., deer, elk, rabbit, bird, and fox populations), taking on the designation "wild game and fish" and addressing animals not only as food but also as sport. These interests introduced a tension into the literature that remains to this day: What are the best ways to protect and propagate local game? It is in these early days of game culture—a "game culture" seemingly far different from what game scholars today usually mean—that the soon-to-be infrangible connections among community, entertainment, and a kind of Foucauldian biopolitics were forged.

When these turn-of-the-century publications lit up with legal debates over game culture and the need for and wisdom of restrictions on the practices of culturing wild game and fish, game culture enthusiasts of all types (especially hunters and fishers) began to create and shape philosophies not only about best practices for game culturing itself but also about the roles, processes, and obligations of collective stewardship. In 1915, for instance, the Society of American Foresters, a group of individuals philosophically disconnected from game culturers, found themselves embodying a fractured old-guard mindset toward the cultivation of land for human use and enjoyment. Smith Riley wrote: "It surely must be possible to give attention to the game culture as a secondary matter in forest protection, particularly when such time given to game culture does not detract from, but adds to, the standard of forest protection and renders the National Forests of greater value to the States and Nation" (1915, 178).[3] Similarly, Chauncey Hamlin, in his report to the National Conference on Outdoor Recreation, cited a survey that documented not only how individuals were using public lands for game hunting but also how businesses and organizations were doing so, extending as part of their member benefits access to private lands for hunting purposes. He specifically cited (among others) the Institute of Small Arms and Ammunition Manufacturers' "game culture," which at the time oversaw the management of 400,000 acres of stocked and huntable land (1928, 85). It is clear from these and similar developments that as the United States renewed its domestic agenda after World War I, the politics driving the practices of resource management were becoming increasingly scientized, industrialized, and militarized.[4]

Given this development of game culture from an informal to an eagerly optimized set of practices, it is no surprise that natural resources were quickly brought to a precipice that is still being negotiated today. Forestry and game culturing as professional and scientific occupations were certainly profitable, but most of the developmental capital was being expended on the resources to be managed and increased, whereas considerations of how this work was actually done were largely ignored. By the time World War II was on the horizon, the term *game culture* had taken another turn, one that was more sensitive to the management of available resources and, newly, to the management of the people whose job it was to steward those sport- and leisure-oriented resources. In the proceedings of the American Game Association Conference of 1930, J. G. Burr argued that the time had come for those taking advantage of the spoils of game culture to learn that such bounty is the result of a laborious enterprise—one that should be rewarded: "In many cases [the hunter's] supply of game is the result of careful attention to its welfare and the laborer is worthy of [the hunter's] hire. The fee paid for the privilege of hunting is a reward for past services and should be accepted as a pledge of [the laborer's] future devotion to game culture. It is just another step in economic evolution" ([1930] 1932, 28). Embedded within this call to further build out the "economic evolution" of game culture was the "pay to play" mandate that today is found everywhere from state park entrance fees to downloadable content for the latest gaming consoles. The recognition that play, like everything else in the capitalist hegemon, is a commodity that requires a complex set of production processes signifies an important cultural shift, one that (if it is to be profitable) requires not only the management of resources but also the ongoing production of interested consumers.

After World War II, another significant shift in the literature suggests a change in the way Americans especially understood the term *game culture*. Although the term maintained its reference to the stewardship of wild game for hunting, a new meaning emerged from the field of social psychology. Within this context, *game culture* signifies a culture that develops either designedly (all players agree on it) or spontaneously within a simulation (a game). In group-dynamics research, interest focused on the formulation of culture as a set of understood or learned social rules of behavior. The words *game* and *culture* were used nearly interchangeably in work by Anita Yourglich as she advised that when entering the game of baseball, for example, one should seek to interact with those who are already a part of the game culture so one can "define the situation, play a role, and earn status in the group of players" (1955, 7). Several years later Harold Guetzkow conducted an experiment in his classroom in which students became citizens within a classroom game. In this simulation, students constructed a culture of international relations. At the end of the experiment, Guetzkow reflected that "game culture takes time to appear. It is likely, then, that it was real culture that we saw in the foreign student run of the simulation, for the run operated less than one full day" (1963, 101). In other words, Guetzkow posited that there can be a culture born inside of a game in which the simulation takes on meaning to the players, charging them to formulate a culture to define and preserve what is acceptable within that simulated world.

Elsewhere in social science research, the term *game culture* was used to describe the ways in which some patients are able to form a simulation within their own minds, one that fosters calm, engagement, and physical comfort. In this sense, *game culture* refers to a kind of treatment for physical and mental maladies. Edgar Schein and Warren Bennis contended, for example, that it might be possible to reach some psychologically distant patients by having them cast their attention away from the session at hand and create an inner world that fostered more ease. Some patients, they noted, do this more naturally than others, choosing "game culture as a defense if what [they hear] is too threatening. It may help therefore for the staff to support the fiction of the game in the early parts of the laboratory. The fact that the laboratory is temporary and has unreal 'as-if' qualities makes it easier for norms of mutual acceptance to be established" (1965, 302). Here again, the dialectic of the term *game culture* is pronounced and important: what began as resource management for the purposes of food security becomes the production of leisure, which becomes the practices associated with that leisure, which becomes the utilization of the feelings and phantasms associated with leisure practices to mitigate various physical and psychological torments. Each iteration of the term differs from previous ones yet also contains traces of them.

In the late 1960s, the term *game culture* had aggregated another important meaning, one that precipitated its most common meanings today. In 1968, Gary Gygax organized Gen Con, one of the first large-scale meetings of game players. With that event, Gygax, who later developed *Dungeons & Dragons* (1974) with Dave Arneson, created a phenomenon that grew rapidly and internationally, eventually popularizing not only tabletop games but also the desire to design, play, discuss, and critique games as a community. Within ten years, this set of communal practices had been operationalized in *MUD*, the first online multiuser dungeon game, developed by Roy Trubshaw and Richard Bartle. Running first on a Digital Equipment Corporation Programmed Data Processor-10, *MUD* and its successors allowed gamers both to play and to socialize together online and in real time. Gary Alan Fine observed in these early days that it seemed to be precisely this kind of shared experience—playing and collaboratively constructing knowledge about a game—that established "a unique group culture": "This culture combines the players' previous information (latent culture) with the game events (manifest culture). This game culture becomes a central mechanism by which group interactions are organized as it provides for a set of shared experiences and common referents" (1982, 21). Fine further observed that these groups' gaming cultures develop in a "particularly intense" manner and speculated that they will tend to expand because all future engagements among these gamers (including the cultural exchanges within their groups) are based on their past interactions (ibid.). History and presence, in other words, had become signature facets of cultural inclusion within the gaming community.

Throughout the 1980s (thanks to the proliferation of arcades) and into the 1990s (thanks to the rise of ubiquitous computing and the World Wide Web), game culture accrued adherents, critics, and (perhaps most important) capital exponentially. By the end of the millennium, the term *game culture* not only had acquired a hegemonic signification relating to the video game industry specifically, but also had spawned numerous subterms and subcultures. In 1999, for example, journalist Monty DiPietro helped popularize both the *otaku* (おたく) subculture—an emerging Japanese trend involving youth who obsessively read and discuss comics and play video games—and the rise of an extreme version of this subculture known as "*hikikomori* [ひきこもり], or social withdrawal" (36). DiPietro misleadingly characterizes *hikikomori* as "an increasingly serious problem" among "video-game fanatics"—there are actually many forms of this particular type of social isolation, only one of which involves an obsession with video games—but the representation stuck and soon became a cautionary exemplum—along with school shootings—of what could go wrong if game culture were allowed to reach a malignant apogee. David Bell characterizes this development of game culture (or at least its mass media representation) well when he observes, "In this kind of reading, 'immersive' gets rewritten as 'addictive,' and 'interactive' as 'antisocial,' since kids were assumed to interact with machines rather than [with] each other" (2001, 46). This agonistic arena is where the term *game culture* largely resides today—that is, as a locution that simultaneously spans disciplines and politics and masks a connotative cornucopia with (to paraphrase *Webster's Dictionary* definition of "culture") a seemingly intuitive denotation: "the collection of values, conventions, and social practices" that circulate around video games (online ed., 2015).

Game Culture as Speech Act

The scope of the term *game culture* is massive and dynamic, which can make it seem inaccessible and even unusable. It is the vitality of the term, however, that makes it particularly salient for game studies because it exemplifies how important it is for scholars to reflect on not only what a term means but also what it does. In the case of *game culture*, consider the work the term has done in its time: propagated the ideas that wild animals are calculable resources, that a labor force should be fielded and an economic infrastructure built to control these resources, that leisure costs dearly, that games and labor are inextricable, and that play and psychosis may go hand in hand. Surely there is other work it can do. In any case, as important as it is to understand what *game culture* means in any given situation, giving attention to the consequences of its use is equally necessary. J. L. Austin observes that "by saying something we do something" (1962, 109). If this is true, then ultimately perhaps that most important meaning of the term *game culture* is that it reproduces itself.

Notes

1. See, for example, the work of Quintilian (2010); Roger Caillois ([1958] 2001); Brian Sutton-Smith ([1997] 2001); Blaise Pascal ([1669] 1995); Michel de Montaigne ([1580] 1993); Raymond Williams ([1958] 1983); Clifford Geertz (1977); Matthew Arnold ([1869] 1966); Leo Lowenthal (1961); Johan Huizinga ([1935] 1955); Theodor Adorno (1951); and Aristotle (1932), to list but a few.

2. Our research focused on English-language texts, including translations. A more expansive study to include such terms as *la culture du jeu*, *Spielkultur*, and ゲーム文化 is needed but beyond the scope of this chapter.

3. We are grateful to Henry Lowood for reminding us that the German term *Forstkultur* (forest culture) dates to the late eighteenth century. The new German forestry science of that time influenced later resource-management movements well into the twentieth century, and Riley came out of that tradition.

4. The inextricably connected (and rapidly expanding) leisure and entertainment culture (e.g., theater, vaudeville, film, recreational camping) predictably bore similar markings. See Aron (1999) for a particularly insightful analysis of the rise of vacation culture, which was significantly impacted by a scientized work–play rhetoric in this same period.

Works Cited

Adorno, Theodor. 1951. *Minima Moralia: Reflections from a Damaged Life*. London: Verso Press.

Aristotle. 1932. *Politics*. Cambridge, MA: Harvard University Press.

Arnold, Matthew. [1869] 1966. *Culture and Anarchy*. Ed. J. Dover Wilson. Cambridge: Cambridge University Press.

Aron, Cindy S. 1999. *Working at Play: A History of Vacations in the United States*. Oxford: Oxford University Press.

Austin, J. L. 1962. *How to Do Things with Words*. Cambridge, MA: Harvard University Press.

Bell, David. 2001. *An Introduction to Cyberculture*. London: Routledge.

Burr, J. G. [1930] 1932. "Does Game Increase When the Landowner Has a Share in the Game Crop?" In *Transactions of the Seventeenth American Game Conference*, 25–33. Washington, DC: American Game Association.

Caillois, Roger. [1958] 2001. *Man, Play, and Games*. Trans. Meyer Barash. Urbana: University of Illinois Press.

DiPietro, Monty. 1999. "Coming of Age, Virtually." *East* 35 (4): 34–46.

Fine, Gary Alan. 1982. "Legendary Creatures and Small Group Culture: Medieval Lore in a Contemporary Role-Playing Game." *Keystone Folklore: The Journal of the Pennsylvania Folklore Society* 1 (1): 11–27.

Geertz, Clifford. 1977. *The Interpretation of Cultures*. New York: Basic Books.

Guetzkow, Harold Steere. 1963. *Simulations in International Relations: Developments for Research and Teaching*. Englewood Cliffs, NJ: Prentice-Hall.

Hamlin, Chauncey. 1928. *National Conference on Outdoor Recreation: A Report Epitomizing the Results of Major Fact-Finding Surveys and Projects Which Have Been Undertaken under the Auspices of the National Conference on Outdoor Recreation*. Washington, DC: US Government Printing Office.

Huizinga, Johan. [1935] 1955. *Homo Ludens: A Study of the Play Element in Culture*. Boston: Beacon Press.

Lowenthal, Leo. 1961. *Literature, Popular Culture, and Society*. Palo Alto, CA: Pacific Books.

Montaigne, Michel de. [1580] 1993. *Michel de Montaigne: The Complete Essays*. Ed. and trans. M. A. Screech. London: Penguin Books.

Pascal, Blaise. [1669] 1995. *Pensées*. Ed. and trans. A. J. Krailsheimer. London: Penguin Classics.

Quintilian. 2010. *Quintilian's Institutes of Oratory: Or, Education of an Orator*. Charleston, SC: Nabu Press.

Riley, Smith. 1915–1925. "Game on National Forests." In *Proceedings of the Society of American Foresters*, 10 vols. Vol. 2: 175–182.Washington, DC: Press of Judd & Detweller.

Schein, Edgar, and Warren Bennis. 1965. *Personal and Organizational Change through Group Methods: The Laboratory Approach*. New York: Wiley.

Sutton-Smith, Brian. [1997] 2001. *The Ambiguity of Play*. Cambridge, MA: Harvard University Press.

Williams, Raymond. [1958] 1983. *Culture and Society: 1780–1950*. New York: Columbia University Press.

Yourglich, Anita. 1955. *The Dynamics of Social Interaction*. Washington, DC: Public Affairs Press.

24 GAME DEVELOPMENT

Katie Salen Tekinbaş

The fourth step in game development is quantifying the relationship between decisions and results.
—Richard Marvin Hill, *Marketing Concepts in Changing Times*

At first glance, the term *game development* presents itself as a bit of a simpleton, a self-descriptive term that means just what it says. Games are developed via a software-development process that merges the design of gameplay, art, and audio with the technical constraints of platform, be it console, mobile device, the Web, or some combination thereof. Development most often occurs in teams, although this was not always the case historically. Today the process involves a fairly standardized set of iterative stages, from preproduction to the design of a proof-of-concept prototype through production, testing, launch, and, in some cases, ongoing revision. Members of a game-development team most often play a specific role in the game's development, although a single person might take on multiple roles. The team can include game programmers or engineers, game designers, level designers, artists (two dimensional, three dimensional, animation, cinematics), sound engineers, producers, writers, and testers. Although development teams today range in number from 5 to 60 or more, as late as the 1980s a team could be as small as one person.

On the surface, this explanation of the term *game development* seems pretty straightforward, but a second look reveals it to be far more interesting. From its origins in analog and tabletop gaming through university and military application to the birth of console gaming and the rise of the software-development industry, the meaning of *game development* has changed in subtle but interesting ways.

Before we dig into that rich history, it is important to note that *game development* is not a stand-alone term: it is tightly coupled with the concept of and term *game developer*, the individual, team, or company undertaking the development of the game, and it's difficult to say which came first. To most current practitioners, including those who create and those who publish games, the terms simply go hand in hand—game developers are the entities that do game development.

Each of these terms also appears to have had independent meanings at various times and places. For example, the term *game developer* has a slightly different meaning in the United Kingdom and the United States. In the United Kingdom, a game developer can be a job title for an individual or a company; when it refers to an individual, there is the definite implication that programming will be involved. In the United States, however, designers and artists who never touch code are also referred to as game developers.[1] This linking of the game developer and the programmer can be traced to the early days of video games: "In the early days, all designers were programmers, because that was the only way they could get their games made. It's a common misconception to say that all early computer games were designed by programmers: actually, many (if not most) were programmed by designers."[2] Designers programmed in order to develop games. But they weren't the first game developers, as we shall soon see.

Emergence of the Term

I would bet [the term game development] goes back to analog games. When I joined Sirtech in the early '80s, it was a known term.
—Brenda Romero[3]

Game development seems to have first emerged as a term descriptive of the work done by the designers of board games and board war games, such as those published by Parker Brothers, Avalon Hill, and Simulations Publications, Inc. (SPI). As a result, the term gets tangled up with *game designer*, a term that most certainly predates *game developer*. The developer in the context of analog-game creation was someone who took the draft of the rules provided by the designer and saw to development, testing, and production of the game. Avalon Hill began giving credits for rule development in 1974, but by the second edition of the game *Third Reich* in 1976 and the release of *Rail Baron* in 1977, there was an actual game-development credit in the rules (Don Greenwood for the former and Randy Reed for the latter).[4] SPI published development credits in 1973 for both *Sniper!* and *Fall of Rome*. And in 1979 SPI's game *John Carter Warlord of Mars* credited Eric Goldberg as the game developer right on the box.

Further, issue 22 of *Moves*, a wargame magazine popular in the 1970s, published the game-"development" schedule for *Firefight*, a tactical-level simulation set in 1970s West Germany pitting Soviet and American forces against one other. Designed by James F. Dunnigan and Irad B. Hardy, *Firefight* was funded by the US Army as a way to develop simulation-based training materials for US ground troops assigned to the European theater during the Cold War ("SPI,

Firefight" 1999, 1). Note that in this context the term *development* refers to the sequence of levels to be designed rather than to the overall process of the game's creation, which is its standard usage today. The development of *Firefight* was to proceed as follows:

FIREFIGHT 1: DEVELOPMENT STAGES

Stage I—The basic introductory level game to include the rules covering movement, combat, sequence of play, victory conditions, as well as description of the playing map and the playing pieces. The map itself will be 8" x 10" using 16 millimeter hexes; the scale will be assumed to be companies of leg infantry, two kilometer hexes and twelve hour turns.
Stage II—This stage will introduce all the rules required for ZOC [zones of control]. These Zones will be assumed to be semi-active rigid.
Stage III—This stage will include the introduction of stacking rules. Stacking will be two high.
Stage IV—This stage will include the introduction of effects of terrain on movement and combat. In Stage I, of course, terrain either allowed you to go into a hex or not to enter it.
Stage V—This stage will introduce ranged fire. This will also include opportunity fire, as well as yet to be developed rules for suppressive. ("SPI, *Firefight*" 1999, 1)

In 1977, the staff of *Strategy & Tactics Magazine* coauthored the study *The History, Production, and Use of Conflict Simulation Games*. One of the chapters, "Game Design and Development: A Case Study" (Berg 1977, 41), cemented the use of the term across the industry.

Publishers such as Avalon Hill and SPI employed game developers, who acted like editors or producers, to "develop" games beyond their initial design. Even today in the board-game world, a developer is usually an employee of the publisher who refines the product and gets it ready for prime time.[5] This structure is in distinct contrast to the way most commercial video game development occurs today, whereby large game publishers fund game-development companies to create games, which they then market and distribute. For example, game publishers such as THQ Wireless, EA Mobile, and UIEvolution (a former subsidiary of Square Enix) fund the game-development studio 5th Cell to produce games such as *Scribblenauts* (2009) and *Drawn to Life* (2007). This is one interesting distinction in how the term *game development* is used today in the contexts of analog and digital game development.

Digital Game Development

If the predictions of Albrecht and other innovative leaders in computer game development are accurate, computer skills will be as common as riding a bicycle.
—Dennie Van Tassel, *The Compleat Computer*

We know from the many written histories of video games that the first video games were developed in the 1950s by computer scientists, required mainframe computers to play, and as a result were not available to the general public. This form of game development in the 1960s and 1970s—digital game development—was occurring simultaneously with board-game and tabletop wargame development. Individuals such as Don Daglow and Kelton Flinn, both prolific game designers in the 1970s, were graduate students and graduate instructors with years of free access to campus computers. They used this access to develop many games. "In the 1971–80 university context," explains Daglow, "we referred to 'game development' as a parallel to the standard term 'software development,' but did not think of it as a trade term, just as an alternative, more descriptive phrase to refer to our hobby (since there was no money in it in the early '70s)."[6] This intermingling with and distinction from software development marks the use of the term *game development* in production of digital games from the very beginning. Daglow was eventually hired in 1980 as one of the original five in-house Intellivision programmers at Mattel: "I know we used the term at Intellivision in 1980–83, because our official group title was 'Intellivision Applications Development,' and we never used that term but always changed it to 'Game Development' internally."[7]

Games such as *Tennis for Two* (Higinbotham, 1958) and *Spacewar!* (Russell/MIT, 1962) are well documented within game studies literature; less well documented are the numerous simulation games, master's theses, and other game-development experiments that took place in marketing, urban planning, military, and engineering schools in the 1960s and 1970s, all of which contributed to the ongoing use of the term *game development*. From *CIDY* (*Communicate! I Dare You!*), a communications game submitted in partial fulfillment of a master's in urban planning at the University of Michigan, we get this game-development feedback gem: "Their questions and criticisms regarding CIDY showed that they [the players] were learning something of game development themselves" (Baldwin 1970, 47). *Collegiate News and Views*, a Pennsylvania State University publication, records that "a 1964 survey by the University of Texas indicated that AACSB schools were using 30 functional games and 27 general management games. Of these games, 29 were noncomputerized. The period of time can be classified as 'The Simulation Period'" (Keys 1974, 17). Further, a report from the University of Michigan president to the Board of Regents for 1975–1976 discussed "a growing interest in games and simulations as facilitators of understanding, dialogue, planning and decision-making, together with a modest increase in the number of organizations requesting services, have made the past year a time of expansion and creativity for the Extension Gaming Service. ... The number of formal gaming 'programs'—workshops, game runs, consultations, demonstrations, game development projects—increased by almost 25 percent over the previous year, from 38–47" (University of Michigan 1975, 18).

So while Dunnigan and Hardy were hard at work developing tabletop simulation games such as *Firefight* for the US government, universities were exploring computer-based simulation-game development for a remarkable range of uses, including education. As a result, both analog and digital game development from this era served as a training ground for many of the designers, programs, artists, and others who went on to find employment in companies such as Atari (founded in 1972) and Electronic Arts (founded in 1982)—companies that became icons in the game-development industry as it is known today. They included designers such as Don Daglow, mentioned earlier, as well as Redmond Simonsen, Greg Costikyan, Warren Robinett, and Chris Crawford, to name but a few.

The final distinction in the evolution of the term game *development* can be made in the change from development handled by a single individual to development handled by a team. Because of the limitations of early computer and video game graphical displays, almost all aspects of early game development were handled by a single developer—programming, game design, art, and audio. We see this focus on the single developer beautifully captured in an advertisement from a November 1982 issue of *Creative Computing Magazine*, which introduces readers to The Frob, "the hardware/software system that converts your Apple II into a sophisticated Atari 2600 VCS game development workstation." Due to the ever-increasing processing and graphical capabilities of arcade, console, and computer products, along with an increase in player expectations, game development soon moved beyond the scope of a single developer (Bethke 2003). This shift marked the beginning of team-based game development, the standard form followed today.

A last bit of history: when Chris Crawford launched the Computer Game Developers' Conference in 1988, he originally wanted to call it "Computer Game Designers' Conference," but "one of our Board members—I think it was Stephen Friedman—pointed out that 'Developers' was a more inclusive term covering people such as artists and writers who made important contributions. After some discussion, we agreed that 'Developers' was a better word than 'Designers.'"[8]

Notes

Many thanks to Earnest Adams, Richard Bartle, Ian Bogost, Richard Garriott de Cayeux, Chris Crawford, Don Daglow, Mark DeLoura, Robert Gehorsam, Raph Koster, Brenda Romero, Warren Spector, and Johnny Wilson for their help (and attics) in researching this chapter.

1. Richard Bartle to Katie Salen Tekinbaş, email, January 5, 2014.

2. Ibid.

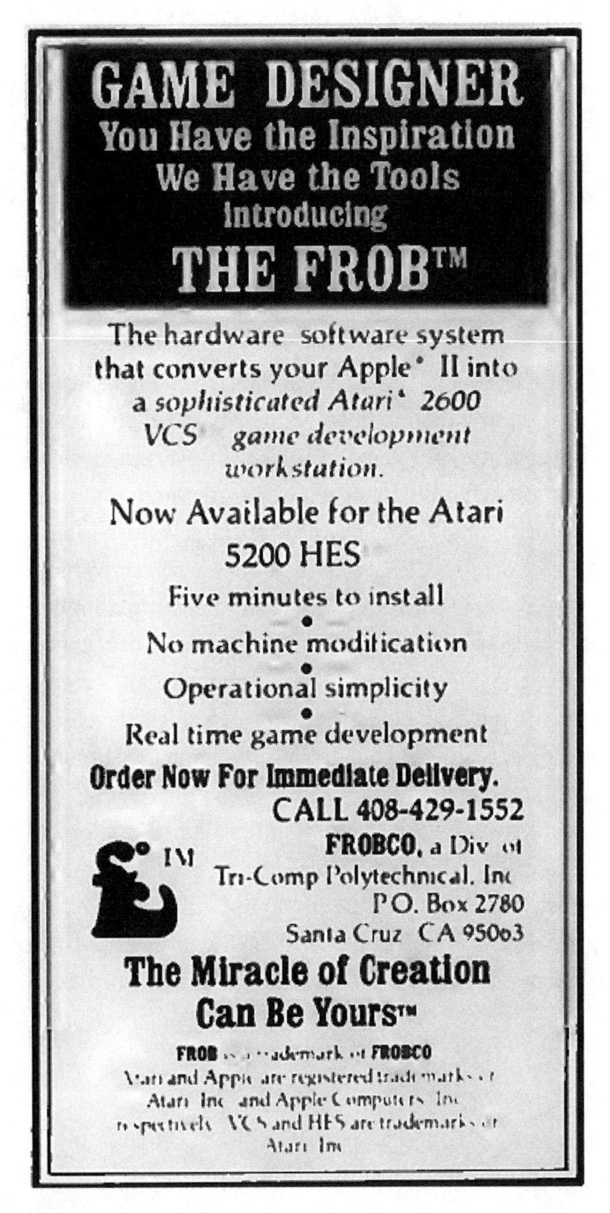

Figure 24.1

Ad for The Frob in *Creative Computing Magazine*, January 1983.

3. Brenda Romero to Katie Salen Tekinbaş, email, January 4, 2014.
4. Johnny Wilson to Katie Salen Tekinbaş, email, January 7, 2014.
5. Ibid.
6. Don Daglow to Katie Salen Tekinbaş, email, January 4, 2014.
7. Ibid.
8. Chris Crawford to Katie Salen Tekinbaş, email, January 5, 2014.

Works Cited

Advertisement for The Frob™. 1983. *Creative Computing Magazine* 9 (1): 234.

Baldwin, Bruce R. 1970. "CIDY (Communicate! I Dare You!): Pilot Development of a Communications Game." University of Michigan, Ann Arbor.

Berg, Richard. 1977. "Game Design and Development: A Case Study." *The History, Production, and Use of Conflict Simulation Games.* Strategy & Tactics Staff Study no. 2. New York: Simulations.

Bethke, Erik. 2003. *Game Development and Production.* Plano, TX: Wordware.

Hill, Richard Marvin. 1960. *Marketing Concepts in Changing Times.* New York: American Marketing Association.

Keys, Bernard. 1974. "The Lecture vs. the Game." *Collegiate News and Views* 28: 17–21. Nashville, TN: South-Western.

"SPI, Firefight." 1976. *Map and Counters*, blog, June. http://mapandcounters.blogspot.com/2009/06/spi-firefight-1976.html. Accessed October 17, 2013.

Strategy & Tactics Magazine Staff. 1977. *The History, Production, and Use of Conflict Simulation Games.* Strategy & Tactics Staff Study no. 2. New York: Simulations.

University of Michigan. 1975. *Financial Statement for the Fiscal Year: The President's Report to the Board of Regents for the Academic Year 1975–1976.* Vol. 2. Ann Arbor: University of Michigan.

Van Tassel, Dennie. 1976. *The Compleat Computer: Being Compendium of Tales of the Amazing & Marvelous Poetry, Informative News Items, Articles for Edification and Enjoyment, Cartoons Plus Many Other Illustrations with a Special Section of Splendiferous Science Fiction Art in Full Color.* Chicago: Science Research Associates.

25 GAME ENGINE

Henry Lowood

Before the release of its upcoming game *DOOM* in 1993, id Software issued a news release. Promising that *DOOM* would "push back the boundaries of what was thought possible" on computers, this press release summarized stunning innovations in technology, gameplay, distribution, and content creation. It also introduced a new term, the *DOOM engine*, to describe the technology under the hood of id's latest game software. The news release predicted that id's technology would change the game industry, with *DOOM* being the first example of a new kind of "open game." The moment of *DOOM*'s release thus merges histories of game terminology, technology, and game design (id Software 1993).

As the term has developed since the early 1990s, *game engine* encompasses the fundamental software components of a computer game. These components typically include program code that defines a game's essential "core" functions, such as graphics rendering, audio, physics, and artificial intelligence, although the components vary considerably from one game to another. Usage of the term in the textual corpus generated by Google Books seems to confirm id's bravado about its technology. The term first appeared in print during the year after *DOOM*'s launch in December 1993 (figure 25.1). We find it in André LaMothe's book *Teach Yourself Game Programming in 21 Days*, in articles in *PC Magazine* and the inaugural issue of *Game Developer*, and—no surprise—in *The Official DOOM Survivor's Strategies and Secrets* by Jonathan Mendoza, all published in 1994. Google Books identifies random citations before that year, but they are unrelated to game development or false hits. These data indicate that the new term *game engine* kept pace with another term, the increasingly prevalent *game software*, during the 1990s, while the older term *game program*—which appeared on Atari game cartridges such as *Yars' Revenge* (1982)—declined as a description of the code underlying computer games. This brief analysis suggests that *game engine* was a neologism of the early 1990s and that the invention of this game technology was a discrete historical event of that decade.

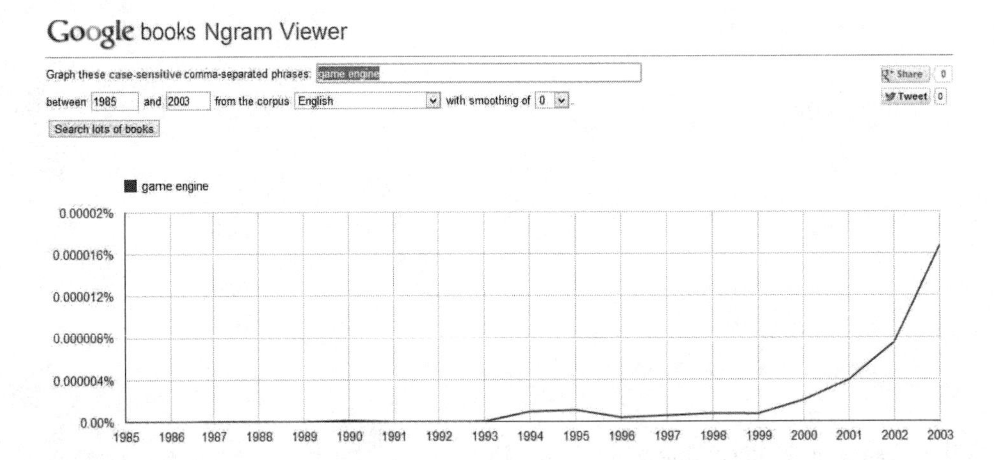

Figure 25.1
"Game Engine," Google Books, 1985–2003.

Focusing on the term itself thus awakens the temptation to treat the history of the game engine as an invention narrative. Telling the history this way would lead us to game technology, beginning with a definition of *game engine* as a discrete technology or set of technologies and then asking questions about circumstances and events: When, where, and by whom was it invented? What was the technical, business, or cultural context for the invention? And so on. The problem with this approach to the history of game engines is that John Carmack, the lead programmer at id, did not just create a new kind of software, as if that were not enough. At its heart, the game engine is also a particular way of organizing the structure of computer game software; this structure separates execution of core functionality by the game engine from the creative assets that define the play space or "content" of a specific game title. Jason Gregory writes in his book on game engines, "*DOOM* was architected [*sic*] with a relatively well-defined separation between its core software components (such as the three-dimensional graphics rendering system) and the arts assets, game worlds, and rules of play that comprised the player's gaming experience" (2009, 11). This way of organizing game software leads us from technology to game design and from there to the "open game" and its impact on the game industry.

Before we circle back to historiography, we need to delve into the chain of events that produced the game engine and the decision to package assets separately. Why was the term *game engine* coined to describe id's game technology? The answer to this question begins with the state of personal computer (PC) gaming circa 1990 and its invigoration by id. It is fair to say that as the 1990s were beginning, the PC was not where the action was. Video game consoles dominated, and the popular 8-bit home computers of the 1980s were on the last legs of a phenomenal run. The PC circa 1990 was neither the dominant target platform for new games nor at the leading edge for the development of game technology. When id released *DOOM* on one of the University of Wisconsin's File Transfer Protocol servers in late 1993, the decade of the PC game was about to begin; by the end of the 1990s, both games and technologies frequently moved from PC to console, led by id's first-person shooters. The release of *DOOM* was a significant moment for this chronicle of events.

In 1990, John Carmack and John Romero were the key figures in a team at Softdisk tasked with producing content for a bimonthly game disk magazine titled *Gamer's Edge*. They had come to the realization that following the release of MS-DOS 3.3 in 1987 and the maturing of the hardware architecture based on Intel's x86 microprocessor family, advanced PCs running DOS were poised to become an interesting platform for next-generation games; they would need to "come up with new game ideas that ... suit the hardware."[1] Romero and Carmack had cut their teeth as young programmers on the Apple II. They had perfected their coding skills by learning and working alone. As they thought about future projects at Softdisk, they realized that they would need to make software as a team. The plan was to divide and conquer. Carmack would focus on graphics and architecture, Romero on tools and design. Their first project was *Slordax* (1991), a *Xevious*-like vertical scroller. During development of this game, Carmack showed how to produce smooth vertical scrolling on the PC. Romero was modestly impressed, but he "wasn't blown away yet." He challenged Carmack to solve a more difficult problem: *horizontal* scrolling, as seen in the Nintendo Entertainment System.

Carmack's response has been well documented (Romero n.d.; Kushner 2003). With help from Tom Hall, Carmack programmed *Dangerous Dave in Copyright Infringement* (1988), acknowledging with the title that it was a blatant copy of Nintendo's *Super Mario Brothers* (1985). Now Romero *was* blown away. "I was like, 'Totally did it.' And I guess the extra great thing is that he used Mario as the example, which is what we were trying to do, right, to make a Mario game on the PC." Horizontal scrolling was the team's ticket out of Softdisk. They realized, as Romero recalls, "We can totally make some unbelievable games with this stuff. We need to get out of here." Yet the separation from Softdisk was gradual. Their new company, id Software, continued to produce games for their former employer. While this was happening, Romero came into contact with Scott Miller of Apogee Software, a successful shareware publisher. After showing the *Dangerous Dave* demo to Miller, id began to produce games for him, too. These overlapping commitments produced a brutal production schedule, even without taking into account independent projects and technology development. The company met this schedule in part by producing games in series.

The first series of *Commander Keen* games, *Invasion of the Vorticons*, was published between late 1990 and mid-1991. It featured Carmack's smooth horizontal scrolling. He used other projects to make progress with three-dimensional (3D) rendering, which seemed like a promising technology for a second *Keen* trilogy. During this period, as Carmack and Romero began thinking about their next series of games, they began to call the shared codebase for these three games the "Keen engine." The engine at that time was a single piece of software that produced common functionality for multiple games. The idea of licensing such an engine as a stand-alone product to other companies emerged quickly, and id briefly tested the idea by offering a "summer seminar" in 1991 to potential customers for the Keen engine. Carmack and Romero demonstrated the design of a *Pac Man*–like game during the workshop and waited for orders to pour in. They received only one, from Apogee. Romero recalls, "So they get the engine, which means they get all the source code to use it." Apogee made one game with id's engine, *Bio Menace* (1993), then made its own engine after gaining access to id's code. "And then," as Romero put it, "[Apogee] didn't license any more tech from us."

Although the trial balloon of the licensing concept was a failure, the game engine stuck as a way of designating a reusable platform for efficiently developing several games. According to Romero, "I don't remember, at that point, hearing of an engine like, you know, *Ultima*'s engine because I guess a lot of games were written from scratch." In other words, game programs had previously been put together for one game at a time. But why call this piece of game software a "game *engine*"? Both Carmack and Romero were automobile enthusiasts, and, as Romero explains, the engine "is the heart of the car, this is the heart of the game; it's the thing that powers it ... and it kind of feels like it's the engine, and all the art and stuff is the body of the car." We can now make a precise entry in our chronicle: id Software invented the game engine around 1991 and revealed the concept no later than the *DOOM* press release in early 1993.

DOOM was the technological tour de force that heralded the "technical revolution" proclaimed in id's news release, issued nearly a year before the game appeared. *DOOM* showcased novel game technology and design: a superior graphics engine that took advantage of a 256-color Video Graphics Array, peer-to-peer networking for multiplayer gaming, and the mode of competitive play that Romero named "death match." It established the first-person shooter and the PC as its cutting-edge platform, even though *DOOM* had actually been developed on NeXT machines and cross-compiled for DOS execution. Last but not least, *DOOM* introduced Carmack's separation of the game engine from "assets" accessible to players and thereby revealed a new paradigm for game design on the PC platform.

DOOM's game engine is a significant event in the history of game software. Carmack would remark later that it was a "really significant inflection point for things, because all of a sudden the [game] world was vivid enough that normal people could look at a game and understand what computer gamers were excited about" (2003). This game engine was not only a technical achievement but also the springboard for changes in perception and play: networked player communities, modifiable content, fascination with the sights and sounds of games, and concerns about hyperrealistic depictions of violence and gore. Most of us would agree with Carmack about the inflection point in computer gaming around 1993 and its aftereffects. A specific example close to my heart is the history of machinima. Before machinima, there were "*Quake* movies." This earlier expression tied game-based movie making to Carmack's separation of the game engine from assets in *Quake* (id Software, 1996, the successor to *DOOM*). *Quake* demo files were a kind of asset file that had been introduced by *DOOM*. A few players figured out how to change these files and produce player-created movies that the game engine could then play back. Although id had not anticipated movie making, it was an affordance of its game technology.

Lev Manovich has described the impact of *DOOM* as nothing less than creating "a new cultural economy" for software production. This cultural economy exploited the full implications of id's software model of separated engine and assets. Already in 1998, Manovich—leaning on Michel de Certeau—described this new economy as a situation in which "producers define the basic structure of an object and release a few examples, as well as tools to allow consumers to build their own versions, to be shared with other consumers." In other words, Carmack and Romero opened up access to their games in a fashion that might be construed as giving up creative control. And yet id's move was not a concession; the two software developers embraced the implications of this approach as the company focused increasingly on technology as a foundation for game development. They encouraged the player community and worked with third-party developers, who modified id's games or made new ones on top of id's engine. Carmack's support for this open-software model can be explained in part by his background as a teenage hacker. He created a robust model of content creation that would allow players to do what he had wanted to do: change games and share the changes with other players. With this accomplished, Carmack's attention shifted easily to improving the technology rather than working on game design.

It is important to acknowledge that users—players, that is—played a significant role in shifting id's focus to the game engine as a content-creation platform. When id released *Wolfenstein 3D* in 1992, dedicated players' efforts to hack the game and insert characters such as Barney the Dinosaur and Beavis & Butthead made an impression on Carmack and Romero. Michael Adcock's "Barneystein 3D" patch and others like it documented players' eagerness to change content, even though the game did not offer an easy way to do this. Romero recalls that *Wolfenstein 3D* demonstrated that players wanted "to modify our game really bad." He and Carmack concluded about their next game that "we should make this game totally open, you know, for people to make it really easy to modify because that would be cool." Carmack's solution then was a response to a perceived demand. Players could now alter assets such as maps, textures, and demo movies without having to hack the engine. These alterations might result in the expression of personal gameplay preferences, artistic statements, independent game production, or offensive content; in other words, opening up access to game assets also opened up issues that the game industry would encounter in the following two decades. The foundation for distribution and sharing of new content was the stability of the engine. Moreover, access to assets encouraged the development of software tools to make new content, which then generated more new modifications, maps, design ideas, and so on. The corporate history of id boasts to this day that after *DOOM* was released, "the mod[ification] community took off, giving the game seemingly eternal life on the Internet" (id Software n.d.).

Manovich points to the implications of the changes introduced in *DOOM* in terms of support for content modifications and reuse of the engine, but he does not say very much about the motivations. In a history of id Software, David Kushner says that Carmack's separation of engine and assets resulted in a "radical idea not only for games, but really for any type of media. ... It was an ideological gesture that empowered players and, in turn, loosened the grip of game makers" (2003, 71–72). At the same time, as Carmack and Romero had predicted, it was also good business. Eric Raymond takes up this theme in his essay "The Magic Cauldron" (2000), bolstering the case for the business value of open-source software by analyzing id's decision to release the *DOOM* source code. The media artist and museum curator Randall Packer (1998) has also noticed that a cultural shift occurred only a few years after the release of *DOOM* and *Quake.* He observes that games had become the exception among interactive arts and entertainment media because game developers did not view the "letting go of authorial control" as a problem. By "game developers," he means id, of course.

My narrative suggests that the logic of game engines and open design led to id becoming the model for a different kind of game company. The PC, as a relatively open and capable platform during the 1990s, was also conducive to this logic. In line with Carmack's focus, id's game became its technology. Several years after *DOOM* was released, Carmack reflected that technology had created the company's value and that by this point there was not much added by game design over what "a few reasonably clued-in players would produce at this point" (2002). In other words, id's creativity was expressed primarily through the game engine, and the provisions for modifying assets opened up possibilities for the kinds of creative expression that came to be called "player-generated" (or "user-generated") content. Of course, the financial return also came from licensing the engine to external developers. That was the formula. This combination of strategic, business, and design innovations was strengthened and emphasized in id's next blockbuster: *Quake,* released in 1996. It proved to be an especially fertile environment for player creativity. Expected areas of engagement such as licensed games and patches were matched by unexpected activities such as *Quake* movies. During the five years from *Keen* to *Quake,* id Software worked out a game engine concept capable of prodding the rapid pace of innovation in PC gaming during the 1990s.

Note

1. I interviewed John Romero at the Computer History Museum in Mountain View, California, in 2011 and 2012; all quotations from Romero in this chapter come from these interviews.

Works Cited

Carmack, John. 2003. "*DOOM 3*: The Legacy." Transcript of video from *The New* DOOM *Website*, http://www.newdoom.com/interviews.php?i=d3video. Accessed December 2003.

Carmack, John. 2002. "Re: Definitions of Terms." *Slashdot*, discussion post, January 2. http://slashdot.org/comments.pl?sid=25551&cid=2775698. Retrieved January 2004.

Gregory, Jason. 2009. *Game Engine Architecture.* Wellesley, MA: A. K. Peters.

id Software. n.d. "id Software Backgrounder." id Software website, c. 2002. https://web.archive.org/web/20121214092527/http://www.idsoftware.com/business/home/history/. Accessed February 2004.

id Software. 1993. *DOOM* press release. *Lee Killough's Legendary DOOM Archive*, January 1. http://web.archive.org/web/20071220155804/http://www.rome.ro/lee_killough/history/doompr3.txt. Accessed December 2003.

Kushner, David. 2003. *Masters of* DOOM: *How Two Guys Created an Empire and Transformed Pop Culture.* New York: Random House.

LaMothe, André. 1994. *Teach Yourself Game Programming in 21 Days.* Indianapolis: Sams Publishing.

Manovich, Lev. 1998. "Navigable Space." http://manovich.net/index.php/projects/navigable-space. Accessed June 2004.

Mendoza, Jonathan. 1994. *The Official DOOM Survivor's Strategies and Secrets.* Berkeley, CA: Sybex Inc.

Packer, Randall. 1998. "Net Art as Theater of the Senses: A HyperTour of Jodi and Grammatron." Beyond Interface website. http://www.archimuse.com/mw98/beyond_interface/bi_fr.html. Accessed April 2004.

Raymond, Eric Steven. 2000. *The Magic Cauldron.* August 2. http://www.catb.org/~esr/writings/cathedral-bazaar/magic-cauldron/. Accessed June 2013.

26 GAME GLITCH

Peter Krapp

According to the *Oxford Dictionary of Slang*, *glitch* is a word "applied to a sudden brief irregularity or malfunction of equipment, originally especially in spacecraft"; however, despite dating it to 1962, the dictionary marks its origin as unknown (Ayto 1998, 418).[1] Probably derived from the German verb *glitschen*, meaning "to slip or slide" (or the related adjective *glitschig*, meaning "slippery"), *glitch* in English has over the past four decades become associated particularly with unpredictable consequences of software errors (Nunes 2010). By contradistinction, the fears of global computing trouble at the turn of the millennium (referred to as "Y2K") helped further popularize the term *software bug*—if the trouble is anticipated, it is thus not classified as a glitch.[2] *Glitch* remains "a precariously vague term, which however captures some of the slipperiness of digital media" (Vanhanen 2001). The word has become particularly common in computer game circles, where it is sometimes synonymous with sudden disruptions—graphic display or sound mistakes, communications errors, failed collision detection, games freezing or crashing, or other malfunctions, often prompted by a particular input combination. Glitches in games can halt gameplay, but in some cases they can lend an unforeseen advantage to players who know how to exploit the situation. Intriguingly, some players seek to re-create the conditions of possibility of a glitch in order to repeat it, sometimes in the hope that the re-created glitch may become an exploit or at least something to brag about.

How does a computing setup that fundamentally does not encourage anything random become the platform for a lively pursuit of the random? What makes a glitch interesting, first of all, is that it simply stands out in a landscape of tightly regulated interactions between software and hardware. If necessity and chance are complementary, then one might expect not only games of chance but also games of necessity. Yet we can see that a fully rationalized game—such as a state-operated lottery that allows for play, minimizes risks, and benefits the commons with profits—is simply a fiscal and moral calculation that recuperates contingency into a legally sanctioned form of gambling that fills the state coffers. Jorge Luis Borges is surely not alone in protesting that the lottery hardly lives up to its promise of an "intensification of chance" because in reality it is a tightly controlled process of finite drawings and final decisions, almost "immune to hazard" (1962, 69–70). It is worth noting that Gilles Deleuze picks up on Borges's suggestion, albeit without making a connection between that digital potential and an ideal game "in which skill and chance are no longer distinguishable" (1990, 60).

Moreover, the common understanding of the glitch is that it has little or nothing to do with operator errors. Computing culture is predicated on communication and control to an extent that can obscure the domain of the error. Between a hermetically rule-bound realm of programmed necessity and efficient management of the totality of the possible, we find a realm of contingency: distortions in the strictest signal-to-noise ratio, glitches and accidents, or moments where the social hierarchy of computer knowledge condescendingly ascribes to "user error" what does not compute. As our digital culture oscillates between the sovereign omnipotence of computing systems and the despairing agency panic of the user, glitches become aestheticized, recuperating mistakes and accidents under the conditions of signal processing: "Glitches can be claimed to be a manifestation of genuine software aesthetics" (Goriunova and Shulgin 2008, 111).

Some game glitches render the situation inoperable not only if they freeze the screen or disconnect from further input but also when they drop the gameplay into bad infinity, as in the example of the *Pokemon* bad egg (in generation III, IV, and V—*Pokemon* 2002 et seq.)—an egg that keeps wanting to hatch with every step the player takes or that hatches further bad eggs. Interestingly, a bad egg can be removed out of one's game with a cheatcode. Not so with the *Pokemon* glitch "MissingNo." wherein a certain sequence of actions in the Nintendo game, as documented in 1999, results in disruption of the game's graphics and potentially in erased games. Players could flee from, fight, or capture MissingNo. yet, despite Nintendo's warnings, several publications provided instructions on how to *arrange* such an encounter (Bainbridge 2007). Any such glitch will break the illusion of an immersive game world and remind the player of the game's software and hardware conditions of possibility.

Due to this extradiegetic dimension, the glitch plays a pivotal role also in alternative play or counterplay—that is, in exploring ways of playing *with* the game rather than just playing the game. As Tom Apperley has it, counterplay explores the limits of digital games, including the cataloging and exploiting of bugs and glitches, but without having to code or hack—it does not interfere with software or hardware at any level except that of the user interface. One example Apperley offers is that "in the original release of *Sid Meier's Civilization* [MicroProse, 1991] a game-winning tactic was quickly developed that involved building large numbers of cities close to each other so that their zones of influence overlapped and then swamping the opposing civilizations with large numbers of low powered units. The use of this strategy meant that the game could easily be over, with the globe dominated by one culture by 400BC" (2010, 142).

This glitch from 1991 is of course no longer a feature of the later installments of the *Civilization* series. Yet there is hardly a game that has not provided a glitch worthy of recording, replicating, and exploiting to its regular players. In the first installment of *Halo* (Bungie, 2001), players were able to make the Covenant Elites and Jackals bleed when they were wounded or killed—even after killing these enemy avatars, players were able to keep hitting them to produce more blue-purple blood. A glitch in the game appeared when the blood did not fade away: "If you kept hitting the body of one dead alien over and over, you could paint the ground in layers and layers of alien blood until the framerate slowed to a crawl as the Xbox tried to deal with all these extra layers of blood" (Keogh 2012, 110). Of course, to discover such a software glitch that actually affects the hardware, the player must first indulge in a cruel excess of alien bashing to accidentally discover the material impact such actions can have on the machine. Like the *Civilization* glitch, this *Halo* glitch raises interesting questions about these games' premises and some basic assumptions held by their players.

Game studies often discusses games in terms of a risky, lawless world versus a safe, riskless world. The fact that there is a veritable gaming culture around the glitch illustrates this tension in game design—what is attractive about the glitch is that it is a risk players incur but also something they seek to harness. Players have found a number of ways to recuperate such moments: "glitching" websites aggregate discoveries in various games and advise how to re-create situations that were discovered accidentally, thus turning exceptions into exploits. Deliberately inducing a glitch can be risky in live play, especially in competitive online situations, but it can also lend an edge to players who successfully arrange it to work in their favor. "When you learn a glitch that can be used in actual gameplay," one glitch website advises, "you may be tempted not to attempt that particular glitch during gameplay (someone might kill you while you're trying to accomplish that glitch.) But if you learn to incorporate that glitch during gameplay I can't tell you how effective that will be."[3]

For example, in *Metroid* (Nintendo, 1986), players aware of a particular glitch were able to access unused map data Nintendo never intended them to see by getting stuck in a blue door, which allowed them to walk through walls and access what some thought were secret worlds inside the game. Similarly, in the Wii game *Mario Kart* (2008), the PlayStation 3 version of *Tom Clancy's Rainbow Six: Vegas* (2007), and the anniversary edition of *Lara Croft: Tomb Raider* (Crystal Dynamics, 2007) a player could turn rapidly to catch a glimpse of what is on the other side of a wall before the in-game camera adjusted its distance behind the player's avatar—a glitch that allowed the player to see through walls.

A more complicated manner of exploiting a glitch is exemplified by the ghost mode in *Call of Duty 3* (Activision, 2006), where a player can have a member of their team go over from the

Allies to the Axis side, get on a motorcycle, meet them by a tank, join them on the motorcycle, and then have a third team member in the tank shoot the motorcycle. If the player has "team damage" turned off, the teammate playing an Axis soldier is killed, but the other player not—they can walk around invisible to their teammates (though not to the enemy). Multiplayer online games must carefully calibrate and balance labor with appeal, both in time sinks and money sinks, to regulate their virtual economy. A classic online game, Lucasfilm's *Habitat* (1986), provided a system of avatar currency (tokens) and stores where items could be purchased. When a glitch in the program produced a virtual money machine, a small group of *Habitat* avatars became virtual millionaires overnight. Over protests from the "poor," the *Habitat* programmers decided to let the finders keep the virtual loot (Dibbell 1998, 70–72). Another early multiplayer example helps to highlight how the glitch can incite new modes of gameplay not foreseen by the game's developers and administrators. During the beta test of *Ultima Online* (Origin Systems, 1997), the game's creator Richard Garriott held an assembly where his avatar, Lord British, addressed a group of players, but one of the players attacked and killed Lord British despite the fact that he was the supreme ruler of the fictional game world who should have been indestructible. A glitch had rendered him vulnerable—a server reboot just before the event (Nitsche 2008, 27)

Strictly speaking, there cannot be a glitch-free computer game or, more generally, a glitch-free computing environment. One of the more interesting insights offered by pioneering computer engineer Severo Ornstein is that synchronization becomes more problematic the faster processing cycles become. Ornstein (who worked at Lincoln Labs, Bolt Beranek & Newman, and Xerox PARC during some of their most influential phases) resolved to debunk any claims about the possibility of a glitch-free circuit and started to propagate the notion that the efforts to remove the glitch should in all honesty just be called "move-the-glitch": "In every case we were able to determine to where, in the design, the ingenious designer had moved the glitch. In each case, the glitch persisted, and the presenter was brought into alignment with this new correct view of the universe. Our fervor approached that of religious zealots. It was our mission to stamp out anti-glitch apostasy wherever we could find it" (2002, 136)

The problem of the glitch, the unavoidable chance factor that introduces mistakes and noise, is thus only too well known among gamers. The interesting question for the history of computing is why—despite steadily rising storage, better and quicker processors, and all kinds of technical advances in software and hardware—a veritable cult of the glitch developed nonetheless, including the cult of 8-bit game sounds. When and how does visual, acoustic, and audiovisual raw material that is way below the technically possible become an aesthetic choice?

It did not take long for the glitch to become aestheticized (Hernandez 2012). Chicago saw several annual GLI.TC/H events,[4] and Oslo mounted the Motherboard Glitch Symposium in 2002. Glitch art, often though not exclusively inspired by computer and game culture, has become acceptable in galleries and museums and has found entry into videos and installations. Early examples include the manipulation of a Bally Astrocade game console for the art video *Digital TV Dinner* (1979) by Raul Zaritsky, Jamie Fenton, and Dick Ainsworth; more recent work by Cory Arcangel and Eddo Stern could also be mentioned in this context, but that would require another whole entry for this book. The machinima clip *Untitled (for David Gatten)* by Phil Solomon and Mark LaPore (2005) exploits the frozen screens, data corruption, and other glitches Solomon and LaPore found in *Lara Croft: Tomb Raider* to create a sequence of "dripping, elongated textures and blotches of green hurtling towards the screen" (Sicinski 2007). The play of aesthetics shows itself in these examples as a reduction of contingency through form, harnessing adventure through aleatory or stochastic management of the event and the surprise. Another example is jodi's game modification *SOD*, a mod of the classic game *Castle Wolfenstein* (Muse Software, 1981) that retains the soundtrack but scrambles the visuals so as to render the screen a black-and-white abstract pattern of glitchy, nonrepresentational patterns, appearing to make the user's computer run amok.[5] A related example is the art game *ROM CHECK FAIL* (2008), an "independent" game predicated on the glitch as an aesthetic tool, shuffling around the elements of a number of classic two-dimensional computer games.[6] As Jarrad Woods describes *ROM CHECK FAIL*, "Every few seconds, there is a glitch internal computer noise, and everything in the current level stays in the same position but changes into something else" (qtd. in Rose 2011, 48). Random switching to alternative rules apparently complicates the game mechanic and allows *ROM CHECK FAIL* to recycle an oddball assortment of arcade and console classics, to be navigated with the arrow and space keys. Of course, this kind of 1980s retro-game generator cashes in on nostalgia and aestheticizes the glitch. Going one step further in that direction is the commercial release *Mega Man 9* (Capcom/Inti Creates, 2008). Though developed for recent consoles including the Nintendo Wii, *Mega Man 9* uses graphics and sounds harking back to the 8-bit era of the original Nintendo Entertainment System. A "legacy mode" emulates the lower frame rates of the game's ancestors, only partially renders sprites, and causes them to flicker when they crowd the screen; this degradation feature is achieved not via an emulator but using a dedicated engine that simulates an outmoded technology. By the same token, it is evident that any glitch becomes more palpable with the advent of higher expectations from audiovisual resolution.

Thus, it follows that "glitch" should also have become a music genre. Clicks and glitches come to undermine the tightly controlled, layered loops of electronic music (Krapp 2011). As Chris Salter observes, "Another movement within the audiovisual coding scene amplified its processual characteristics through the revealing and subsequent use of errors and glitches"

(2010, 178). But in a recursive system where differentiations between data and programs are obsolete, it comes as no surprise that there should also be a Glitch plug-in for audio software that chops up audio files and applies a variety of effects, depending on how much you tweak the controls. With unremarked irony, the website for that Virtual Studio Technology software plug-in warns, "This version of *Glitch* is still just a prototype and contains a few bugs—obviously I am working towards fixing everything as soon as possible."[7] Thus, it becomes clear that even with a software routine that can be directed to apply randomized effects, human–computer interaction remains predicated on an interpretation of control that tends to bar artists from finally taking the place of contingency: "It is from the 'failure' of digital technology that this new work has emerged: glitches, bugs, application errors, system crashes, clipping, aliasing, distortion, quantization noise, and even the noise floor of computer sound cards are the raw materials composers seek to incorporate in their music" (Cascone 2004, 393). Even and especially in the electronic music inspired by game technology, digital tropes of perfect sound copies are abandoned in favor of errors, glitches become aestheticized, and mistakes and accidents are recuperated for art under the conditions of signal processing. The glitch has been almost completely domesticated (Menkman 2011). There is even a game called *Glitch* (2011), a web-based massively multiplayer online puzzle developed by Tiny Speck (led by Flickr cofounder Stewart Butterfield) that sends players into the past to inhabit the minds of 11 giants (Terdiman 2010).

Yet if we allow ourselves to conceive of human–computer interaction as organized around the glitch, it is not to smuggle a covert humanism in through the back door of technological determinism; rather, it is to emphasize what Alexander Galloway calls "the cultural or technical importance of any code that runs counter to the perceived mandates of machinic execution, such as the computer glitch or the software exploit, simply to highlight the fundamentally functional nature of all software (glitch and exploit included)" (2006, 326). One might conclude, however provisionally, that gaming glitches are part of the art form in the same way that brushstrokes are part of painting. Game developers may be tempted to regret this comparison as little consolation to a user who just had a program crash or indeed to a programmer who is trying to debug the system. However, the glitch may be that hairline fissure that can widen on to new vistas in gameplay and game studies—and thus on to better mistakes.

Notes

1. *A Dictionary of Astronomy* defines *glitch* as "an abrupt disturbance in the regular train of pulses from a pulsar, appearing as a sudden change of the pulsar's period and spin-down rate. Glitches tend to occur in younger pulsars whose rate of spin is slowing rapidly, notably the

Crab Pulsar and Vela Pulsar. They are believed to be caused by the sudden release of stress energy either in the crust (a starquake) or between the crust and the superfluid interior" (Ridpath 1997, 191). It is plausible that John Glenn, who used the word *glitch* in the sense of random system error in 1962 (in the book *Into Orbit*, an account of Project Mercury, the United States' first human spaceflight program), would have been aware of the astronomical meaning of the term.

2. Y2K fears anticipated what might happen when year fields switched over from 99 to 00 because memory conservation in COBOL (common business-oriented language) had limited years to two-digit representation. However, the majority of Y2K issues in finance and insurance software were fixed during 1999, and the new millennium began without any major software errors attributable to the Y2K bug.

3. Glitching advice from the now defunct site http://glitchmania.com/ is partially archived by the Internet Archive at https://web.archive.org/web/20110129003547/http://glitchmania.com/Home.html; another repository, also defunct, of game glitch exploits was http://glitchblog.com, partially extant at the Internet Archive at https://web.archive.org/web/20111021025055/http://glitchblog.com/.

4. The archives of three glitch festivals (2010, 2011, and 2012) are found at http://gli.tc/h/.

5. For *SOD*, go to http://sod.jodi.org/.

6. For more on *ROM CHECK FAIL*, seehttp://www.farbs.org/games.html. Also see Joan Leandre's art projects *R/C* and *NostalG* at http://www.retroyou.org and http://runme.org/project/+SOFTSFRAGILE/.

7. The Glitch 2 plugin for VST is available from http://illformed.org/plugins/. The original Illformed glitch plugin is ranked #5 among 100 best free VST plugins; see https://edmranks.com/top-100-free-vst-plugins-for-electronic-music-producers/.

Works Cited

Apperley, Tom. 2010. *Gaming Rhythms: Play and Counterplay from the Situated to the Global.* Amsterdam: Institute of Network Cultures.

Ayto, John. 1998. *Oxford Dictionary of Slang*. New York: Oxford University Press.

Bainbridge, William Sims. 2007. "Creative Uses of Software Errors: Glitches and Cheats." *Social Science Computer Review* 25 (July): 61–77.

Borges, Jorge Luis. 1962. *Ficciones*. New York: Grove.

Cascone, Kim. 2004. "The Aesthetics of Failure: 'Post-digital' Tendencies in Contemporary Computer Music." In *Audio Culture*, ed. Christoph Cox and Daniel Warner, 392–398. London: Continuum.

Deleuze, Gilles. 1990. "Tenth Series of the Ideal Game." In *The Logic of Sense*, 58–65. London: Athlone.

Dibbell, Julian. 1998. *My Tiny Life: Crime and Passion in a Virtual World*. New York: Henry Holt.

Galloway, Alexander. 2006. "Language Wants to Be Overlooked: On Software and Ideology." *Journal of Visual Culture* 5 (3): 315–332.

Glenn, John. 1962. *Into Orbit*. London: Cassell.

Goriunova, Olga, and Alexei Shulgin. 2008. "Glitch." In *Software Studies: A Lexicon*, ed. Matthew Fuller, 110–119. Cambridge, MA: MIT Press.

Hernandez, Patricia. 2012. "It's Not a Glitch, It's a Feature. It's Art. It's Beautiful." *Kotaku,* August 10. http://kotaku.com/5955722/its-not-a-glitch-its-a-feature-its-art-its-beautiful.

Keogh, Brendan. 2012. *Killing is Harmless: A Critical Reading of "Spec Ops: The Line."* Marden, Australia: Stolen Projects.

Krapp, Peter. 2011. *Noise Channels: Glitch and Error in Digital Culture*. Minneapolis: University of Minnesota Press.

Menkman, Rosa. 2011. *The Glitch Moment(um)*. Amsterdam: Institute of Network Cultures.

Nitsche, Michael. 2008. *Video Game Spaces: Image Play and Structure in 3D Worlds*. Cambridge, MA: MIT Press.

Nunes, Mark, ed. 2010. *Error: Glitch, Noise, and Jam in New Media Cultures*. New York: Continuum.

Ornstein, Severo. 2002. *Computing in the Middle Ages. A View from the Trenches 1955–1983*. New York: Authorhouse.

Ridpath, Ian. 1997. *A Dictionary of Astronomy*. New York: Oxford University Press.

Rose, Mike. 2011. *250 Indie Games You Must Play*. New York: CRC Press.

Salter, Chris. 2010. *Entangled: Technology and the Transformation of Performance*. Cambridge, MA: MIT Press.

Sicinski, Michael. 2007. "Phil Solomon Visits San Andreas and Escapes, Not Unscathed: Notes on Two Recent Works." *CinemaScope Magazine* 30. http://www.academichack.net/solomon.htm.

Terdiman, Daniel. 2010. "In Depth with Tiny Speck's Glitch." *CNET News*, February 9. http://news.cnet.com/8301-13772_3-10449721-52.html.

Vanhanen, Janne. 2001. "Loving the Ghost in the Machine: Aesthetics of Interruption." *CTheory* a099. http://www.ctheory.net/articles.aspx?id=312.

27 GAMES AS A MEDIUM

Mary Flanagan

It might be argued that everyone knows a game when they see one. But we should ask ourselves, How does that game come to us? What do players manipulate in order to play? What is the *medium* of a game? Since the dawn of time, humans have played games. Material artifacts, such as the Neolithic game boards found in what is now Jordan, might be carved of stone. Ancient texts such as the Egyptian *Book of the Dead* from the thirteenth century BCE depict wooden Senet boards; indeed, tombs were filled with Senet games (King Tut had at least three) so that deceased pharaohs and queens might play for their fates. For thousands of years, games have helped players pass the time, settle arguments, negotiate for their souls, express themselves, and be entertained, and thus they have formed an integral part of human activity since antiquity. Flash forward to the present: although we still play board games, games on computers have proliferated. The digital game industry has exploded in popularity, and contemporary artworks increasingly reference games or play. Games today are typically referred to by genre, but they are also referred to by *medium*: "console game," "card game," "board game," and so on. Games across various media are indeed different, but "gameness" endures. What is specific to the game if it is truly a "creative medium"? Is there a game medium?

I trace two paths in this essay to help us consider the notion of medium in a game context. One is the link to the arts, which long has used the term *medium* to discuss both the material and social properties of creative works. The other line follows the developments of computing technologies and their influences on computer games. These trajectories intersect to shape notions of games and their media.

The Notion of a Medium in Art: Material and Social Aspects

During the 1980s, critic Frederic Jameson was exploring a relatively new medium of art—video—when he declared that an art medium consists of three distinct "signals": (1) that of "an artistic mode" or "form of aesthetic creation"; (2) that of a "specific technology, generally organized around a central apparatus or machine"; and (3) that of a "social institution." All three signals lead to a definition of an artistic "medium" ([1986] 1991, 67). Let's unpack the idea of a medium along Jameson's lines, for these useful dimensions cross any art format or genre. First, the notion of a medium creates a type of art category. A painting is one class of objects, and a print is quite another. Both share properties of flatness, color, and so on, but they do not share other properties (e.g., paintings are generally "one-offs," whereas prints are multiples). Like any category, there are always exceptions and crossovers (e.g., Andy Warhol).

Second, the notion of a medium in art—the materials used in the creation of an artwork—is an ancient concern. Media have always *influenced* and *delineated* the ways in which the artist expresses himself or herself. Clay, paint, flax linen, and marble are distinctly different materials that require different tools—a knife, a brush, a loom—with which an artist might craft vastly different aesthetic and expressive works with different messages.

Third, particular uses of a medium in a time and a place take on certain conventions that are socially constructed, which artists seek to subvert or maintain. For example, the use of marble in Greek sculpture became nearly synonymous with the rendering of highly detailed human and animal figures. Marble is extremely hard, is expensive, and is a difficult material with which to work. Therefore, mastering the delicate features of the living form was seen as a pinnacle of the artistic use of that material. As a medium, marble can be considered across the "signals" that constitute it.

Marble's challenging material properties make it instantly "classic," and the techniques developed to work with this stone are unique and might not be useful in pursuing, say, calligraphy. These ancillary material properties are technologies and inform the artwork directly. Marble also has societal expectations and connotations. Because the material is expensive to acquire and use, it is typically employed for monumental works with grand themes. Sculptor Louise Bourgeois famously noted that marble was her favorite material. "It permits one to say certain things that cannot obviously be said in other materials. ... Persistence, repetition, the things that drive you toward tenacity" ([1988] 1996, 39). Thus, the material and the social combine to create a set of values associated with the material (permanence, wealth, and tenacity in the case of marble).

An artistic medium, as a general category of creative object, has particular tangible properties. A sculpture can be felt. A film, however, would typically be watched. Paint has adhesive properties, can be saturated with a pigment, and can create a permanent mark on a surface. Yet the idea of "medium" must not be oversimplified to mean *materials* alone: painting as a *medium* takes into account these properties but must in addition encompass the social use of that medium of expression—for instance, cave paintings, Egyptian painted statues,

Renaissance painting, abstract and contemporary painting practices, as well as the category of other work called "paintings." Each art form's medium shapes its conventions, such as how viewers or participants interact with that art form and how the interaction is governed. Agnes Martin noted that "people think that painting is about color; [i]t's mostly composition; [i]t's composition that's the whole thing" ([1973] 1996, 128). Martin shows us here that a medium importantly brings with it a nexus of societal and conceptual debates, such as how paintings are exchanged and valued, who is expected to own a painting, who paints, the future of painting as an art form, the death of painting, and so on. Conceptual debates include the types of representation styles that are in vogue, what is not a legitimate use of the medium, what constitutes mastery, and so on. An artistic medium fosters a constellation of *ideas*. Thus, a medium not only implies a material substance but also carries ideas about social customs, aesthetics, societal debates, valuation, taboos, and subversive interpretations.

From wood to stone to code, games are part of an age-old category of human activity. They are also made of materials, some of which are materials in an ephemeral sense—for example, using the body in sports and performative games such as charades. When we think about games in the classic terms used to discuss artistic media, we must include societal and conceptual debates as part of that medium's constitution. Such debates on games happen in "gaming" culture, but also in pop culture, the art world, gambling, psychology, sports, and so on. Among other questions, we can ask, How do contemporary creations of games made for artistic expression fit into historic games by artists? Do games have to be playable and "winnable" to be good art?

A Technological Medium of Play

Game scholar Jesper Juul noted in 2003, "There is no single game medium, but rather a number of game media, each with its own strengths. The computer is simply the latest game medium to emerge" (30). Indeed, when people think of games in contemporary culture they might immediately think of computer games: aside from entertainment, such games appear in education, health care, civic engagement, and artistic fields. When the public discusses computer games, they tend to conjure images of shooter games. This connection is made based on many factors, such as shooter games' prevalence in advertising and game review sites or perhaps the mainstream media's attraction to shooter games as an easy target when news about youth violence surfaces. There is in the history of computing a strong link to support contemporary interpretations of computer games as at their core competitive and filled with conflict, violence, and heroic victory. Early computers were originally developed to better accomplish advances in war technology that had obsessed thinkers for as far back as the times of Leonardo

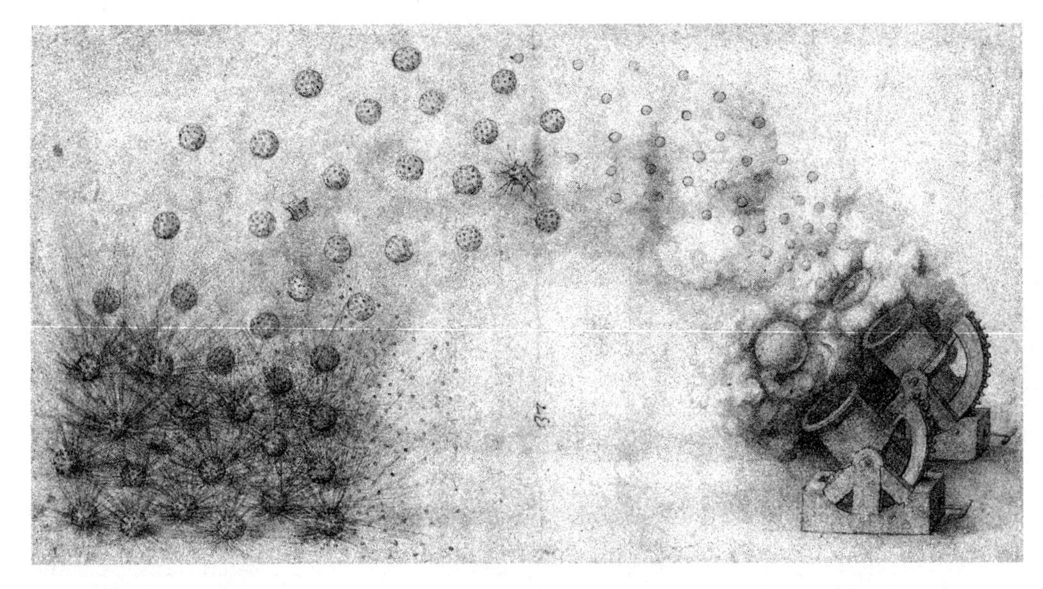

Figure 27.1

Leonardo da Vinci's drawing *Huge Mortars with Explosive Projectiles*, c. 1500–1519. Atlantic Codex (Codex Atlanticus), folio 33 recto. Study of two mortars able to throw explosive bombs. Copyright Veneranda Biblioteca Ambrosiana.

da Vinci, who designed machines from crossbows to canons. His notebooks document myriad proposed empirical tests on trajectories hundreds of years before Sir Isaac Newton developed classical mechanics or the field of ballistics shaped what was to become the field of physics (figure 27.1).

The computer's affinity for mathematics was actually part of its origin story: all early analog precursors to computers were for math, "computing" is a math activity, and computer science as a discipline was born in university math departments in the United States. The math behind many twentieth-century experiments in computation was linked to war, namely World War II from 1939 to 1941. This history of computing has certainly shaped the platform for games. As "mysterious machines of the future," computers became a target of various public-relations efforts during the Cold War era to make the machines less frightening. Games were developed on computers as "demos" that would seem fun and approachable to the public. These demos that were developed around the globe, however, yielded very different experiments regarding play. Computer games developed in the United Kingdom, for instance, focused on puzzles and the computer as an artificial intelligence, such as *The Nimrod* (early 1950s), a mathematical puzzle, an early chess game, and OXO, a "Naughts and Crosses" game (Tic Tac Toe for Americans). Additional playful experiments also had influence: British computer scientist and linguist Christopher Strachey's *Love Letter Generator* (1952) was part poetry and part language play and used the computer to see language in a new way. Thus, British games highlighted brain teasers, code breaking, poetry, problem solving, and artificial intelligence.

Meanwhile, across the pond in the United States, early game inventions were emerging around technological fascinations of a different sort: William Higinbotham's analog computer game *Tennis for Two* (1958) bears the mark of a physicist: although games did not have to have any connection to real-world laws, *Tennis for Two* is a simulation machine for trajectories, velocity, drag, and collision. *Spacewar!* (Russell/MIT, 1962) later modeled human-centric situations of conflict by pitting player against player in a war of annihilation, demonstrating the technological capacity of the time to calculate velocity and collision (other authors in this volume no doubt reference these games in more detail than noted here). These American games highlighted notions of accuracy, strategy, the laws of physics, binary combat models and outcomes, zero-sum conflicts, competition, and efficiency. And it was these values that entered into the conceptual framework of the invention we call the computer and that continue to shape the game industry. By the time of the explosion of commercial games in the 1970s, themes such as shooting projectiles, destroying the enemy, and playing virtual sports had already become embedded in computer games.

The key point to understand, however, is that only *some* of the early computer games and demos were developed based on physics, fighting, and projectiles; other early demos offered other modes of equally compelling play. The US games *Spacewar!* and *Tennis for Two* showcased physics concepts such as gravity and drag, whereas UK demos such as *Naughts and Crosses* allowed for a player versus computer mode and notions of the computer's capacity for artificial intelligence, with Strachey's poetry showcasing imaginative possibilities. The four demo games of this era mentioned here reveal the influence of very different assumptions about the next steps in technology.

These early demo uses of computers are not mere curiosities. History shows how these games' particular values have entered the technical underpinnings of computer-based systems. This spectrum of features is certainly embedded in contemporary computer games through their very medium, but it is easy to see that the US model for computing that focused on conflict, wartime calculations, and physics has been dominant in the past 20 years of gaming and that such games were in fact precursors of what was to become the foundation for computer game engines used in many popular games over the years. War-related influences on the computational medium were significant.

It is important to note that demo games were not built with a well-thought-out scheme for what might be important to the game as a medium over the next 100 years. They were built quickly, and their makers certainly did not expect to shape the game landscape for years to come. Nevertheless, the cultural context and values among those who produced the early demo computer games created a sort of "habit" for thinking about technological interpretations of games produced from then on. This is why although *Spacewar!* was not the first computer game, it is consistently gestured to as the first game—because it contains many of the conventions from US computer science labs that "stuck."

Thus, early games had an infrastructural impact on technological production and set out the ways in which the technology itself was imagined. The demos formed a medium in the classic arts sense, meeting Jameson's criteria for materials, technologies, and cultural context. Software architecture, existing practices, and demos were the way that certain presumptions entered the hardware, code, and technoculture that would soon mass-produce the computer game as a medium. Just think—if game makers harkened back to *Naughts and Crosses*, computational poetry, and NIM (an early computational combinatorial game) as their technological precursors, we would have inherited a very different commercial games industry.

A Transmedia Phenomena

Within the past century, games have been a key reference in contemporary art practice. From Alberto Giacometti's game sculptures to Alison Knowles's nonsensical bean games to Marcel Duchamp's play of chess as a form of expression to critic Dave Hickey's fascination with basketball, the art world has claimed certain types of games as art forms.[1] Surrealist artists used games as generative creative procedures, and games became part of the vocabulary of Fluxus artists as they made provocative commentary on contemporary life, using properties of the game medium to reference and reinforce the absurd. For example, when artist Oyvind Fahlström made "variable paintings"—paintings and interactive board games—they were images as well as games. Fahlström referenced the medium of painting (something hung on a wall, something typically fixed in time) while toying with the essence of painting by introducing elements from games (a game board, interactive parts, suggested turn taking, and role playing). In this way, he was able to take his concept (critiquing atrocities committed by the US Central Intelligence Agency) and look at that concept through several aesthetic lenses (the image and the game). Fahlström said games could be "seen either as realistic models (not descriptions) of a life-span, of the Cold War balance, of the double-code mechanism to push to bomb button—or as freely invented rule-structures. Thus it becomes important to stress relations. ... The necessity of repetition to show that a rule functions—thus the value of space-temporal form, and of variable form. The thrill of tension and resolution, of having both conflict and non-conflict" ([1966] 1996, 194–195).

The inclusion or implication of rules, a delineated play space, and interactive parts all suggest properties of a game. They are the basic elements of a game the way that plot, characters, and sequence are the basic elements of a narrative (no matter what their final material output might be, a novel or a film) or that tones, tempo, and pitch are the basic elements of music (no matter if played live or recorded as digital media). The particular medium within which a game manifests is not the game properties per se, though they can strongly affect the expression of the game. Like music or narrative, elements of games help explain how as a creative form games transcend notions of a "medium" to be an art in their own right. The game incorporates a combination of aesthetic, practical, social, and experiential elements. Societal interpretations of games and the media upon which they manifest continue to come and go. The contradictory nature of a game in a particular medium was demonstrated in the 1980s, when very little evidence existed of public worry about the medium of the video game and its link to violence. Rather, media and discourse of the time attributed violence to players of tabletop role-playing games such as *Dungeons & Dragons* (Gygax and Arneson/Tactical Studies Rules, 1974), not to players of *Pac-Man* (Namco, 1980) or interactive text adventures. Thus, social norms and fears move across media where games are concerned, demonstrating a public fear of fantasy and gameplay itself rather than of the computational medium that the game structure might currently inhabit.[2] This small example among many others demonstrates that societal debates about a medium vary and are formed from a wide spectrum of influences, from forces such as marketing and public relations to stories, from media reports to individual experiences.

This essay situates the game among other creative media, recognizing that the game does not fit as a medium per se but inhabits a conceptual space larger and broader than the definition of a medium might allow. A game *is* its own expression of the creative act, akin to large creative phenomena on grander scales such as "narrative" or "image making" and is expressed through a medium that adds far more than mere material concerns. A game manifests simultaneously a material phenomenon as well as a series of processes, social practices, histories, expectations, and experiences.

Here I have explored the fundamental properties of the game as a high-level cultural phenomenon that can manifest in various traditional art media as well as computational media through key characteristics that constitute a medium. The synthesis between the material and sociocultural delivery of a game and its technological influences produces a game medium. In the end, I argue that a game is not a medium at all but a higher-level creative concept, like "music," "narrative," "image making," and so on. Yet due to their dynamic nature, games are quite heavily influenced by their material representations and properties. Like any art media, both the game and the medium carry with them their own social debates and practices. A game is a larger art form phenomenon expressed through transmedia.

Notes

1. On Knowles's work, see Higgins 2002; on Duchamp's chess playing, see Naumann, Bailey, and Shahade 2009; and on Hickey and the beauty in basketball, see Hickey 1997.

2. As video games began to be marketed for their more realistic violence with the rise of the games *DOOM* (id Software, 1993) and *Quake* (id Software, 1996), the violence debate shifted to ideas of "realistic violence." But, for some, games as vehicles for agency and fantasy were and continue to be a real threat.

Works Cited

Bourgeois, Louise. [1988] 1996. "Interview with Donald Kuspit." Reprinted in *Theories and Documents of Contemporary Art: A Sourcebook of Artists' Writing*, ed. Kristine Stiles and Peter Selz, 38–41. Berkeley: University of California Press.

Fahlström, Oyvind. [1966] 1996. "Take Care of the World." In *Manifestos*, Great Bear Pamphlets, 9–13. New York: Something Else Press. Reprinted in *Theories and Documents of Contemporary Art: A Sourcebook of Artists' Writing*, ed. Kristine Stiles and Peter Selz, 304–305. Berkeley: University of California Press.

Hickey, Dave. 1997. "The Heresy of Zone Defense." In *Air Guitar: Essays on Art & Democracy*, 155–163. Los Angeles: Art Issues Press.

Higgins, Hannah. 2002. *Fluxus Experience*. Berkeley: University of California Press.

Jameson, Frederic. [1986] 1991. "Reading without Interpretation: Postmodernism and the Video-text." Reprinted as "Surrealism without the Unconscious" in *Postmodernism: Or the Cultural Logic of Late Capitalism,* 67–96. Durham, NC: Duke University Press.

Juul, Jesper. 2003. "The Game, the Player, the World: Looking for a Heart of Gameness." In *Level Up: Digital Games Research Conference Proceedings*, ed. Marinka Copier and Joost Raessens, 30–45. Utrecht: Utrecht University.

Martin, Agnes, with Ann Wilson. [1973] 1996. "The Untroubled Mind." *Flash Art* 41 (June): 6–8. Reprinted in *Theories and Documents of Contemporary Art: A Sourcebook of Artists' Writing*, ed. Kristine Stiles and Peter Selz, 128–138. Berkeley: University of California Press.

Naumann, Francis M., Bradley Bailey, and Jennifer Shahade. 2009. *Marcel Duchamp: The Art of Chess*. New Haven, CT: Readymade Press.

28 GENRE

Mark J. P. Wolf

If, as Rune Klevjer has noted, "there is a curious lack of genre studies" in video game studies (2005), it is certainly not surprising, considering what a mare's nest video game genre classifications have now become. Genres are used to refer to multiple domains, for example, to describe the technology used for gaming (what we might more properly refer to as "form factors," such as arcade games, console games, computer games, online games, and mobile games—which is almost synonymous, but not quite, with the older term handheld games); the number of players who can play (such as single-player games, multiplayer games, and massively multiplayer online games); hardware used (vector games, raster games); software and graphical capabilities (two-dimensional [2D] games, 2.5D games, 3D games); the functions or purposes for which games are used (such as advergames [games that promote particular brands or products], educational games, training games, and the overarching "serious games," which can include all of the others); the level of commitment required (such as casual games and hardcore games); perspective (first-person, second-person, and third-person games); and, finally, the content of the games themselves (such as adventure games, shooting games, strategy games, fighting games, and so on).

Individual games can obviously appear in multiple form factors (*Pac-Man* [Namco, 1980], for example, has appeared as an arcade game, a console game, a handheld game, a computer game, an online game, and a mobile game), and a particular game might be classified in multiple generic domains at the same time (for example, *Chase the Chuck Wagon* [Purina, 1983] was simultaneously a console game, advergame, maze game, and single-player game). To further complicate the situation, a genre can also reference multiple domains (for example, a massively multiplayer online role-playing game is a multiplayer game, an online game, and a role-playing game all at the same time). Within individual domains, there is much overlap between genres as well—some console games are played online, online games are played on computers and on mobile devices, and so on; and media convergence will require even more revised terminology. Video games vary so widely when it comes to game content and types of player experiences that genres are still struggling to convey these things. This can be seen in the variety of systems of generic classification in use, which include scholarly, journalistic, industrial, commercial, and vernacular ones, among which there is great inconsistency in the terminology, the degree of precision, the boundaries demarcating generic divisions, and the genres themselves. Yet in all of these cases genre is used to convey a sense of player experience and frames the way games are considered and discussed, placing more importance on player experience than on iconography or narrative.

The concept of genre began in antiquity with Plato and Aristotle, where it distinguished between the various kinds of literary works (such as comedy and tragedy), giving an audience some idea of what to expect from the work. Experience with works of a particular genre also led to genre conventions, understood by both authors and audiences, through the use of recognizable elements repeated from one work to another. The *Oxford English Dictionary* (2nd ed., 1989) informs us that the word *genre* (French for "kind," "sort," or "style," from the Latin *genus* and the Greek *genos*) was applied to literary works as early as the 1770s and that during the 1800s it was applied to painting and music as well. As works from these media were adapted into other media during the early twentieth century, the notion of generic classifications followed (such as Western novels becoming Western films or stage musicals becoming film musicals). As is evident from their content—their iconography, action, characters, storylines, and narrative situations—the design of video games was certainly influenced by these genres as well as by the genres developing in the area of electromechanical games (such as shooting games and racing games), which video games would join in the arcade.

By the 1960s, arcade games and electromechanical games had already been organized within their industry into genres based on the action they offered, such as pinball, shooting, racing, and so forth, emphasizing the type of player activity rather than the games' iconography, suggesting that companies felt the former was more important to the player than the latter. Early arcade video games, made by many of the same companies that produced electromechanical games (such as Bally, Midway, Gottlieb, Williams, and others), would inherit these genres while at the same time being touted as something new—namely, "video games." By the time video games appeared, then, the term *genre* was being used in various media to indicate literary qualities and story content, iconography, and player interactivity; and all of these uses would be adopted (to varying degrees) by video game classifications, further complicating the notion of genre.

From their very inception, video games have been subject to the influence of generic classification; Thomas T. Goldsmith Jr. and Estle Ray Mann's Cathode Ray Tube Amusement Device of the late 1940s was conceived as a shooting game with airplane targets in mind, and *Spacewar!*

(1962), developed at MIT, was designed around spaceships and inspired by E. E. "Doc" Smith's *Lensmen* series of novels in the realm of science fiction. A few years later, inventor Ralph Baer wrote down his initial ideas for games that could be played on a television set: "I meticulously documented everything I did, starting with a 4-page paper I wrote September 1, 1966, in which I laid out the whole idea of playing games on a TV set and defined many specific game categories" (2005, 9). In Baer's four-page paper, the second section, "Some Classes of Games Considered," lists action games, board (skill) games, artistic games, instructional games, board chance games, card games, and sports games (2005, 21–22). Unknown to Baer at the time, Roger Caillois's classifications of play into *agôn*, *alea*, mimicry, and *ilinx* in 1961 were also action based, but Baer's classifications were closer to genres in that some of them hinted at game content; reference to content—in particular gameplay and player experience—is what separate genres from mere classifications. In this sense, Baer's classifications were arguably not genres in a fully developed sense, but they began the sorting and classifying of games that would lead to generic classification.

Early arcade video games extended some of the activities of electromechanical games into the virtual realm (such as shooting, racing, and pinball), and paddle-based games were popular in both arcade and home systems. Arcade video games were distinct from other types of arcade games and were collectively referred to as "electronic games" and "computer games" in the *Reader's Guide to Periodical Literature*; it was not until the March 1973–February 1974 issue of the *Reader's Guide* that the term *video games* was listed. Thus, the terms *video games* and *computer games* were used to distinguish these games from other types of arcade games or home games, but the relatively small number of such games at the time meant that no internal system of classification was needed. Around the same time, however, other types of games were being classified in more detail; *The Study of Games*, edited by Elliot Avedon and Brian Sutton-Smith (1971), encouraged the academic study of games, setting the stage as well as a precedent for the academic study of video games that would one day follow, and many of its essays also classified games according to the activities they involved.

As the video game industry spread and companies' game catalogs expanded during the late 1970s, the growing number of games created a greater need for generic classification. The *Atari Catalog* in 1978 had a page or two dedicated to each game and a page organizing games by which controller they used (but gave no overall names to these categories) (Atari 1978). By 1981, the *Atari Catalog* listed 45 games and grouped them into categories based on game content, many of which indicated the type of player activity present (such as "Combat Zone" or "Adventure Territory") (Atari 1981). Although the term *genre* was never used in these catalogs, such categories functioned increasingly like genres, organizing games by content and player experience; they would also help indicate the presence of particular game conventions, which were becoming codified as more games appeared. These conventions in turn arose out of the technical limitations imposed by hardware and software, which influenced the shape that games took more than literary conventions or literary genres.

Game journalists, whose reporting and analyses helped to bridge the gap between industry and academia, also began using terms that hinted at groupings of similar games and the formation of genres. For example, in 1978 Kris Jensen wrote in *Popular Electronics*, "National Semiconductor's *Adversary 370*—introduced last year as a tennis–hockey–handball game—has been joined by the company's new *Adversary 600*, which has 12 action fields and 23 games that include *Pinball* and *Wipeout* (with 240 stationary targets) games as well as some of the traditional paddle-ball games" (33–34). Here the descriptions used ("tennis–hockey–handball game," "Pinball" games, and "Wipeout" games) refer to games that are similar to particular games, with the game and its variations or imitators being grouped together. Two years later, in what is probably the first recounting of the history of video games, Jerry Eimbinder and Eric Eimbinder wrote about such groups in the first part of a multipart essay in *Popular Electronics*; for example, "The *Pong* paddle game survived and prospered. (Eventually, space war games, in simplified versions, would also capture the public's fancy)" (1980, 54). This comment demonstrates how the names of generic classifications (paddle games and space-war games) became more formalized and functioned as genres for some time before the term *genre* was used (at the time, *genre* was more of an academic term and not one found in common parlance).

The term *genre* would finally be applied to video games when they began to receive academic attention a few years later. The third chapter of Chris Crawford's book *The Art of Computer Game Design* (1984), "A Taxonomy of Computer Games," used the word *genre* and opened with three paragraphs that described the difficulties of creating a taxonomy, thus already recognizing some of the problems to come. Although the categorizations described earlier were motivated by commercial concerns, Crawford's divisions reflected a more scholarly approach aimed at analyzing games themselves. Crawford justified the project, writing, "A taxonomy would illuminate the common factors that link families of games, while revealing critical differences between families and between members of families. A well-constructed taxonomy will often suggest previously unexplored areas of game design. Most important, a taxonomy reveals underlying principles of game design" (Chapter 3).

Crawford organized games into two broad categories, each of which was further broken up into genres: "Skill-and-Action Games" (combat games, maze games, sports games, paddle games, race games, and miscellaneous games), and "Strategy Games" (adventures, *Dungeons & Dragons*–type games, wargames, games of chance, educational and children's games, and interpersonal games). Crawford's two broad categories reflect the two different modes of play that still apply today (action-based games requiring skills and fast reflexes and contemplation-based games requiring strategy and puzzle-solving skills), and his genres are based mostly on gameplay action, though some (e.g., miscellaneous games and interpersonal games) refer to other domains.

Some academic authors applied genre terminology from other media; for example, Marsha Kinder wrote in her book *Playing with Power in Movies, Television, and Video Games* (1991), "This structure is apparent in games designed in the genre of romance, like the *Super Mario Bros.* and *Zelda* series, or of non-stop warfare, like *Teenage Mutant Ninja Turtles, Contra, Double Dragon*, and *Renegade*, but also in sporting games like *Skate or Die*" (110–111). Though today a game such as *Super Mario Bros.* (Nintendo, 1985) would be considered a platform game, Kinder's inclusion of the game in the literary genre of romance reveals her focus on narrative elements (Mario and Luigi's goal is to save the princess) carried over from literature. These early writings used a variety of generic systems and distinctions to discuss the shared traits of groups of games, attempting a more formal analysis of what constitutes each type of game, though they varied as to what aspects were important to their analyses. Yet certain genres, such as sports games, racing games, and strategy games, appeared frequently enough in generic systems that they seem to have gained some legitimacy as enduring generic conceptions, despite having lines of demarcation that differed from one usage to another.

Games have evolved so much since Kinder's analysis that one can add more genres or divide the current genres into series of subgenres (see Wolf 2005, 2001). For example, shooting games, (or "shooters," as they are commonly called today) could be further divided into such subgenres as first-person shooters, second-person shooters (as in *Second Person Shooter Zato* [Kongregate, 2011]), third-person shooters, rail shooters, shooting galleries, tactical shooters, and so forth. One might even argue that first-person shooters are deserving of their own separate genre instead of being merely a subgenre; when enough games exist within a genre, distinctions between variations become the basis for further divisions and subclassifications, allowing subgenres to emerge or even to become new genres. Although video game genres can be discussed in an academic context, the genre categorizations arise mainly from popular usage and discussion, which attempt to demarcate the various player experiences or game mechanics so as to indicate their presence within a group of games. Marketing departments may pick up on these genre names as well, but the names have historically emerged largely from player communities. All this only serves to demonstrate how genres and even their names are continually in flux and are not likely to be immutably codified, yet such change is slow enough to make these classifications useful to gaming communities.

The generic classification of video games is further complicated by video games' presence in transmedial franchises, which are composed of related works in a variety of media, each of which may have its own generic classification system and criteria. The various works may also appear in different genres, and adaptations may even hybridize genres—for example, the *LEGO Star Wars* video games, which are science fiction like the films on which they are based but also comedy due to cutscenes (scenes that break into gameplay) that parody scenes from the movies. Single games can also contain different sections, each of which may contain different forms of interactivity (each based on some part of the franchise) and thus elements of different genres. For example, *M*A*S*H* (1983) for the Atari 2600 (a game based on a TV show based on a movie based on a book) begins with a collecting game in which competing helicopters collect wounded soldiers from the field while avoiding projectiles fired from an enemy tank at the bottom of the screen; next, players must operate on a body, removing bullets without touching them to the sides of the body onscreen (much like the Milton Bradley board game *Operation* [1965]). Other games on the same cartridge involve catching bodies falling from a plane. Although such games can, of course, exist by themselves without being a part of a transmedial franchise, games that are a part of one are more likely to adapt different objects, iconography, and interactive situations from works in other media, resulting in hybridized activities that cross genres and further complicate generic classifications. When cinematic and literary genres become involved, video games may also be classified by their iconography and representational content, narrative themes, and other content that overlaps with other media. (Even games outside of such franchises can be affected by the generic systems of other media; for example, the more cinematic games appear, the more they may be classified by cinematic genre classifications.)

As franchises grow larger and their amount of narrative material increases, their scope broadens and is more likely to cross generic boundaries; the larger the franchise, the less useful generic classifications are (for example, the *Star Wars* franchise contains war, romance, monsters, comedy, drama, science fiction elements, and so forth). Likewise, even individual video games can contain such a wide variety of player activities and game mechanics that older generic distinctions weaken and are no longer useful in characterizing them, creating the need for newer terminology. Sandbox games (also known as "open-world games"), for example, did not arise until the second half of the 1980s and can be traced back to *Elite* (Acornsoft, 1984) ("Complete History" 2008); such games provided a game mechanic and player activity that required a new term to describe them.

So as useful as genres may be, their use is still often ad hoc and haphazard, crossing multiple domains and varying in definition (which is perhaps not surprising, considering that there is still only limited consensus regarding even the presentation of the name of the medium itself, which is still given as both *video games* and *videogames*). Other discussions, such as those asking how many genres are necessary (and how many are too few or too many) or how narrow and specific genres should be, depend on such things as the degree of accuracy in describing games' experiences versus the degree of simplicity required for their organization. No matter how genres divide up the gaming landscape, hybridizations will always be possible, resulting in overlaps, and genre membership will always be a matter of degree. Even the compromises and balances that are struck in the devising of generic systems will vary with their functions because academic, journalistic, industrial, and common usages will never agree completely.

Yet the notion of genre continues to be a useful way of talking about video games and seems likely to remain so in a variety of areas, including academia, marketing, player-forum discussions, and game design, even as each of these areas may divide up the field differently due to the differing criteria that they consider important. At the same time, transmedial franchises may have lessened the usefulness of genre at least in the area of marketing, where games are positioned to highlight the intellectual property they feature rather than their genre.

Despite these tensions and perhaps because of them, new genres and even new domains that organize genres continue to appear, resulting in a more hierarchical approach to genre in which subgenres play a larger role (for example, in the shooting game genre described earlier). Just as video games' widely ranging form factors, appearances, and uses have called into question what exactly constitutes a "medium," they also reveal the difficulties in devising taxonomies and generic classification systems because video games have so many aspects that simultaneously influence user experience. And the ways genres call attention to different aspects of games can inspire designers to deliberately devise games that fill gaps and cross generic boundaries—hybrid games and game technologies that further lead to new forms of games and gaming. In this sense, generic classification is more than merely an organizational scheme; rather, it is an important influence shaping the future of video games and gaming itself.

Works Cited

Atari, Inc. 1981. *Atari Catalog*. CO16725-Rev. D. Sunnyvale, CA: Atari, Inc. http://atariage.com/catalog_thumbs.html?CatalogID=32.

Atari, Inc. 1978. *Atari Catalog*. CO-12737. Sunnyvale, CA: Atari, Inc. http://atariage.com/catalog_thumbs.html?CatalogID=55.

Avedon, Elliot, and Brian Sutton-Smith, eds. 1971. *The Study of Games*. New York: Wiley.

Baer, Ralph H. 2005. *Videogames: In the Beginning*. Springfield, NJ: Rolenta Press.

"The Complete History of Open-World Games (Part 1): Feature: From Elite to Elder Scrolls." 2008. *PC Zone*, May 24. http://web.archive.org/web/20141211152956/http://www.computerandvideogames.com/189591/features/the-complete-history-of-open-world-games-part-1//.

Crawford, Chris. 1984. *The Art of Computer Game Design*. Berkeley, CA: McGraw-Hill and Osbourne Media.

Eimbinder, Jerry, and Eric Eimbinder. 1980. "Electronic Games: Space-Age Leisure Activity, Part 1." *Popular Electronics* 18 (4): 53–59.

Jensen, Kris. 1978. "New 1978 Electronic Games." *Popular Electronics*, January.

Kinder, Marsha. 1991. *Playing with Power in Movies, Television, and Video Games: From Muppet Babies to Teenage Mutant Ninja Turtles*. Berkeley: University of California Press.

Klevjer, Rune. 2005. "Genre Blindness." Hard Core column no.11, Digital Games Resarch Association, December 28. http://www.digra.org/hc11-rune-klevjer-genre-blindness/.

Wolf, Mark J. P. 2005. "Genre and the Video Game." In *Handbook of Video Game Studies*, ed. Joost Raessens and Jeffrey Goldstein, 193–204. Cambridge, MA: MIT Press.

Wolf, Mark J. P. 2001. "Genre and the Video Game." In *The Medium of the Video Game*, ed. Mark J. P. Wolf, 113–134. Austin: University of Texas Press.

29 IDENTITIES

Carly A. Kocurek

A young man sits at a computer. His shoulders are hunched, his hand hovers on the mouse. He chews his bottom lip as he stares intensely at the screen. Noise bleeds from his headphones. Explosions. Shouting. A moment later he whoops and throws a fist in the air. Victory. The young man is a gamer. And because he is a gamer, we can know some things about him or at least can assume them. Some of those assumptions are positive—he may be assumed to be technologically competent and intelligent, for example. Others are less so—perhaps he is socially inept and isolated, maybe even misanthropic or violent. Over the decades since video games first achieved commercial success, the gamer has become an archetype, a readily recognizable cultural trope invoked by critics, companies, and fans, as well as a means for people to express their individuality or their group allegiance—in short, an *identity*.

To invoke gamer identity is to further reify narrow ideas of who gamers can or should be. Game companies do it when they produce games for and advertise to an idealized hardcore demographic. The press does it when it presents gamers as a monolithic category or worries over the impact of games on a generation of bright young men—the only gamers that matter. And self-identified gamers take an active role in this reification when they aggressively police the boundaries of who can and should be gamers. Gamer identity may be recognizable, but it is also fraught. The iconic nature of the gamer and the presupposition of narrow limits to gamer identity have also provided a generative arena for resistance and playful engagement. That the rise of gamer identity coincides with the rise of the term *identity politics*, as first used to refer to the political organization and activism around social categories in the 1970s, is also worth noting. Indeed, a number of initiatives aimed at diversifying gaming culture, both among players and among producers, have been organized around categories of identity and extend or echo concerns raised in various movements formed around identity politics.

Identification around and through gaming has a long history. The earliest advertisements for coin-operated video games offered a broader vision of who players might be. Whereas the advertisement for *Computer Space* (Nutting Associates, 1971) featured a female model to showcase the machine, as was common in ads for boats, cars, and luxury goods, ads for games just a few years later often depicted both men and women playing games, thus highlighting the games' potential as social bar amusements, or showed families playing games together, with daughters and mothers participating alongside fathers and sons (see figure 29.1). As games became more popular, however, they also came to occupy a more limited position in the popular imagination. Games were for youths, but not just all youths, only boys. Through advertising and public events, the industry actively participated in shaping public perceptions of gamers. The coin-op industry, in particular, looked to the crop of young men drawn to video games as an opportunity to make claims of respectability.

The longevity of Penny Arcade Expo's ongoing "dickwolves" controversy, which began with a rape joke made in the *Penny Arcade* comic in 2011, suggests the tightrope many gaming or gaming-adjacent companies try to walk. In the dickwolves case, the resultant backlash, which included eloquent critiques by rape survivors, was mocked on the Penny Arcade forums and in a flippant *Penny Arcade* comic. Penny Arcade then began selling "Team Dickwolves" (which translates roughly to "Team Rapists") merchandise, further incensing critics. When the company pulled the merchandise, the controversy abated, but then in 2013 Penny Arcade artist Mike Krahulik publicly stated at Penny Arcade Expo Prime that he regretted pulling the merchandise. Those speaking out against Krahulik's sentiment were met with death and rape threats by Penny Arcade loyalists. Penny Arcade has subsequently found itself in the unenviable position of having either to defend its fan base, even as those same fans threaten people with bodily harm and defend the rightness of rape jokes, or deliberately to criticize that same fan base.[1]

Relationships between gamers and game companies are often fraught and are also enmeshed in a complex cultural politics. Contemporary images of gamers have been decades in the making. One of the most widely distributed images of gamers from the 1980s specifically celebrated the achievements of coin-op players. In January 1983, *LIFE* magazine published a now iconic image: a group of teenage boys and young men posed behind a row of arcade cabinets placed in the center of Main Street in Ottumwa, Iowa, with a team of cheerleaders in the foreground. The photograph, part of *LIFE*'s annual "year in pictures" issue, was intended to represent the growing significance of gaming during the previous year. The players pictured, gathered by Twin Galaxies founder Walter Day, were all world-record holders, embodying, as Day would have argued, the best and brightest that video gaming had to offer. In the years since its first publication, the image has become a touchstone of arcade culture, occasionally found on display behind counters in arcades and the subject of a documentary, *Chasing Ghosts* (Lincoln Ruchti, 2007).

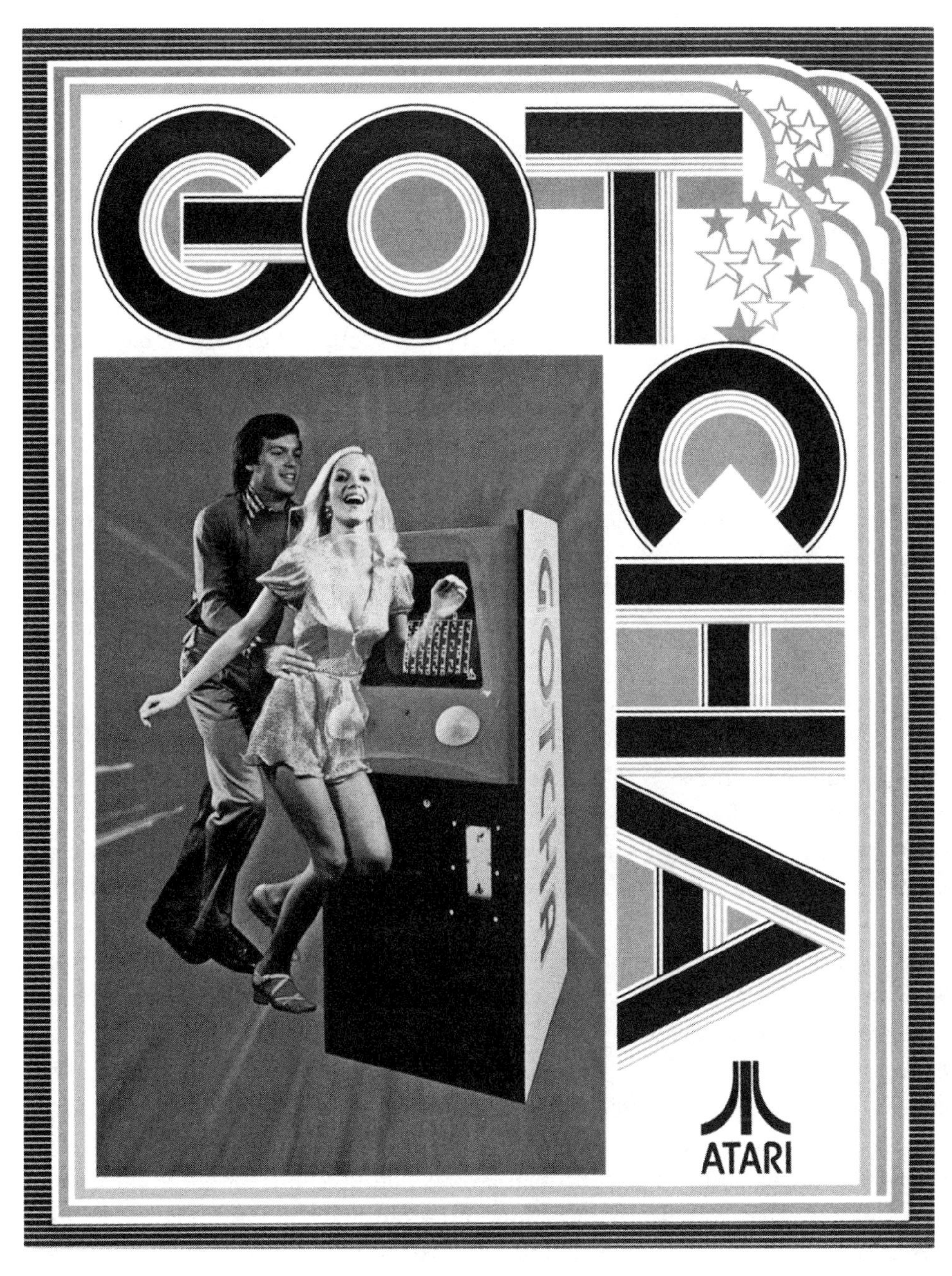

Figure 29.1
The flyer for Atari's game *Gotcha* (1973) features a man and woman, both of whom seem to be hustling past a cabinet conveniently located in a night club.

The resonance of this image, now decades after its initial publication, speaks to the importance of the moment in which it was captured. But it also suggests the extent to which the iconography developed around and through such presentations functions as a specific type of mythology. The boys and young men shown posed in the street in Ottumwa were among some of the first to be identified by and through their video gaming; although they were on occasional derisively called "vidiots" or "joystick jockeys," they are today known and recognized as gamers. They are identifiable not only for their visual association with the games displayed alongside them but also for other characteristics. This iconic image and the intervening decades of news articles, films, and advertisements on gamers support a relatively narrow view of who gamers can or should be. Gamers are young, the myth tells us, they are male, they are competitive, they are white, they are at least middle class, and they are technologically brilliant.

The components of this myth are powerful and widely invoked both in popular discourse and in industrial maneuverings. Although early advertisements for arcade games and consoles often depicted families or heterosexual couples playing together, over time marketing efforts became strongly gendered and more heavily focused on an idealized male gamer. Today, even as player demographics have shifted, the marketplace for gaming personal computers (PCs) and other console systems continues to identify and target that idealized demographic.[2]

An advertisement for the PlayStation 4 in 2013, for example, shows players as actors in iconic gaming landscapes set to Lou Reed's song "Perfect Day." In the ad, Reed's haunting anthem of fleeting joy is recast as celebratory. The gamers are punch-drunk and exhilarated, singing through a sword fight, a high-speed automotive race, and a futuristic battle scene. The ad, which ends with the phrase "Greatness awaits," relies in part on the juxtaposition of these modern young men—and all of them *are* young men—with these fantastical landscapes. It also relies on the recognizable mythologized image of the gamer as it invokes an idealized sales demographic of affluent young men. We all, the ad reminds viewers, know what gamers look like and should be able to readily recognize them even as they wield maces and swords in a computer-generated wonderland.

When gamers are so easily recognized as young, straight, white, able-bodied men, it is unsurprising that the game industry might overtly cater to this perceived market. Entire practices have in fact developed from these assumptions and in turn have shaped the industry. For example, the wide employment of "booth babes" at industry conferences also underscores long-standing and widespread beliefs about who gamers are and who designs games. In 2013, the hiring of scantily clad models at an International Game Developers Association party prompted an outraged response among the organization's membership, at least two of whom resigned board positions, leaving the party's sponsor, a game company, scrambling to defend the models as gamers who just so happened to be models. The party staffing decision, like the use of booth babes and other models at industry events, speaks volumes about gaming companies' assumed audience both within the industry and among consumers. That attractive models, many of whom know little about the products they are representing, are utilized as a standard branding and sales strategy reveals a normative vision of gamers that presumes not only maleness but also heterosexuality and standardized heteronormative desire. They are seductive window dressing, a wink and a nod for insiders, not unlike the *Hot Coffee* minigame embedded in *Grand Theft Auto: San Andreas* (Rockstar North, 2004)—initially hidden in the larger game's code, inaccessible to players, but later exposed. The exposure launched a short-lived scandal, but the content of the minigame differed only by degrees from that of the broader *Grand Theft Auto* universe.

A few companies and organizations have pushed back against these practices. Ubisoft, for example, recruits and sponsors teams of women gamers to promote their products. The first of these teams, the US-based Frag Dolls was formed in 2004. The women were so successful as a branding strategy that the company sponsored teams in the United Kingdom and France as well; although, perhaps reflecting changing tides in gaming, the last of the teams was disbanded in 2015. Although the Frag Dolls may have *looked* like booth babes to the untrained eye, each of the Frag Dolls was herself a competitive gamer, and all of the women were known by their gamer handles. In 2006, the Frag Dolls placed first in a *Rainbow Six: Vegas 2* (Ubisoft, 2007) tournament held by the Cyberathlete Professional League, becoming the first all-women team to win a professional circuit video game tournament. The Frag Dolls presented a conscious effort to present women not only as a lure for men who identify as hardcore gamers but also as skilled gamers in their own right.

While Ubisoft manipulated notions of gamer identity through the sponsorship of women as professional gamers—even as the company refuses to add female characters to its popular *Assassin's Creed* franchise—other companies work to find purchase among consumers outside the hardcore demographic. The positioning of Nintendo's Wii console as a disruption of the existing console market also speaks to the narrowing and calcification of gaming communities. Nintendo's code name for the project, *revolution*, represented an effort not only to reclaim Nintendo's significance in the console market but also to rethink who the industry should appeal to. When the console was launched in 2006, Wii's roster of games reimagined gamers as families, senior citizens, small children, and dabblers—a significant departure from an overt focus in the console and PC game market on the vaunted hardcore-gamer audience. Nintendo had also worked to market its portable Nintendo DS to an audience beyond the hardcore; advertisements for the console's release in 2004 targeted women and featured notable female celebrities such as Gwyneth Paltrow and America Ferrera. The strategy has been effective: the Wii was so popular that consumers waited in long lines to snap up the consoles, and the Nintendo DS holds the title of top-selling handheld console. Worth noting, of course, is Nintendo's sometimes problematic emphasis on "family values." In 2014, the company responded to player demand for same-sex romances in the life-simulation game *Tomodachi Life* with an official announcement reinforcing the choice to exclude those relationships. The announcement sent a clear message not only about how Nintendo imagines its audience but also about how the company defines and defends its "family-friendly" image.

Video gaming has produced a number of avenues for the formation and discussion of identity; in-game avatars and player choice of avatar, for example, have proven a generative area for scholarship, and the increased professionalization of game production has led to a growing number of career paths and attendant professional identities within the field. Further, the wide adoption of video gaming means that the number of people who play games has increased while the demographic profile of those people has become less focused. The Entertainment Software Association's industry report for 2013 notes that 58 percent of Americans play video games, 45 percent of those who play video games are female, and that the average player is 30 years old (2–3). These industry statistics counter popular perceptions of who video gamers are or should be. The real gamers of 2013 were a diverse group in terms of age and gender, but this diversification has proven contentious.

Although the industry regularly deploys *gamer* as a type of catch-all term for those who play video games, some players have proven themselves less willing to allow for the flexible use of this identity. What it means to be a gamer and who has the right to deploy this identity are a source of ongoing debate within gaming culture and are in some ways at the heart of the #GamerGate hashtag campaign, initially begun in the fall of 2014. That gaming has its own gatekeepers who are unwilling to recognize those participants deemed less properly authenticated is unsurprising given the increased cultural significance of gaming—gaming has become an arena with its own cultural and social capital, and the effort to limit who can identify as a gamer is also an effort to limit access to that capital. Identity in the case of gamer identity is a source of cultural and social capital accrued through a process of authentication. High scores and other displays of skill, for example, are part of a validation process; a broad knowledge of games, a state that can be likened to being "well read," is another way of articulating mastery and authenticity. That the cultural and social capital of gaming are often accrued through a consumer process that requires the expenditure of both money—coins dropped into arcade machines or dollars sunk into the purchase of games and consoles—and time—as in hours spent playing and mastering games, reading reviews and other games coverage, and even waiting in line to purchase games—is worth noting because it supports the narrowing of access and presents real limits to meaningful participation.

Because gamer identity functions as a kind of myth, the myth may persist even in the presence of factual evidence that counters it. The industry's own numbers reveal that the iconic image this chapter opened with is far from representative, and yet popular discourse on gamers, including advertisements like those launching the much-anticipated PlayStation 4, is quick to reinforce long-held beliefs. The prevalence of these popular beliefs necessitates further research into gaming practice, in particular research that does not proceed from limited assumptions about who gamers are.

A number of groups and programs have addressed the marginalization of women in gaming. Game designers have historically worked to engage girls in gaming by diversifying approaches to design. The "games for girls" movement of the 1990s, for example, saw the launch of several design studios, most notably Purple Moon. And ongoing efforts to diversify the industrial workforce and support the diversity of players have taken a number of forms. For example, Girl Gamer, which bills itself as "the definitive social network & video game site for female gamers," has a readership in the thousands. The more than 4,000 members of Women in Games International advocate for the inclusion and advancement of women in the game industry. The Toronto-based organization Dames Making Games offers workshops and other professional opportunities for women interested in game development. And individual women have undertaken efforts to build community and raise awareness. Jenny Haniver, who runs the blog *Not in the Kitchen Anymore*, has playfully documented the in-game harassment she faces for daring to game while being female.

Although the organization and activism of women gamers have reached a relatively high level of visibility, women gamers are not alone in organizing for a more inclusive gaming culture, nor are they alone in recognizing and reacting to incidents of marginalization and exclusion. Members of the lesbian, gay, bisexual, transgender, and queer community, for example, have both highlighted issues of representation and inclusion and established events and organizations to support their participation in gaming both as players and as designers. GaymerX, a convention aimed at creating a safe space for gay, lesbian, bisexual, transgender, and queer gamers attracted 2,300 attendees when first held in 2013. Racial representation in games is also a problem, as is racial diversity in the industry. Again, there have been organizational efforts to diversify through identity-based activism. Blacks in Gaming (BIG), for example, provides opportunities for networking, mentoring, and education for people of color interested in working in the game industry.

Disabled gamers face unique challenges in part because of the perceived difficulty in providing accessible gaming interfaces. The AbleGamers Foundation, based in the United States and founded in 2004, advocates for game accessibility and has collaborated with game companies on improving hardware and software accessibility. The organization also gives small grants to gamers who need specialized equipment to be able to access games and reviews games for their accessibility. In the United Kingdom, SpecialEffect also provides gaming devices and other support to disabled gamers.

The lack of diversity in the game industry—in terms of gender, race, sexuality, and so on—helps propagate a lack of diversity on screens and in the gaming community. The assumed whiteness, the assumed maleness, the assumed heterosexuality, the assumed youth, and the assumed physical ability of gamers is broadcast through a web of cultural touchstones: the imagery of avatars, the identities of visible industry leaders, the language deployed in gaming forums both official and unofficial, the interfaces and controllers that demand advanced manual dexterity and clear vision, and the deluge of popular media that represent gamers through a lens that doesn't question the established gaming hegemony.

Gamers and game industry professionals marginalized because of gender, race, sexuality, or ability have distinct concerns and face distinct challenges, and the range of identity-based initiatives aimed at diversifying game culture speaks to an oppressive homogeneity. The hostility that greets many who seek to engage in gaming culture or to pursue professional opportunities in the industry derives from a narrow sense of who gaming is for—a narrow sense supported by popular discourse and historical representations that privilege certain kinds of images.

That photograph from *LIFE*, for years tacked, yellowing and curling, to arcade walls, is not just a pop-culture oddity; it is now an image made myth, a touchstone, an iconography of meaning easily invoked even years later, even when the gamers photographed have been lost to history or have passed quietly into middle age somewhere in the Midwest. What organizations such as BIG, the AbleGamers Foundation, and GaymerX are demanding may not seem radical. After all, if most Americans play video games, then surely the audience for those games is necessarily diverse. But in demanding inclusion and representation in a community that has long been defined by an origin myth that heralds the achievements of a select few, they are storming the flickering walls of a pixilated castle defended by boy kings as fierce as dragons. And in identifying as gamers, those who do diverge from long-held assumptions about what gamers look like and do are subtly assaulting the gamer identities of those who fit more neatly into boxes.

Identity is an important facet of the history of video gaming not only for what it tells about who is left out but also for what it can tell us about who is included and why. Is gaming culture, for example, merely mirroring broader social inequities? And when gamers and industry insiders defend the culture's status quo, are they assuming a defensive posture, or are they aggressively constructing an exclusive enclave? In either case, what is being defended, and who does that defense benefit? The stakes of these questions are high because they concern the ownership of and access to a major cultural form; they also involve the legitimacy of those working to broaden that access.

Gamer identity is a contentious arena, but it is also a potentially rich avenue for research. Investigations into how the gamer operates as myth and icon versus the reality of how gaming is currently practiced among the 58 percent of Americans who play games would provide further insight into the subtleties of an increasingly diverse community. Further, considerations of identity are integral to investigations of gaming's past as assumptions about gaming and gamers shape both representation and the historical record. As researchers work from archival sources documenting the development of gaming, we would do well to remember who might have been left out of that record and why.

Notes

1. Given the visibility and influence of Penny Arcade, incidents like this present not minor media kerfuffles, but instead serious discussions about gaming culture's fraught gender politics.

2. For detailed gamer demographic information, see Entertainment Software Association annual report *Essential Facts about the Computer and Video Game Industry*.

Works Cited

Entertainment Software Association. 2013. *Essential Facts about the Computer and Video Game Industry*. Washington, DC: Entertainment Software Association.

30 IMMERSION

Brooke Belisle

Step into the Game

Unless you are under water, the term *immersion* is a metaphor: a plunge into an all-enveloping experience, a sudden shift into another atmosphere. In the context of modern media, it describes a state of engagement in which viewers or users feel transported into and absorbed by the world of a representation. Games are said to be immersive when play is so compelling that it completely captivates the player, and the game world so fully elaborated that it seems to overflow the limits of representation. By affording and rewarding investments of attention, intention, and agency, an immersive game seems to temporarily operate as the player's lived world. Rather than offering a complete etymology of the term or history of this idea, this chapter asks what *immersion* means in contemporary media discourse and what may be at stake in the very contingency of its metaphor.

As of 2014, the next wave of immersive media was eagerly anticipated in the arena of digital gaming—consumer virtual reality (VR). A VR headset with a sci-fi name, Oculus Rift, was crowdfunded with almost $2.5 million in 2012 and then purchased by Facebook for $2 billion in 2014. Oculus Rift uses a combination of motion tracking, wrap-around imaging, and stereoscopic 3D to let players "step into the game" ("Oculus Rift" n.d.). Oculus's 21-year-old founder, Palmer Luckey, claims he was inspired as a child by popular ideas of VR that animated the 1990s (Beilinson 2014). The release of Oculus Rift is heralded as finally achieving the immersive potentials that were promised by early VR but never achieved.

Describing media as "immersive" and imagining specifically digital forms of "immersion" seem to correlate with the rise of the Internet and emergence of VR in the 1980s and 1990s. Early instances of the term *immersion* begin to appear in connection with digital media around 1983, when Jaron Lanier, the purported popularizer of the term *virtual reality*, first started selling VR goggles. The following year the fantasy of digital immersion was influentially expressed in William Gibson's novel *Neuromancer*. The novel's main character "jacked in" to "cyberspace" by networking his neurological functioning with the electrical signals of machines.

In some ways, the goal of digital immersion aspires toward more embodied and visceral experiences than earlier media have offered—something more physical than semiotics. Fantasies of digital immersion approach a vanishing point, however, where the body may disappear. For example, in the early 1990s sci-fi horror film *The Lawnmower Man* (Brett Leonard, 1992), a character uses VR games as training exercises to develop superhuman abilities; he eventually leaves his body behind as his mind is transubstantiated into "pure energy" and absorbed into a computer mainframe. Today this endpoint is eagerly awaited as "the Singularity"; proponents of this endpoint, including Google founder Sergey Brin, expect that human consciousness will eventually merge with digital forms of artificial intelligence in ways that undermine any clear distinction between actual and virtual or human and machine (Vance 2010).

Although some may still anticipate hyperbolic possibilities of digital immersion, the assumptions informing such possibilities have been critiqued since the early days of VR. During the 1990s, media artists pioneering the medium of digital interactivity experimented with VR to challenge any simple relationship between the virtual and the visceral. For example, in Char Davies's installation *Osmose* (1995), participants wore not only a head-mounted VR display but also a vest that sensed their breathing and balance. Using breath and balance as navigational inputs tied a heightened awareness of actual, embodied sensation to participants' experience inside a digital, virtual world. Another influential installation, *The Legible City* (1995) by Jeffrey Shaw, was reprised in 1999 as an installation titled *The Distributed Legible City*: participants wearing stereoscopic head-mounted displays rode stationary bikes to navigate through a virtual scene of three-dimensional words that Shaw described as an "urban textual landscape" (see Shaw n.d.). Shaw's experiment suggested that experience is already partly virtual—cities are conceptual landscapes, language an architecture of thought.

Following the popular imagination of VR and the prominence of interactive, digital art, scholars in the late 1990s and early 2000s theorized the cultural impact of digital immersion. Janet Murray imagined the potentials for new forms of immersive storytelling and performance in her book *Hamlet on the Holodeck* (1997). In *Virtual Art: From Illusion to Immersion* (2003), media historian Oliver Grau positioned "immersion" as a goal that was slowly advanced through successive formats—from frescoes through panoramas to IMAX screens—but achieved only with the advent of digital technology. Media theorist Mark B. N. Hansen argued in *New Philosophy for New Media* (2004) that the close coupling between human and technical processes offered by digital VR expresses what Henri Bergson (1998) called "creative evolution" and articulates the latest phase of human "technogenesis."

Media scholars enthusiastic about digital technology saw the immersive potentials of computational media as marking an unprecedented aesthetic and cultural opportunity. Scholars less invested in new media, however, were more critical. The film scholar Vivian Sobchack, for example, warned that our investments in digital forms of immersion "tend to *diffuse* and *disembody* the lived body's material and moral gravity," inviting us to float free of ethical coordinates and embodied consequences (2004, 158). Such suspicions that immersion may be immoral continue a critique that has attended every aesthetic medium extending back to religious bans on graven images and Plato's expulsion of the poets in the *Republic*. Immersion suggests the draws and the dangers of visceral, sensual experience. From the romantic novels that corrupted Emma Bovary to the purported contagion of violent games, immersion tends to be associated with "low" cultural forms deemed too simple to be challenging and with indulgent forms of aesthetic encounter seen to lack the "critical distance" needed for thoughtful response.

In the first decade of the twenty-first century, attention shifted away from VR. With the rise of "Web 2.0" and the revolution in mobile, "social" media, interest in immersion moved toward new ideas of distributed participation and networked interaction. The shift to networked and mobile media, however, actually helped make VR more viable as a consumer technology. One reason VR never became popular was that it required expensive, heavy helmets and data gloves strung with tangles of wires. Today, immersive gaming platforms such as Oculus Rift can take advantage of the smaller, lighter, cheaper, and more stable components that were developed for hand-held devices. Facebook's $2 billion purchase of Oculus Rift signaled a "resurrection" of VR through a new synergy with mobile and networked media (Dredge 2014; Parkin 2014; Rubin 2014). Recent films such as *Transcendence* (Wally Pfister, 2014) seem also to resurrect the cultural imagination of VR, inflecting it with newer influences—such as the idea of "the Singularity"—that update the fantasy of immersion.

As Vannevar Bush once wrote, "The world has arrived at an age of cheap complex devices of great reliability; and something is bound to come of it" ([1945] 2003, 38). Bush, a pioneer of early computing, made this remark in his essay "As We May Think," published in the *Atlantic Monthly* in 1945. At the end of that essay, he speculated that because both the human nervous system and our media machines rely on electrical signals, "some day the path might be established more directly" (47). Fifty years later this idea remained speculative, elaborated in fictions such as *Neuromancer* and *The Lawnmower Man*. Today, after another fifty years, forging a more direct link between our bodies and digital technologies forms a significant research area called brain–machine interface or brain–computer interface. Companies such as Emotiv have developed electrode-filled headsets that allow users to control digital devices through "brain waves." On blogs, gamers imagine the next horizon of immersion arriving when platforms such as Emotiv and Oculus Rift fuse. Thirty years after *Neuromancer*, it seems we are still attempting to transport ourselves into virtual worlds—waiting, on couches everywhere, for new ways to "jack in."

In well-known work by Vannevar Bush and, later, Marshall McLuhan, the idea of digitally enabled immersion seems to appear years before digital computing was widespread and before the term *immersion* gained traction. Increased use of the word *immersion* in the 1990s may signal how the human–machine encounter was being rethought in the shift from electronic to digital culture. In the electrical era, potentials of human–machine interaction were often understood through metaphors of signal transmission and analogies between electrical circuitry and the nervous system—metaphors and analogies that electrical technology both reflected and helped inspire. With the rise of digital computing, VR, and the Internet, the word *immersion* suited a shifted set of more spatialized metaphors and analogies. As the names of early web browsers such as Netscape Navigator and Internet Explorer suggest, we began to imagine "cyberspace" as a place we could navigate, a virtual world we could explore.

Other Immersions

If we let go of the term *immersion* and instead attempt to tell a history of the idea that immersion seems to name, we can begin this history as far back as we choose: mid-twentieth-century cinematic experiments such as Smell-O-Rama and Sensorama, elaborate painted frescos, or even the shadows in Plato's allegory of the cave. Long before immersive, digital media, we imagined stepping into other forms of representation. We find alternate vocabularies for immersion in literary notions of "the reader in the text" or in the "absorption" that Walter Benjamin and Michael Fried ascribe to painting. It would be a mistake to imagine immersion as an affordance or even invention unique to digital media.

Tracing immersion back in time, we might organize all of media history in a trajectory of "precursors." We might view Oculus Rift, for example, as a digital reinvention of the nineteenth-century visual technologies of panoramic painting and stereoscopic photography; it certainly repeats the claim that both these formats once made of enabling spectators to step inside virtual worlds (see Silverman 1999; "The Stereoscope, Pseudoscope, and Solid Daguerreotypes"1852). Promising edgeless views, Oculus Rift invokes how panoramic paintings extended representation beyond the boundary of the frame and the limits of peripheral vision. As a stereo headset prompting perceptual three dimensionality, Oculus Rift traces a line of inheritance through the Viewmaster that debuted at the New York World's Fair in 1939 and back to the boxy Brewster stereoscope Queen Victoria first encountered at the Great Exhibition of 1851.

Figure 30.1
A: Back view of the Oculus Rift headset. From the developer kit in 2013 funded by a Kickstarter campaign in 2012. High-resolution image available at http://upload.wikimedia.org/wikipedia/commons/2/26/Oculus_Rift_-_Developer_Version_-_Back.jpg. B: *Viewmaster* (2008) from "Objects of Affection" series by Karen Kurycki; watercolor 8x10". Copyright Karen Kurycki. Used by permission of author. C: Brewster Stereoscope, *Popular Science Monthly* 21 (1882).

The panorama and the stereoscope, in particular, are often invoked as steps in a technological progression toward digital forms of immersion. Oliver Grau (2003) and Lev Manovich (2001) position the panorama as setting a precedent for VR. Mark B. N. Hansen (2004) and Jonathan Crary (1992) discuss how the stereoscope's interleaving of body and machine anticipated forms of digital interactivity. Although it would be a mistake to define immersion as a new potential of digital media, it would also be a mistake to assume immersion is a universal concept that operates across changing technologies and distinct historical contexts. Each instance we look back upon as a "precursor" of digital immersion might, if untethered from this teleology, suggest how changing modes of mediation are bound up with changing ideas of what it would feel like or mean to be "immersed."

To understand what is at stake in the idea of immersion, we do not need a definition as much as a genealogy, a way of asking how this term in itself seems to stabilize a set of ideas

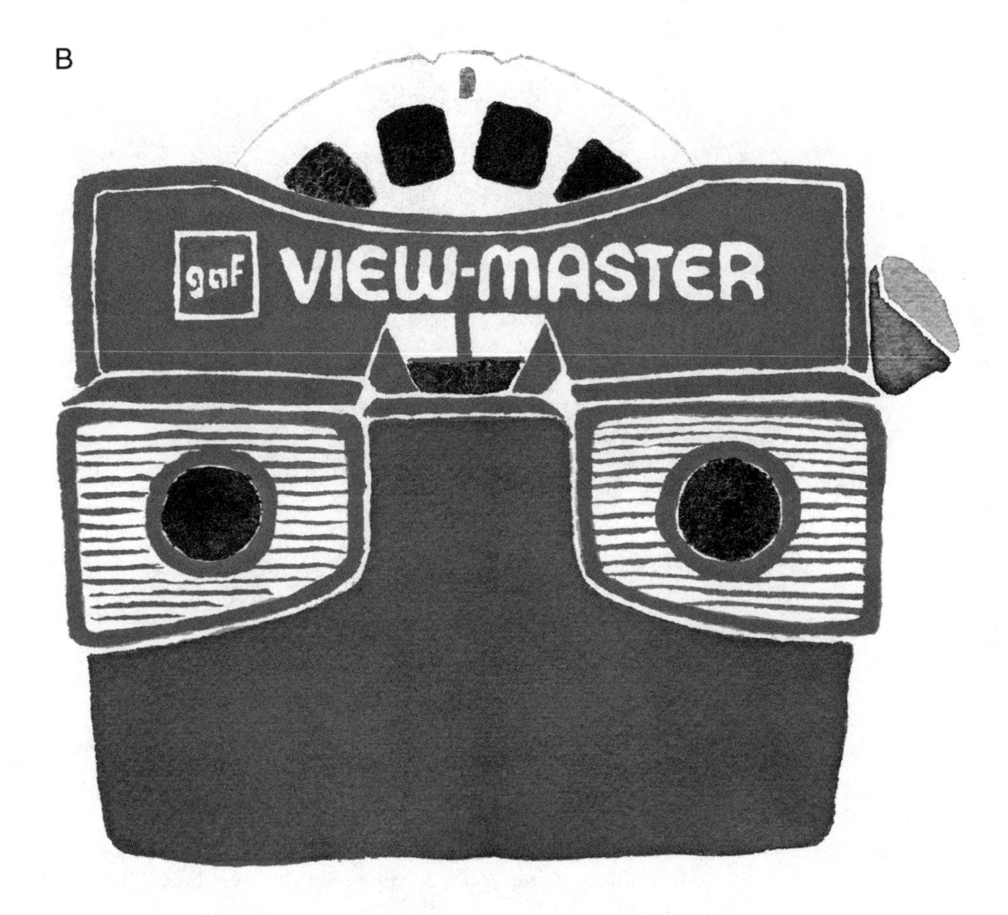

Figure 30.1 (continued)

that are not at all static. The panorama and stereoscope, for example, did not articulate more primitive versions of the digitally mediated experience we would later understand as immersion; each articulated a distinct understanding of what it might feel like and might mean to be fully absorbed by a perceptually enveloping representation. The forms of immersion that they offered were bound up with broader perceptual habits, aesthetic conventions, and technical strategies that cannot be simply transposed to other historical contexts and that were not directly inherited by later technologies.

In the nineteenth century, wrap-around panoramas were spectacles that viewers really could step into (see Oetterman 1997; Oleksijczuk 2011). First installed in major European cities such as London and Paris, early panoramas depicted primarily scenes of cultural and historical significance, such as "foreign" landscapes and colonial battles. The panorama's wrap-around

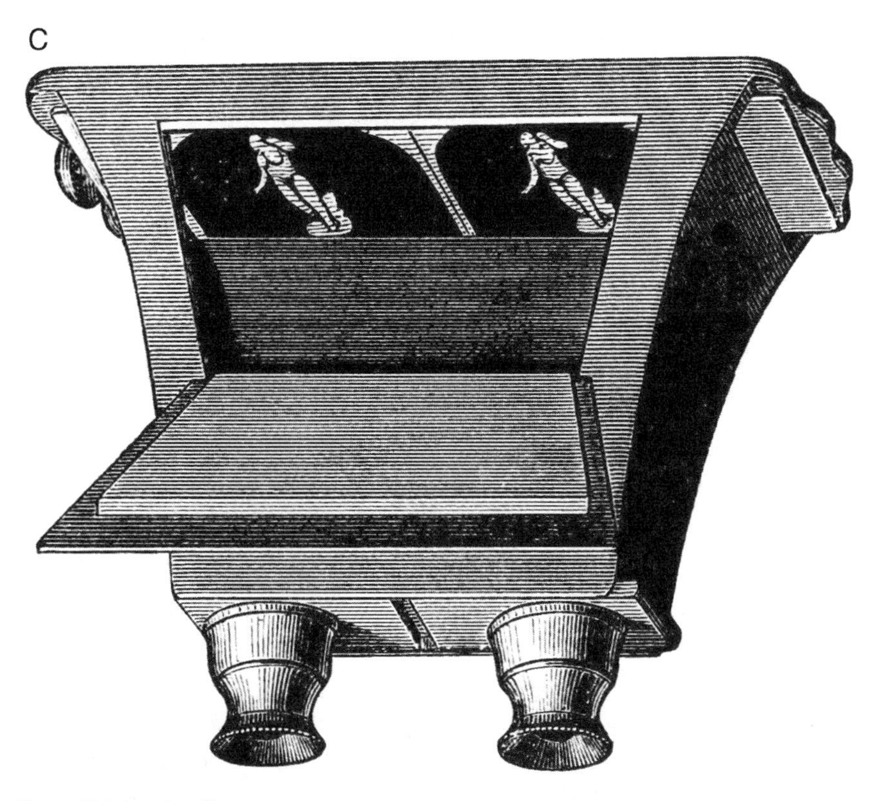

Figure 30.1 (continued)

image was made by affixing a linked series of enormous, painted canvases around the inner circumference of buildings designed for this purpose. Using careful strategies of light, scale, and perspectival distortion, a panorama concealed its status as a composite; it asserted its apparent seamlessness and expansiveness as the signatures of visual reality.

The sense of immersion that a wrap-around panorama produced was not grounded in the precise, optical verisimilitude of its image. A wrap-around panorama instead effected immersion as an act of perceptual integration; each spectator seemed to stand at the organizing center of the visual experience and to receive it in the same manner that an unmediated view might be received. The patent that coined the term *panorama* in 1787 emphasized its ability to present "nature as a glance," uniting multiple perspectives into one look. The patent instructs the panorama artist to "delineate correctly and connectedly every object which presents itself to his view as he turns round" in the place he intends to depict. This ensures that later the "observer turning quite round" inside the panorama can reenact the artist's look, reconstituting the correct coordination of every object and integrating multiple canvases into the perceptual unity of an embodied "view."[1]

Figure 30.2
Cross-section illustration of Robert Barker's first circular panorama. Reprinted from Robert Mitchell, *Plans and Views in Perspective, with Descriptions, in England and Scotland*, London: Wilson and Co., 1801.

The experience of immersion that the panorama offered both presumed and articulated the idea that the panorama's strategy of coherence modeled the coherence of nature itself. In other words, panoramic immersion both relied on and helped advance a broader, cultural logic in which the actual world seemed to be organized according to the principle of panoramic coordination. In retrospect, wrap-around panoramas mapped the ideological coordinates that also structured colonialism, tourism, and capitalism in the early nineteenth century. Panoramas structured visual experience not only as anchored by the subject but as ordered by his gaze and as displayed for his enjoyment; relationships that appeared external, between things themselves, were triangulated from the viewer's position. The way that the panorama seemed to immerse urban, European spectators within "other" landscapes suggested that the visible world, with all its diversity, might be organized around and understood from a privileged, central point of view.

The stereoscope is often positioned as the next major step, after the panorama, in a technological progression of immersion. After emerging in the second half of the nineteenth century, stereoscopic representation was arguably the most popular new form of mass, global, visual media until the advent of cinema. In an influential essay published in the *Atlantic Monthly* in 1859, Oliver Wendell Holmes claimed that when one looks through a stereoscope, "the mind feels its way into the very depths of the picture. The scraggy branches of a tree in the foreground run out at us as if they would scratch our eyes out. ... There is such a frightful amount of detail, that we have the same sense of infinite complexity which nature gives us" ([1859] 1980, 57–58). This sense of visceral presence was produced more by viewers than by the stereoscope. Interpreting two images according to the spacing of their two eyes, viewers perceived a relationship of depth anchored by their own embodied look.

Taking advantage of the new medium of photography and scientific research into perception, stereoscopic immersion relied on the precise detail of photographic images and the binocular depth effect created by presenting slightly different images to each eye. It also relied on and furthered analogies between eyes and cameras, sight and photography, things themselves and their photographs. Expanding forms of industrialism and colonialism that required control by proxy relied on precisely such equivalencies between perception and representation, things and images, bodies and machines.

As Jonathan Crary (1992) has argued, the stereoscope's widespread use aligned with ways bodies and machines were being geared together by forces of global capital. The mobility and reproducibility of stereo images fit with a culture increasingly defined by mass production and international trade. The idea of presence that stereoscopic images seemed to recapitulate relied on a calculated relationship of seer to seen: it modeled vision as a form of virtual grasp in which every spectacle became specimen. The experience of immersion that the stereoscope offered helped shore up the idea that almost anything could be captured, collected, stored, and exchanged as a portable commodity.

The values that determine how we judge media experience as "immersive" articulate our assumptions about the how the world "really" presents itself or should present itself, how embodied perception "naturally" operates or should, and how technological mediation might recapitulate, model, or stand in for these processes. Rather than tracing how every new technology gets closer to simulating the way we are "immersed" in the real world, it is more important to ask how changing concepts and constructs of immersion help us interrogate whatever unmediated relationships of presence immersion aims to replicate. In our ideas of immersion, we see volatile attitudes about the proper relationship between perception and representation, "actual" experience and "aesthetic" experience, subjects and objects, human and not human.

The term *immersion* names the limit conditions of mediation, the horizon at which the apparatus or interface seems to disappear and a mediated experience feels immediate. It thus points to expectations refueled with every new technology and presses a boundary always being redrawn between human and machine. At the same time, the felt experience of immersion also describes a very real albeit highly contingent coordination. It expresses a temporary but powerful alignment between the technical conventions of a particular medium, the aesthetic form of a particular representation, the cultural logic of a particular historical moment, and the perceptual framework of a particular participant's embodied experience. The changing terms of immersion articulate shifts much broader than any simple story of progress.

The digitally inflected notion of immersion has named one constellation of ideas about the potentials and threats that digital media pose. It has correlated not only with fantasies about "virtual reality" but also with debates about the "posthuman" and questions about how the rise of networked, computational media impact relationships of technical mediation and embodied experience. Analogies that governed ideas of immersion from the early days of computing through the rise of the Internet—brain waves as electrical signals and the nervous system as a network—may be giving way to newer analogies that will govern new ideas of immersion in the emerging era of ubiquitous computing, ambient technology, and Big Data. As other words displace *immersion*, we are likely to retrace the history of media through them.

Note

1. Robert Barker's patent application, June 19, 1787, is archived in *Repertory of Arts and Manufactures 4* (London, 1796), 165–167, and reprinted in Oetterman (1997), 358.

Works Cited

Beilinson, Jerry. 2014. "Palmer Luckey and the Virtual Reality Resurrection." *Popular Mechanics*, May 28. http://www.popularmechanics.com/technology/gadgets/news/palmer-luckey-and-the-virtual-reality-resurrection-16834760.

Bergson, Henri. 1998. *Creative Evolution*. Trans. Arthur Mitchell. Mineola, NY: Dover.

Bush, Vannevar. [1945] 2003. "As We May Think." *Atlantic Monthly*, July. Reprinted in *The New Media Reader*, ed. Noah Wardrip-Fruin and Nick Montfort, 35–47. Cambridge, MA: MIT Press.

Crary, Jonathan. 1992. *Techniques of the Observer: On Vision and Modernity in the Nineteenth Century*. Cambridge, MA: MIT Press.

Dredge, Stuart. 2014. "Oculus Rift—10 Reasons Why All Eyes Are Back on Virtual Reality." *Guardian*, March 29. http://www.theguardian.com/technology/2014/mar/31/oculus-rift-facebook-virtual-reality.

Gibson, William. 1984. *Neuromancer*. New York: Ace Books.

Grau, Oliver. 2003. *Virtual Art: From Illusion to Immersion*. Cambridge, MA: MIT Press.

Hansen, Mark B. N. 2004. *New Philosophy for New Media*. Cambridge, MA: MIT Press.

Holmes, Oliver Wendell. [1859] 1980. "The Stereoscope and the Stereograph." *Atlantic Monthly*, June. Reprinted in *Photography: Essays and Images*, ed. Beaumont Newhall, 57–58. New York: Museum of Modern Art.

Manovich, Lev. 2001. *The Language of New Media*. Cambridge, MA: MIT Press.

Murray, Janet. 1998. *Hamlet on the Holodeck: The Future of Narrative in Cyberspace*. Cambridge, MA: MIT Press.

"Oculus Rift." n.d. Oculus Rift corporate website. http://www.oculus.com/rift/. Accessed September 18, 2014.

Oetterman, Stephen. 1997. *The Panorama: History of a Mass Medium*. New York: Zone Books.

Oleksijczuk, Denise Blake. 2011. *The First Panoramas: Visions of British Imperialism*. Minneapolis: University of Minnesota Press.

Parkin, Simon. 2014. "Oculus Rift: Thirty Years after Virtual-Reality Goggles and Immersive Virtual Worlds Made Their Debut, the Technology Finally Seems Poised for Widespread Use." *MIT: Technology Review* 117 (3). http://www.technologyreview.com/featuredstory/526531/oculus-rift/. Accessed September 18, 2014.

Rubin, Peter. 2014. "The Inside Story of Oculus Rift and How Virtual Reality Became Reality" *Wired*, May 20. http://www.wired.com/2014/05/oculus-rift-4/.

Shaw, Jeffrey. n.d. Descriptions and images of artworks on the artist's website. http://www.jeffrey-shaw.net/. Accessed September 18, 2014.

Silverman, Robert. 1999. "The Giant Eyes of Science: The Stereoscope and Photographic Depiction in the Nineteenth Century." In *Instruments and the Imagination*, ed. Thomas L. Hankins and Robert J. Silverman, 148–177. Princeton, NJ: Princeton University Press.

Sobchack, Vivian. 2004. "The Scene of the Screen: Envisioning Photographic, Cinematic, and Electronic Presence." In *Carnal Thoughts: Embodiment and Moving Image Culture*, 135–162. Berkeley: University of California Press.

"The Stereoscope, Pseudoscope, and Solid Daguerreotypes." 1852. *Illustrated London News*, January 24.

Vance, Ashlee. 2010. "The Singularity Movement: Merely Human? That's so Yesterday." *New York Times*, July 10.

31 INDEPENDENT GAMES

John Sharp

What Is an Indie Game?

Depending on who you ask, the term *independent games* or *indie games*, as it is often phrased, refers to video games made by amateurs, video games produced and distributed without the involvement of large companies, video games published and played outside traditional markets, video games made with avant-garde intentions, and even card games and board games made outside the toy and hobby game industry. People look to indie games to fill all sorts of purposes and niches, too. Greg Costikyan (2005) views independent games as the only way to save the game industry, and anna anthropy (2012) views them as an expressive medium available to everyone, in particular marginalized voices. Jeff Vogel (2014) sees indie games as a new cycle of business in the video game industry. Celia Pearce and Akira Thompson (2013) see them as the space within which games can become broader and more inclusive.

As Bennett Foddy (2014) has noted, there are endless examples of games and game makers and game-making practices from the past 30 years that can be considered indie: public-domain game distribution via magazines and bulletin board systems; the shareware-distribution model used to great success by id Software; Activision's formation as the first third-party development studio; the home-brew computing community; the release of level editors and game-making software such as the Adventure Game Construction Set.

The more closely examined the term *indie* gets, the more the question "What is an indie game?" is asked.

Ron Gilbert (2014), a game designer whose career reaches back to the 1980s, has teased out many of the problems with the term *indie*. He raises questions about independent developers and their relation to capital, game genre, and distribution method, among other factors. Indeed, who and what qualifies as "indie" is deeply contested as a marker of status and authenticity even as more and more game developers self-identify as "indie devs." *Indie* has come to simultaneously mean an alternative to the mainstream game industry, a progressive mindset for game makers, a loose marketing concept, and a genre.

In 2002, well before the independent games phenomenon, game designer Eric Zimmerman explored the economic, technical, and cultural potential of independent games to help make sense of them. He looked to independent cinema for comparison in the absence of any useful models from inside the games industry. The term *independent film* speaks to both the economic and production infrastructures and an aesthetic sensibility. From an economic perspective, independent films are produced by individuals or small companies operating outside the large film studios such as Fox, Sony, and Columbia. The budgets tend to be much smaller, with crews as small as a handful of individuals and scaling up to several dozen or more people (King, Molloy, and Tzioumakis 2012). In part because of the smaller scale and risk, independent cinema was the locus of much of the innovation in film. For decades, artists and filmmakers explored and expanded storytelling, cinematography, subject matter, and many other aspects of film outside the main film industry.

In the early 1990s, however, *independent film* became more a category or marketing concept than a term grounded in concrete criteria relating to production and distribution or formal and aesthetic considerations. For example, major studios (Sony Classics, Fox Searchlight, etc.) created divisions for distributing independent films, and large theater chains in American cities are as likely to show independent films as are arthouse theaters. The independent-film phenomenon came to represent a category of film under which a diverse group of filmmakers are collected—for example, Darren Aronofsky to Jane Campion to David Lynch—whose work share few traits.

This description of independent film very much speaks to the current state of indie games. There is no one indie game, just as there never was a single independent film.

Music is another useful model for thinking about the indie game phenomenon. *College music* or *indie rock* were terms used to describe the rise of certain styles of music in the mid-to late 1980s. These names did not signal the start of music making outside major labels; instead, they were attempts to codify and package a wide range of already-existing musical styles, music scenes, and musicians. Bands from local communities such as Athens, Georgia; Boston; Los Angeles; and Seattle produced music as part of their local communities. Supporting infrastructures took shape: independent record labels such as Dischord, SST, and Slash released the music, college and regional radio stations played the music, indie record stores sold the music, and clubs permitted national and international touring. Press outlets formed, further expanding awareness while creating consumer categories for grouping the disparate bands and scenes. Major labels began signing these bands and even started quasi-independent labels of their own.

Although there were commonalities in the music grouped under the banners *college music* and *indie rock*—much of it was guitar driven, for example—there was little connecting the musicians' intentions, their audience, and other important factors. Few musicians referred to themselves as "independent musicians" or their output as "college music." They might have instead talked about their genre—hardcore, postpunk, grunge, and so on—or referred to their local scene or community. So the term *indie rock* was really used for consumer-focused codification and promotion purposes rather than derived from or used by the musicians and their communities.

In the same way independent film and music came to represent modes of production, artistic sensibilities, and marketing categories, indie games have come to represent these things in the gaming milieu. With indie games, there are similar communities of like-minded practitioners—what Eric Zimmerman (2002) once called "garage band game studios." And, as Jesper Juul (2014) has noted, a predilection for retro-inspired pixel graphics pervades certain communities within the indie games landscape. Some of these practitioners are geographically bound in cities such as Austin (Juegos Rancheros, Fantastic Arcade), Los Angeles (Glitch City, IndieCade, the University of Southern California's Interactive Media and Games Division, and the related Game Innovation Lab), New York (Babycastles, New York University's Game Center, and Parsons School of Design's Design & Technology program), and Toronto (Dames Making Games, the Hand Eye Society, Sheridan College). A diverse range of games are also grouped under the indie banner—from platformers to e-sports to art games to text-based games to puzzle games and beyond.

Just like there was not one independent film or music outside marketing and press, there is no single indie game. The seemingly out-of-the-blue success of games such as thatgamecompany's *flOw* (2007), Number None Inc.'s *Braid* (2008), and 2D Boy's *World of Goo* (2008) coalesced the concepts "indie games" and "indie development" into idealistic if not completely accurate caricatures. By 2008, there was a clearly visible, if not easily definable, category "indie games," with hundreds of game developers self-identifying as "indie devs" and hundreds of thousands of gamers playing "indie games."

What led to this upswell in attention some thirty years after the first independent games and just six years after it was unclear if indie games existed at all?

Factors in the Rise of Independent Games

Nine factors created the conditions from which the indie phenomenon emerged: indie-focused game websites, indie-operated download websites, digital distribution services, independent game studios, Internet-based communication channels, a focus on experimentation, academic game programs and research groups, indie game festivals and conferences, and increased media attention.

Indie-focused game websites were the first factor in the establishment of indie games. Sites such as Newgrounds (founded 1995) hosted Flash-based games alongside animations and short videos, which gave animators, comic artists, and game developers a platform for showcasing their work. As these websites gained in popularity, tightly curated and managed portal sites such as Shockwave.com, Yahoo.com, and Microsoft's Gaming Zone began making deals with indie developers. They were poor options for developers, however, because the content was tightly curated, thus limiting experimentation, and the only revenue stream was ads, which were not yet a sustainable income source.

With a new outlet established for playing indie games, the second factor took shape: indie download services. The developer-publisher-distributor model, in which shelf space and advertising dollars were tightly controlled, put traditional distribution out of the reach of independent game developers. Downloadable game services such as Greg Costikyan and Johnny L. Wilson's Manifesto Games (founded 2005) allowed smaller studios and even individuals a way to release their games outside the AAA distribution system. When Manifesto Games launched its website in 2005, digital downloads of games were unusual—game products were still shrink-wrapped and either disc or cartridge based. Costikyan and Wilson's vision hinted at things to come once high-speed Internet became more widely adopted and consumers became more comfortable with downloaded purchases. Services such as the Humble Bundles and Gumroad have since flourished, giving developers a more accessible means of selling their games.

When the downloadable-content market solidified, it was predominantly the domain of large corporations such as Sony, Microsoft, Apple, and Valve. More than the indie download websites, these services gave smaller developers access to much broader audiences and thus expanded the potential audience for their games. This was the third factor in the rise of indie games.

Particularly important was the launch of Sony's PlayStation Network (2006). Few mainstream industry developers were willing or able to produce games for Sony's new download service. Thanks to the success of *Cloud* (2006), a game by students in the University of Southern California's Interactive Media Division, Tracy Fullerton was able to help thatgamecompany secure a three-game deal that resulted in the release of *flOw*, *Flower* (2009), and *Journey* (2012).[1] The success of *flOw*, followed by even better sales of *Flower*, opened the door for other indie studios to receive development deals.

Soon after the launch of *flOw*, Microsoft began publishing indie-produced games through its Xbox Live Arcade service. Though this service was launched in 2004, indie games didn't begin to find success on it until 2008. That summer, Number None, Inc.'s game *Braid* and The

Behemoth's *Castle Crashers* were released and promoted as part of the XBLA Summer of Arcade promotion.

The console platforms were tightly controlled, with only a handful of indie developers given the opportunity to release games. Valve launched Steam, its personal computer game download service in 2003, though it did not see indie game success until the release of Introversion Software's *Darwinia* and Qi Studios' *Rag Doll Kung Fu* in 2005.The Apple App Store (launched in 2008) provided a more open platform. Early indie successes included Semisecret Software's *Canabalt* (2009) and Tiger Style's *Spider: The Secret of Bryce Manor* (2009). Together, the successes on the download services brought a great deal of attention to indie games.

The fourth factor in the rise of indie games was independent game studios, which produced the games distributed through these new channels. The typical approach used by studios such as the New York–based Gamelab (founded in 2000) and Toronto-based Capybara Games (founded in 2003) was to produce games for clients to fund the studio's original games released on portal sites or through mobile phone service providers. It was a hard path, with many studios spending most of their time making games for others and for poor financial return. Many had to close their doors, as was the case with Gamelab in 2009. This approach sometimes succeeded in getting a studio to a place where it could sustain the development of original games, as happened to Capybara Games, to this day a thriving indie studio.

The pioneer independent studios learned the lessons the hard way and shared their successes and failures in both public ways (e.g., blogs and conference talks) and private ways (e.g., closed email lists and online forums). This leads to the fifth factor: Internet-based communication as a central source of community. Where geographic proximity was once a precondition for communities of practice, the Internet enabled dispersed communities to connect. Private email lists such as Indie Biz and Artgame List functioned as a way for members of the indie games community (at least those invited to participate) to share best practices for development, business, marketing, and other topics. Forums such as those found on TIG Source (founded in 2005 by Jordan Magnuson and soon thereafter taken over by Derek Yu) gave developers a place to discuss development, business, and marketing and more generally to talk about games.

A culture of experimentation was the sixth factor in the rise of indie games. Had independent games simply been smaller-scale versions of the kinds of games published by large publishers, the indie phenomenon would not likely have grown to such a degree. Greg Costikyan's "Scratchware Manifesto" (2000), in addition to being a call to arms around the economics of video games, emphasized the need for a new aesthetic that eschewed visual realism and spatial navigation. A similar call came from Paolo Pedercini in his essay "Indie Developers Do It Better" (2006). With their "Mechanics, Dynamics, Aesthetics" methodology, Robin Hunicke, Marc Leblanc, and Robert Zubek (2004) sought to develop tools to strengthen game designers' ability to deliver a wider range of play experiences. The Ludica collective, made up of scholars and game designers Celia Pearce, Tracy Fullerton, Janine Fron, and Jacki Mori, looked to the New Games movement for paths to make video games more inclusive and aesthetically diverse (Ludica 2005). A particularly visible outlet for experimentation was the Experimental Games Workshop, an annual session at the Game Developers Conference since 2003. Originally organized by Jonathan Blow, Doug Church, and Chris Hecker, the session showcased experimental games, providing developers with a platform for sharing their attempts to expand the possibilities of games.

The seventh factor influencing the emergence of the indie games phenomenon involved academic programs and research groups, which fostered exploration outside the constraints of the marketplace. Programs such as Parsons School of Design's MFA Design and Technology Program (founded in 1997), Carnegie Mellon University's Entertainment Technology Center (1998), Georgia Institute of Technology's Digital Media and Computational Media Programs (founded in 1998 and 2008, respectively), and the University of Southern California's Interactive Media and Games Division (founded in 2002) embraced experimental games. These programs often had research labs attached to them. Mary Flanagan's Tiltfactor (2003), Tracy Fullerton's Game Innovation Lab (2004), Colleen Macklin's PETLab (2006), and Celia Pearce's Emergent Game Group (2006) contributed to a new way for game makers to think about what games could be. The most crossover with other approaches to game making happened inside academia—from the New Games movement to artist's games to folk games to board and card games to serious games and even sports.

Besides sending students out with expanded notions of what games could be, these programs and labs produced a number of important early independent games, including—in addition to *flOw* and *World of Goo*, mentioned earlier—Lima Ship's *Doodle Jump* (Parsons, 2009), and the Copenhagen Game Collective's *B.U.T.T.O.N.* (IT University of Copenhagen, 2010).

The eighth factor in the rise of indie games was the formation of indie festivals and conferences. Events such as the Independent Games Festival (started 1998), IndieCade (started 2005), the SlamDance Guerilla Games Festival (2006), and the Independent Games Summit (started 2007) further fueled the budding indie games scene. These events gave developers a place to showcase their games, to meet like-minded developers, and to see themselves as part of a community.

The burgeoning game press covered these events and their award shows, giving indie games and their developers greater legitimacy and attention. The ninth and final factor in the development of the indie games phenomenon was the rise of an indie games press and increased attention from mainstream media. Developers and players alike could go to *Indie Games the Weblog* (2005), *GameSetWatch* (2005), *Rock Paper Shotgun* (2007), *Offworld* (2008), and other blogs and sites for news and reviews. The mainstream press began to pay closer attention to video games in general and to indie games in particular. The *New York Times*, for example, began to review games, and the *Atlantic Monthly* and the *New Yorker* wrote about indie game developers from time to time. All this attention further increased the visibility and credibility of the growing indie games phenomenon.

The watershed moment in media attention was the release of *Indie Games: The Movie* (James Swirsky and Lisanne Pajot, 2012, filmed in 2008–2010). The film chronicled the lives of Edmund McMillen, Tommy Refesnes (Team Meat), Phil Fish (*Fez* [Polytron, 2012]), and Jonathan Blow (*Braid*). The film propelled the four developers and indie games as a whole further into the popular imagination. For many, this small circle of developers became the face of independent games. However, although these four men were an important part of the rise of indie games, the film's focus on white male developers left the people of color, women, and nonheteronormative developers essential to the rise of indie games out of the most visible indie origin myth.

Conclusions

These nine factors together created the conditions from which the indie game phenomenon emerged. As with indie film and music, there is no one indie game and no single moment that catalyzed the phenomenon. Instead of one "indie game," there are innumerable overlapping communities—from the TIG Source developers to the New York–based indie/academic scene to the polished experimentation of the Los Angeles scene to the queer games community to the indie scenes found throughout Africa, Asia, Australia, Europe, and South America. There is no single way to access and play indie games; there are instead multiple ways: consoles, galleries, festivals, smartphones, web browsers, coffee tables, and on and on. And there is no single kind of indie game. Instead, there are platformers, puzzle games, tabletop role-playing games, card games, not-games, shoot 'em ups, e-sports, art games, text-based games, and more. There is no single intention behind the games, either: activism, expression, entertainment, financial gain, and so on.

A couple of things are clear with respect to indie games: because of them, the games community is a more inclusive and varied space, and players have a wider range of play experiences at their disposal than ever before. As the indie phenomenon closes in on a decade of existence, it should be noted that as ubiquitous as the term has become, it has lost much of its power and allure within the game development communities. This is in part due to its transition from a term used by portions of the community into a market concept and genre category. And in the same way that *college rock* and *independent film* are now markers for a time in the past, so too will indie games be in the not so distant future.

Notes

Thanks to Tracy Fullerton, Jesper Juul, and Eric Zimmerman for feedback on drafts of this essay.

1. Tracy Fullerton to John Sharp, email, August 24, 2014.

Works Cited

anthropy, anna. 2012. *Rise of the Videogame Zinesters: How Freaks, Normals, Amateurs, Artists, Dreamers, Drop-outs, Queers, Housewives, and People Like You Are Taking Back an Art Form*. New York: Seven Stories Press.

Costikyan, Greg. 2005. "Death to the Games Industry, Part I." *The Escapist*, August 30. http://www.escapistmagazine.com/articles/view/video-games/issues/issue_8/50-Death-to-the-Games-Industry-Part-I. Accessed September 1, 2014.

Costikyan, Greg. 2000. "Scratchware Manifesto." http://www.homeoftheunderdogs.net/scratch.php. Accessed August 22, 2014.

Foddy, Bennett. 2014. "State of the Union." IndieCade East keynote, October 10, Culver City, California. https://www.youtube.com/watch?v=7XfCT3jhEC0. Accessed August 31, 2014.

Gilbert, Ron. 2014. "What Is an Indie Developer?" *Grumpy Gamer*, April 9. http://grumpygamer.com/1312686. Accessed April 11, 2014.

Hunicke, Robin, Marc Leblanc, and Robert Zubek. 2004. "MDA: A Formal Approach to Game Design and Game Research." In *Proceedings of the Challenges in Games AI Workshop, Nineteenth National Conference of Artificial Intelligence*. San José, California: American Association for Artificial Intelligence. http://www.aaai.org/Papers/Workshops/2004/WS-04-04/WS04-04-001.pdf.

Juul, Jesper. 2014. "High-Tech Low-Tech Authenticity: The Creation of Independent Style at the Independent Games Festival." In *Proceedings of the 9th International Conference on the Foundations of Digital Games*. http://www.fdg2014.org/proceedings.html.

King, Geoff, Claire Molloy, and Vannis Tzioumakis. 2012. *Independent American Cinema: Indie, Indiewood, and Beyond*. New York: Routledge.

Ludica. 2005. "Sustainable Play: Towards a New Games Movement for the Digital Age." http://www.ludica.org.uk/DAC_SustainablePlay.pdf. Accessed August 29, 2014.

Pearce, Celia, and Akira Thompson. 2013. "Be the Solution: Creating an Inclusive Video Game Industry: A Proactive Response to #1reasonwhy." *Be the Solution*. http://be-the-solution.tumblr.com/post/36891883562/be-the-solution. Accessed August 29, 2014.

Pedercini, Paolo. 2006. "Focus: Indie Developers Do It Better." *Videoludica*, May 8. http://www.videoludica.com/news.php?news=257. Accessed August 22, 2014.

Vogel, Jeff. 2014. "The Indie Bubble Is Popping." *Bottom Feeder*, May 21. http://jeff-vogel.blogspot.de/2014/05/the-indie-bubble-is-popping.html. Accessed May 22, 2014.

Zimmerman, Eric. 2002. "Do Independent Games Exist?" In *Game On: The History and Culture of Videogames*, ed. Lucien King, 120–129. London: Laurence King.

32 INTELLECTUAL PROPERTY

Jas Purewal

What Is Intellectual Property?

The term *intellectual property* (IP), according to the World Intellectual Property Organization (WIPO), "refers to creations of the mind, such as inventions; literary and artistic works; designs; and symbols, names and images used in commerce. IP is protected in law by, for example, patents, copyright and trademarks, which enable people to enable recognition or financial benefit from what they invent or create" (n.d.). Put another way, *intellectual property* is an umbrella term for a wide range of property rights over both intangible and tangible aspects of valuable assets such as software, films, music, and video games. In turn, *IP law* is a broad term covering a series of related but separate sets of laws that define and protect these different types of IP. From an economic perspective, IP law is about protecting and allocating the value in these property rights: there can be competing demands on IP between rival interested parties or between society and an owner, so one of IP law's primary aims from its inception was to police that boundary between the creators profiting from their inventions and society's right to benefit from the innovations.

Understanding IP is key to understanding video games because they essentially are just a bundle of IP rights with no physical existence (apart from any storage medium) and so rely on the law to delineate and enforce them. Video games are also a fast-growing and legally uncertain industry, where IP law is helpful in striking a balance between competing factions and wider society. In this chapter, I explore what IP is, how the understanding of the term has changed, and how it applies to video games.

Copyright

Copyright is the branch of IP law that protects the creative expression of an idea: a book or an artwork or a game soundtrack, for example. It is different from trademarks or patents, which, as discussed later, are more focused on protecting brand value or industrial innovation. For a video game, copyright law protects computer code, graphics, audio, textual materials, user interface, and sometimes even gameplay mechanics and the "look and feel" of a game, among other things, as separate copyrighted works bundled together.

Copyright law varies from country to country, although there are core common international principles.[1] The most basic commonality is that copyright law gives a creator rights focused around the ability to stop someone from exploiting the creator's work without authorization. These rights generally exist automatically: unlike with trademarks or patents, there is no registration *requirement* (though registration provides certain legal advantages in the United States). However, copyright is not a monopoly: the rest of society is free to create a similar or competing work or both *provided* that the new work meets certain legal tests. In particular, most countries have a test of substantiality between the two works: How similar are they to each other? The exact formulation of that test varies; in the United States, for example, it is broadly one of "*substantial similarity*"[2] (although this formulation is a judicial construct that has become very complicated and debated over the years and remains a "*sine qua non* of every copyright infringement determination [but] remains one of the most elusive in copyright law" [Lippman 2013, 515]). In the United Kingdom, the test is whether the new work copies "all or a substantial part" of the original work.[3] Even if there has been such copying, the new work will evade liability if it falls under one of the applicable defenses to copyright infringement. For example, in the United States the principle of "fair use" provides a wide-ranging copyright-infringement defense (although quite how this defense applies to video games is still being established). However, most other countries have a significantly more restrictive approach. For example, in the European Union, generally speaking the use of copyright materials for research or education purposes or potentially for private and noncommercial use can constitute one of the few defenses to copyright infringement. Copyright has a limited duration: although, again, the copyright period varies from country to country and the kind of work involved, in the West it lasts for 70 years after the author's death,[4] and during that time the author (or his or her estate) can control the work.[5]

Of course, exactly how all these aspects of copyright apply in any given situation can be debated. For example, to what extent can or should rival creators claim the ability to explore the idea of a puzzle game in which a sequence of tetramino shapes cascade down a screen and must be removed lest they fill the screen and end the game? The Tetris Company does not have a monopoly on such an idea, but copyright law does give it rights regarding its particular interpretation of the idea, which it is entitled to enforce against rival creators in court (and which it has in fact done recently against a rival mobile game clone of *Tetris*).[6] Copyright law continues to evolve rapidly in response to technological changes. One of the most (in)famous legal developments in the past decade was the advent of the US Digital Millennium Copyright Act (DMCA) of 1998 and other legislation worldwide to balance the competing demands of owners of online content such as online films and music as well as Internet service providers regarding piracy and other online IP infringement.[7] On the whole, these statutes serve this balancing act reasonably well, although there is frequent popular debate online regarding to what extent the DMCA and similar legislation stifle the free distribution of online content and potentially even free speech (Electronic Frontier Foundation n.d.).

Trademarks

Trademarks are the branch of IP law that protects brands and goodwill, such as company or product names, logos, graphics, and phrases. Such things are valuable: the Apple logo is immediately recognizable, and consumers have feelings associated with that brand. Protecting a games brand is what makes trademarks the second most important part of video games IP (after copyright). Again, the precise legal tests vary from country to country, but certain core principles hold internationally. Like copyright, trademarks give the owner the right to restrain third-party exploitation of the work but not a monopoly right: Activision cannot stop *all* third-party usages of the phrase "call of duty," nor can Sega control all use of the term *sonic*. The basic test is whether a new product or service is similar or identical to a trademarked product or service and whether there is a "likelihood of confusion" for consumers. Unlike copyright, trademarks need to be registered in order to be fully enforceable, must be renewed periodically, and must be enforced (failures can mean that the trademark will lapse or become invalid).

Patents

Patents protect industrial innovation and inventions. Unlike copyright and trademarks, they are a monopoly right but last for only 20 years. Patents traditionally were awarded to physical innovations, such as console controllers or DVD drives. Software *content* is typically not itself patentable, but in some countries (notably the United States) software functionality is patentable. As discussed later, in the early days of the industry, when hardware and software were effectively one and the same thing, patents were much less important than they are today, when software dominates over hardware.

Other IP Rights

There is a wide range of secondary IP rights, including moral rights, publicity rights, design rights, and database rights, which can be important to some games but on the whole are invoked less often than copyright, trademark, or patents.

When Did IP Law Start?

IP law can trace its roots to the rise of sixteenth-century European mercantilism. In particular, during that period it became common for states to grant traders increasingly valuable monopolies and other preferential trading arrangements, particularly in England under the reign of Queen Elizabeth I (r. 1533–1603). There inevitably came increasing legal scrutiny in such trade arrangements, resulting in England's Statute of Monopolies in 1624,[8] the precursor to modern patent law, and in the Statute of Queen Anne in 1709, now regarded as the world's first copyright statute.[9] Neither statute, however, used the term *intellectual property* or, indeed, *patents* or *copyright*, even though they were the foundational texts of IP, patent, and copyright laws. Nor in other major legal documents of the time, including the original US Constitution, is there any reference to these terms.[10] The terminology of modern intellectual property did not develop until much later.

When, exactly? Well, it is not yet entirely clear, and there is surprisingly little historical analysis of the development of IP law. Even by the late nineteenth century, there was no clear consensus on what IP consisted of or what its subsidiary branches might be. The Berne Convention of 1886, a seminal world convention on copyright law and still one of the world's key IP statutes, in its original version made no reference to intellectual property,[11] although the term *copyright* was coming into use by then.[12] The parallel Paris Convention for the Protection of Industrial Property of 1883 (a key source of international patent law) took a similar approach. This rundown is of course by no means an exhaustive analysis of historical references to IP law, but the point is that until the twentieth century the major international sources of what we now refer to as IP law did not really use the term *intellectual property*.

But let us jump forward to 1967 and the formation of WIPO,[13] now part of the United Nations. One of WIPO's stated main objectives is "to promote the protection of intellectual property worldwide." Since WIPO's formation, IP laws around the world have mushroomed, including more than 25 multinational treaties as well as regional legislation[14] and national legislation throughout the world.[15]

It can reasonably be argued, therefore, that the modern understanding of the term *intellectual property* came to the fore in the 1960s and thereafter, which is convenient for my main purpose here—to clarify the use of the term with respect to video games—because that is when the video games industry was born.

IP and Video Games

Possibly the oldest IP claim regarding video games was from two employees at the US television company Dumont, who filed for a patent over their Cathode Ray Amusement Device in

January 1947.[16] Although that technology was ultimately not pursued, this patent did begin a pattern of US patent law being used to protect the early video game innovations. Ralph Baer, one of the founding fathers of the video games industry and inventor of the Brown Box (the "*first fully-programmable, multi-player video game unit*" [Baer 1997, emphasis added]) filed a series of patents regarding video games technology from 1968 on, which were assigned to his employer Sanders Associates Inc. and in turn licensed to a company called Magnavox, which went on to release the seminal Magnavox Odyssey video games console (Lowood 2009). These patents formed the nucleus of IP law in video games, just as the first commercial successes took place in the industry—above all the rise of Atari's hit game *Pong* (1972)—which inevitably gave rise to the first video games IP lawsuits. In particular, the year 1976 saw a landmark lawsuit in which Sanders and Magnavox successfully sued Atari and other companies over allegations that the defendants' *Pong* or *Pong*-type games infringed these early Baer/Sanders video games patents (Baer 1997). Meanwhile, on other IP fronts, in 1972 Atari had filed for its first trademark (over its own name), a practice later followed by virtually all other games companies, including Activision in 1980 and Electronic Arts in 1982.[17] Regular use of copyright registration in the United States also followed.[18] All of this took place a scant few years following the founding of the WIPO and the modern international IP law regime it heralded.

Litigation inevitably expanded beyond patent matters into other forms of IP. In particular, in 1982 Atari sued Phillips over Phillips's game *KC Munchkin* (1981), which Atari claimed was a clone of—and therefore infringed copyright in—Atari's game *Pac-Man* (1980).[19] This lawsuit proved to be a seminal video games case and came at a time when there was intense competition in the video games industry that in part resulted in rampant copying of any successful game—most of all the worldwide hit *Pac-Man*, which alone was copied dozens of times by rivals. In this atmosphere, the *KC Munchkin* case was the first time that legal guidance was given on how copyright-infringement tests apply to a video game (or, indeed, to software generally), and the principles that it (and its successor cases) established continue to be applied today in the United States.

IP in Video Games in the Twenty-First Century

Into the twenty-first century, IP law is now firmly at the legal center of the video games industry. Registrable forms of IP law—trademarks in particular—have proliferated across the games industry and are becoming matter of course. New devices are patented as early as possible. Games companies use copyright law to defend themselves against piracy, copying of their games, and other unauthorized third-party use. These battles are not necessarily between rival games publishers, as they were originally in the games industry, but can be between developers and publishers,[20] solutions providers and developers,[21] even consumers and the games industry.[22]

New IP legal issues continue to arise regularly, but three types of issues apply to video games in particular. First, if a game unauthorizedly features digitized college athletes, who owns the IP, and can the athletes sue? The extent to which a real person can be replicated in a video game raises complex issues of copyright infringement, parody, fair use, publicity rights, and constitutional free speech, which have lain dormant for some years but are now being addressed in a series of US lawsuits by athletes and celebrities against major games publishers, including Activision and Electronic Arts.[23] Second, if someone misappropriates IP from within a game, is that theft? This question, which is in part a matter of IP law and in part a matter of the nascent law of virtual property, has received almost no authoritative legal attention to date, though there is some slowly developing case law. For example, in the English case *R v. Ashley Mitchell*, an individual "stole" virtual Zynga Poker chips and was given a criminal sentence for misappropriating that Zynga IP.[24] Third, if a video is made of sequences in a video game, who owns the footage legally? This was largely an academic legal question for some years, but the rise of digital video platforms such as YouTube and Twitch have brought it to the fore and added considerable commercial stakes to the balance. As yet there has been no authoritative legal decision on the matter. However, it is likely that when that moment comes, there will initially be rulings in favor of the game creator, which will over time give way to a more nuanced analysis attempting to balance the rights of the game creator, who wishes control his or her or its work creatively and financially; of the video creator, who has invested skill and labor in the video; of consumers, who want access to the video and may not care particularly about the legalities; and, finally, wider society's interest in rewarding creators but also protecting freedom of expression.

The same fundamental reasons underlying the original rise of IP law with respect to video games at the time of the Magnavox Odyssey II remain as true now in this era of iPhones and the World Wide Web. IP law exists not only to ensure that the creator of a work can be associated with the fruits of his or her labor and profit from it but also not to scare off other creators and ensure that society eventually receives benefit from the work too, so that the virtuous cycle continues. IP law is the police officer of innovation, and it is needed especially in industries as young and fast growing as the video games industry. So as video games continue to develop in complexity, maturity, and wealth-generating potential, we can expect more and more IP issues to arise in the future and therefore for an understanding of IP law in the video games industry to remain as critical as it is now.

Notes

1. See, for example, the Berne Convention at http://www.wipo.int/treaties/en/text.jsp?file_id=283698 and other copyright treaties at http://www.wipo.int/treaties/en/.

2. See *Apple Computer, Inc. v. Microsoft Corp.*, 35 F.3d 1435, 1442 (9th Cir.1994).

3. United Kingdom's Copyright Designs and Patents Act of 1988, sec. 16(3).

4. The US Copyright Office has published a guide offering more information about copyright duration in the United States (see http://www.copyright.gov/circs/circ15a.pdf).

5. In practice, problems can arise when the author can no longer be contacted or identified, which gives rise to the issue of "orphan works"—an issue that is growing over time in the video games industry as it becomes older, authors disappear, and rights fragment.

6. *Tetris Holding, LLC v. Xio Interactive, Inc.* (NJ DC, Action No. 09-6115).

7. The most significant international counterpart to the DMCA is the European Union's E-Commerce Directive (Directive 2000/31/EC of the European Parliament and of the Council of 8 June 2000 on Certain Legal Aspects of Information Society Services, in Particular Electronic Commerce, in the Internal Market), which achieved essentially similar aims for the European Union.

8. The text of the Statute of Monopolies is available at http://www.ipmall.info/hosted_resources/lipa/patents/English_Statute1623.pdf.

9. The text of the Statute of Queen Anne is available at http://www.copyrighthistory.com/anne.html. This statute weighed in at a ponderous 293 words (sadly not a characteristic followed by its modern descendants).

10. Article 1, sec. 8, of the US Constitution does state that Congress has the power "to promote the Progress of Science and useful Arts, by securing for limited Times to Authors and Inventors the exclusive Right to their respective Writings and Discoveries." See http://www.archives.gov/exhibits/charters/constitution_transcript.html.

11. The original Berne Convention wording of 1886 is available at http://keionline.org/sites/default/files/1886_Berne_Convention.pdf.

12. Copyright is referred to only twice in the Berne Convention, however, and in the appendix at that (specifically section 3 of the final protocol).

13. For more on WIPO, see its website at http://www.wipo.int/.

14. For an example of such regional legislation, see the European Union's Copyright Directive 2001/29/EC (Directive 2001/29/EC of the European Parliament and of the Council of 22 May

2001 on the Harmonisation of Certain Aspects of Copyright and Related Rights in the Information Society").

15. Such national IP legislation includes the US Copyright Act of 1976, the UK Copyright, Designs, and Patents Act of 1988, and the French Intellectual Property Code of 1992.

16. The original text of the Dumont employees patent is available at http://www.pong-story.com/2455992.pdf.

17. All these trademarks are available for view at the US Patents and Trademark Office website at http://www.uspto.gov/.

18. Unfortunately, there are no easily publicly available records of such copyright registrations because the US Copyright Office does not at present have an online public register.

19. *Atari, Inc. v. North American Philips Consumer Electronics Corp.*, 672 F.2d 607 (7th Cir. 1982).

20. For example, in 2011 Mojang, a darling of the games-development world and maker of *Minecraft* (2009), announced its second game, *Scrolls*. It was promptly sued for trademark infringement by industry publishing giant Bethesda for alleged infringement of its trademarks in its well-known game series *The Elder Scrolls*. The case ultimately settled out of court. For more on this case, see http://arstechnica.com/gaming/2012/03/bethesda-mojang-settle-trademark-dispute-over-scrolls-name/.

21. For example, in 2012 a long-running legal battle concluded between Silicon Knights (a game developer) and Epic Games (maker of the Unreal game engine, which powers a wide series of video games) over allegations regarding infringement of Epic's IP rights in the Unreal engine, among other things. For more on this battle, see http://arstechnica.com/gaming/2012/05/epic-awarded-nearly-4-5-million-in-silicon-knights-lawsuit/.

22. In particular, class-action lawsuits have become increasingly prevalent in the video games industry. At the time of writing, one of the more prominent ongoing lawsuits is against Sega and games developer Gearbox Software over claims that they engaged in misleading advertising and misrepresentations regarding their game *Aliens: Colonial Marines* (2013). For more on this lawsuit, see http://www.wired.co.uk/news/archive/2013-05/1/aliens-colonial-marines-lawsuit.

23. See, for example, the recent legal battle between Electronic Arts and the US National Collegiate Athletic Association at http://espn.go.com/espn/otl/story/_/id/11010455/college-athletes-reach-40-million-settlement-ea-sports-ncaa-licensing-arm.

24. *R v Ashley Mitchell*, Exeter Crown Court 03/02/2011. For more on this case, see http://www.gamerlaw.co.uk/2011/the-first-virtual-currency-crime-hacker-jailed-after-12m-zynga-theft/.

Works Cited

Baer, Ralph H. 1997. "Video Game History." http://www.ralphbaer.com/video_game_history.htm.

Electronic Frontier Foundation. n.d. "No Downtime for Free Speech." https://www.eff.org/issues/ip-and-free-speech.

Lippman, Katherine. 2013. "The Beginning of the End: Preliminary Results of an Empirical Study or Copyright Substantial Similarity Opinions in the US Circuit Courts." *Michigan State Law Review*, 513–565. http://digitalcommons.law.msu.edu/cgi/viewcontent.cgi?article=1010&context=lr.

Lowood, Henry. 2009. "Videogames in Computer Space: The Complex History of *Pong*." *IEEE Annals of the History of Computing* 31: 5–19.

World Intellectual Property Organization (WIPO). n.d. "What Is Intellectual Property?" http://www.wipo.int/about-ip/en/.

33 KRIEGSSPIEL

Matthew Kirschenbaum

The German word *Kriegsspiel* (literally "wargame") translates easily enough, but little else about its usage escapes complication. It denotes several things: a specific set of game rules originating in Prussia in the early nineteenth century; these rules' tangled genealogy of derivatives (typically distinguished as either "free" or "rigid" Kriegsspiel and by national characteristics—that is, British or American Kriegsspiel); *other* specific games that are about war and therefore also titled *Kriegsspiel* but are not at all the Prussian Kriegsspiel or its derivatives (sometimes indeed predating it); *and* the generic class of games about war that are wargames in a purely adjectival sense. The word thus has the obvious potential for misunderstanding whenever it appears, and it doesn't help that *Kriegsspiel* often sheds an *s* in its English orthography. Thus, the "Kriegspiel" that was famously enjoyed by computer pioneer John von Neumann was not the Prussian Kriegsspiel or its descendants, but a blind chess variant wherein each player can see only his or her own pieces. In 1970, the Avalon Hill Game Company published a board wargame called *Kriegspiel*, but it shares nothing except its genre with the nineteenth-century rules. Similarly, when Alexander Galloway coded a digital implementation of Guy Debord and Alice Becker-Ho's game *Le Jeu de la Guerre* (1977), translated as *A Game of War*, he called it "Kriegspiel" (as did Debord himself at times), though all of its concepts and mechanics originated with the Situationist duo (RSG n.d.).

This essay examines the Kriegsspiel that was introduced to the Prussian military establishment first in 1811 by Georg Leopold von Reisswitz and then in 1824 by his son, Georg Heinrich Rudolf Johann von Reisswitz. Above all others, it is the Reisswitz family's Kriegsspiel that occupies a pivotal place in game history. Joining both mathematics and the science of cartography to military theory and instruction, it brought together such concepts as representational terrain, time-based turns, asymmetrical opposing sides, hidden information, umpiring (i.e., the "gamemaster" function), scenario design, and the use of dice-driven probability tables over chesslike deterministic outcomes. Not all of these innovations originated with Reisswitz or his son, but never before had they been so effectively synthesized. Even the familiar red-and-blue color convention for opposing sides on military maps and in field exercises (which had its popular apotheosis in Rooster Teeth Productions' *Red vs. Blue* machinima series) has its origins in Kriegsspiel. The original Kriegsspiel itself is meanwhile still played by enthusiasts, with games organized through clubs and the Internet. My intent here is to offer readers some grounding in the design and conduct of the game (its core concepts and components), its relation to the military contingencies of the day, and its ongoing significance to the history of game design.

The story of Kriegsspiel's initial appearance at the court of King Friedrich Wilhelm III in 1811 is well documented and has been covered in detail by various authorities—notably Peter Perla (1990), Francis McHugh (1966), and Milan Vego (2012)—based on accounts in the *Militär-Wochenblatt*, the news organ of the Prussian officer corps, as translated by the Anglophone world's foremost Kriegsspiel authority, William Leeson. Two other recent accounts by Philipp von Hilgers (2012) and Jon Peterson (2012) are based on fresh examinations of the primary source materials.[1] The elder Reisswitz, a military adviser and strategist, arranged demonstrations of the game first to Prince Friedrich Wilhelm and then the following year to the king himself. Reisswitz had described his creation in a pamphlet tantalizingly titled *Anleitung zu einer mechanischen Vorrichtung um tactische Maneuvers sinnlich darzustellen* (Instructions for a mechanical device that represents military maneuvers to the senses). Although the game was initially played on a sand table, for the royal demonstration he devoted more than a year to the fabrication of a far more elaborate and beautiful wooden cabinet and tabletop, which remains preserved to this day in the Charlottenburg Palace in Berlin. The cabinet housed drawers filled with modular plaster tiles painted to depict fields, forests, villages, rivers, and other elements of the landscape, anticipating exactly the kind of components we see today in games such as *Settlers of Catan* (Klaus Teuber [Franckh-Kosmos], 1995). The playing pieces themselves were small wooden blocks marked with conventional military symbols—a feature that may have originated in a desire for economy but that in fact freed the game from reliance on the sort of figurines that invariably invited comparisons to chess. The game's rules covered primarily the various contingencies of maneuver and reconnaissance. Actual combat was adjudicated with the assistance of tables, an artifact of the eighteenth century's enthusiasm for military mathematics.

Reisswitz's game did not, of course, emerge from a ludic vacuum. Its ancestors include the wide variety of "war chess" variants in vogue in central Europe for much of the seventeenth and eighteenth centuries, including one self-described as "Kriegsspiel" by Johann Christian Ludwig Hellwig and introduced at the court of the duke of Brunswick in 1780. Hellwig's game proved popular among the nobility, and in 1797 it was expanded upon by Johann Georg Julius Venturini, also a Brunswicker, who introduced the "New Wargame" (Neues Kriegsspiel). This game was most typically played on a 60-by-36-inch grid, with each square representing a stylized depiction of the major terrain type in an area on the Franco-Belgian border. Units represented infantry, cavalry, fortresses, and supply trains. (The game was also accompanied by a 230-page rulebook and so in that respect was not dissimilar to the board wargames that would have their heyday almost two centuries later!) Although both Hellwig and Venturini thus figure as immediate precursors to the Prussian Kriegsspiel, both fell short of Reisswitz's key innovation, which was to decouple the grid of the game from the positioning of the pieces. That is, Reisswitz's plaster tiles were strictly a contrivance for configuring the game's terrain—in actual play, the pieces representing troop formations were not confined to the grid of abutted squares. As with the unadorned playing pieces, this innovation helped distance the game from the more stylized war chess tradition. Reisswitz also followed Hellwig and Venturini in encouraging the centrality of the concept of the *scenario* to wargaming, the contingent and temporary configuration of the recombinant board together with specific force allocations and objectives collectively determining the possibility space for the game. Players might be charged with escorting a supply train, for example, or safeguarding a bridgehead or attacking (or defending the flanks of) an advancing troop column. A scenario, in other words, was what leveraged wargames out of an exclusively ludic sphere and into the realm of simulation. Jösef Kostlbauer puts it this way: "This amazing wooden box can be regarded as a kind of analog computer allowing for a great variation of battlefield conditions and tactical moves" (2013, 174).

Though Reisswitz's game was admired and enjoyed by Friedrich Wilhelm III and his immediate circle, the "amazing wooden box" was not practical to reproduce, and so it saw no further development or uptake for more than a decade, until a second major iteration was published in 1824 by Reisswitz's son, then a lieutenant of artillery, as *Anleitung zur Darstellung militairischer Manöver mit dem Apparat des Kriegs-Spieles* (Instructions for the representation of military maneuvers with the wargame apparatus). This is the version of Kriegsspiel that became the basis for the many future variants and that is still available today in modern editions. The rectilinear lead playing pieces again served as indexical representations of troop formations whose physical dimensions, when placed on the map, corresponded to standard tactical deployments—lines, column, square—at scale. But unlike its predecessor, the 1824 game was played on actual 1:8,000 military topographic maps, themselves a critical intervention in the conduct of war (the importance of such maps in Napoleon's then just-concluded campaigns has been widely documented). This meant that a Kriegsspiel could be arranged to rehearse operations over practically any real-world terrain—a capability that completed the transformation of the game from the realm of abstract symbolic representation to a working model, with space and time (each piece could advance only over the ground it could cover in two minutes of actual maneuver) unified in a consistent schema. Crucially, Reisswitz the younger's Kriegsspiel also heavily encouraged umpired play conducted in teams, with the participating officers responsible for their individual subordinate commands (usually at the battalion level) and no communications permissible except in writing. Only the impartial umpire had complete knowledge of the forces and their dispositions. The 1824 Kriegsspiel thus brought to the fore so-called soft factors, namely command and control and the proverbial fog of war—the very "friction" that one of its devotees, Carl von Clausewitz, was shortly to write about so influentially.

This emphasis on soft factors is above all what served to distance the 1824 Kriegsspiel from its progenitors by Hellwig, Venturini, and the elder Reisswitz. Although the umpire's authority was absolute (not coincidentally, he was typically the ranking officer), the 1824 rules also introduced greater structure and formality, particularly in the area of combat results. The results were now adjudicated by rolling specially marked dice, which ensured that the effectiveness of an individual volley of fire could deviate by greater or lesser extents from a mean (derived from weapons efficacy as measured by actual field tests), while also covering such contingencies as the terrain and distance to the enemy, the number of ranks the troops were arranged in, and the particulars of their armaments (see figure 33.1). In employing the dice for this function, the younger Reisswitz broke decisively with the determinism of chesslike strategy games while also anticipating the Combat Results Table that would become ubiquitous in the twentieth-century wargaming hobby after its introduction by Charles S. Roberts in 1954—then an innovation sufficiently close to the Monte Carlo mathematics in use in Cold War think tanks that it earned Roberts a summons from the brain trust at RAND, or so goes a widely circulated hobby story. Similar procedures were put into place for resolving artillery fire and hand-to-hand combat.

Kriegsspiel showcased a distinct approach to the conduct of military operations, one focused neither on grand strategic considerations (the theater-level affairs of whole armies and fronts) nor on the tactical minutiae of drills and formations. Rather, the Kriegsspiel emphasized the art of maneuver or what would later be called *Bewegungskrieg*: the finding and fixing of the enemy, the evolution of a battle plan, and the advance to contact. The game was

Figure 33.1

Plate from Georg Heinrich Rudolf Johann von Reisswitz's manual *Anleitung zur Darstellung militairischer Manöver mit dem Apparat des Kriegs-Spieles* (1824), displaying the mechanism for resolution of combat. Results for five different 6-sided dice (numbered I–V) are illustrated—the choice of die was made by the umpire based on his determination of the type and likely success of the combat. Each face of each die had multiple columns of results representing different ranges and other circumstances. Die I, for example, was used for firing by formed infantry and skirmishers, but it was also used for hand-to-hand combat when the odds were deemed about equal for both sides as well as to see if a fire broke out when cannon were employed against a village. *Source*: http://web.archive.org/web/20140112010022/http://bitzkrieg.net/archive/872, scanned from the copy in the University Library in Bayreuth.

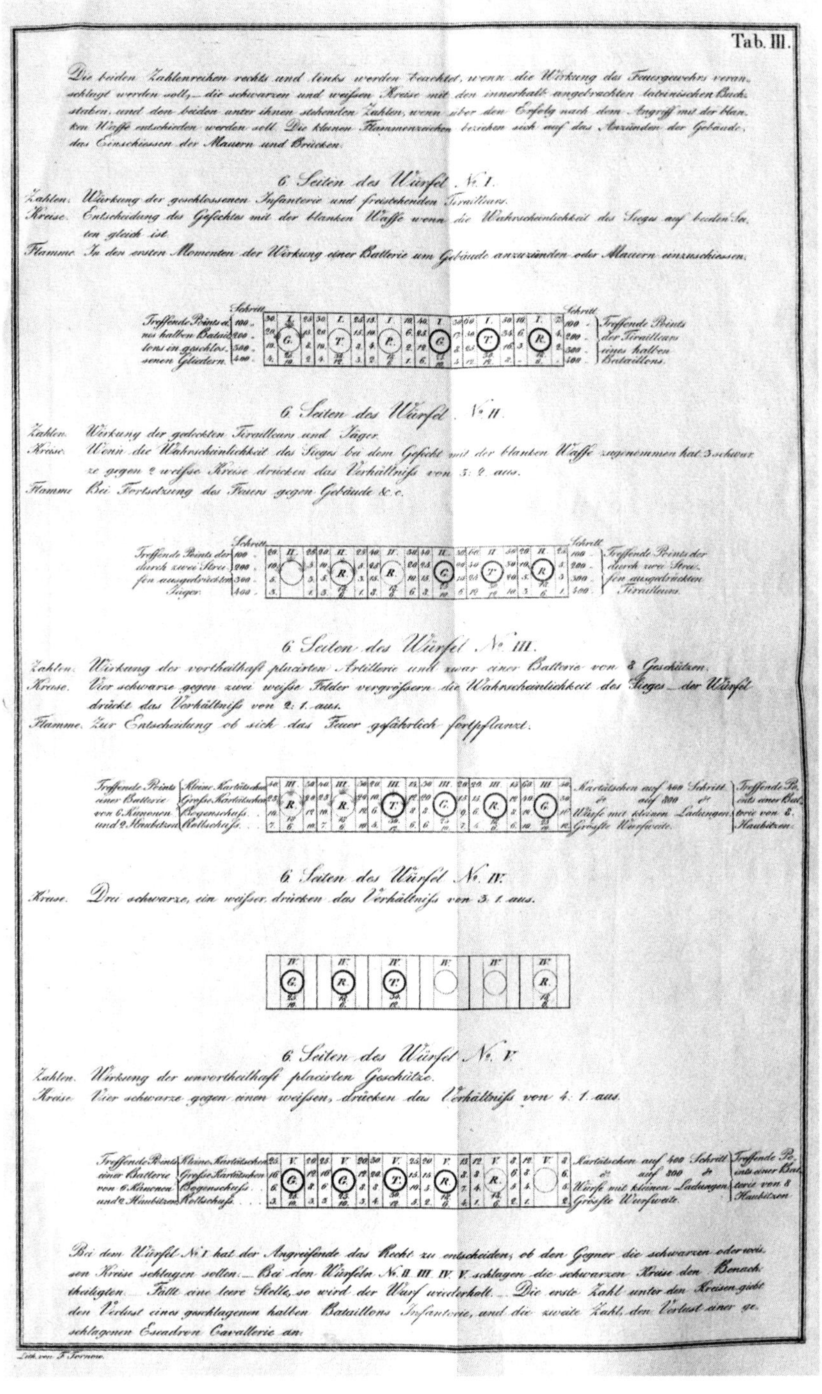

Tab. III.

Die beiden Zahlenreihen rechts und links werden beachtet, wenn die Würkung des Feuergewehrs veranschlagt werden soll, – die schwarzen und weißen Kreise mit den innerhalb angebrachten lateinischen Buchstaben, und den beiden unter ihnen stehenden Zahlen, wenn über den Erfolg nach dem Angriff mit der blanken Waffe entschieden werden soll. Die kleinen Flammenzeichen beziehen sich auf das Anzünden der Gebäude, das Einschiessen der Mauern und Brücken.

6. Seiten des Würfel № I.

Zahlen: Würkung der geschlossenen Infanterie und freistehenden Tirailleurs.
Kreise. Entscheidung des Gefechtes mit der blanken Waffe wenn die Wahrscheinlichkeit des Sieges auf beiden Seiten gleich ist.
Flamme. In den ersten Momenten der Würkung einer Batterie um Gebäude anzuzünden oder Mauern einzuschiessen.

Links: Treffende Points eines halben Bataillons in geschlossenen Gliedern. Rechts: Treffende Points der Tirailleurs eines halben Bataillons.

Schritt	I. (G.)		I. (T.)		I. (P.)		I (G.)		I. (T.)		I. (R.)		Schritt
100	30	25	30	25	15	10	40	30	60	50	10	7	100
200	20	15	20	15	10	6	25	17	30	34	6	4	200
300	10	8	10	8	4	2	12	8	25	16	3	2	300
400	4	2	4	3	2	1	6	4	12	8	–	–	400
Kreis	25/10		35/12		15/6		25/10		30/12		15/6		

6. Seiten des Würfel № II.

Zahlen. Würkung der gedeckten Tirailleurs und Jäger.
Kreise. Wenn die Wahrscheinlichkeit des Sieges bei dem Gefecht mit der blanken Waffe zugenommen hat. 3 schwarze gegen 2 weiße Kreise drücken das Verhältniß von 3: 2. aus.
Flamme. Bei Fortsetzung des Feuers gegen Gebäude &c.

Links: Treffende Points der durch zwei Streifen ausgedrückten Jäger. Rechts: Treffende Points der durch zwei Streifen ausgedrückten Tirailleurs.

Schritt	II. ( )		II. (R.)		II. (R.)		II. (G.)		II (T.)		II. (R.)		Schritt
100	20	25	20	25	40	30	50	30	60	50	20	25	100
200	10	5	10	5	25	20	25	20	30	50	10	5	200
300	5	5	5	3	15	10	15	15	25	40	5	3	300
400	3	1	3	1	8	6	3	5	12	10	3	1	400
Kreis			15/6		15/6		25/10		30/12		15/6		

6. Seiten des Würfel № III.

Zahlen. Würkung der vortheilhaft placirten Artillerie und zwar einer Batterie von 8 Geschützen.
Kreise. Vier schwarze gegen zwei weiße Felder vergrößern die Wahrscheinlichkeit des Sieges – der Würfel drückt das Verhältniß von 2: 1. aus.
Flamme. Zur Entscheidung ob sich das Feuer gefährlich fortpflanzt.

Links: Treffende Points einer Batterie von 6 Kanonen und 2 Haubitzen. Rechts: Treffende Points einer Batterie von 8 Haubitzen.

	III. (R.)		III. (R.)		III. (T.)		III. (G.)		III. (R.)		III (G.)		
Kleine Kartätschen	40	30	40	30	20	15	30	20	20	15	60	50	Kartätschen auf 400 Schritt
Große Kartätschen	25	20	25	24	10	12	20	15	15	12	40	30	do auf 800 do
Bogenschuß	16	12	10	12	6	8	8	9	6	8	10	10	Würfe mit kleinen Ladungen
Rollschuß	7	10	7	10	4	6	6	7	4	6	10	10	Grösste Wurfweite
Kreis	15/6		15/6		30/12		25/10		15/6		25/10		

6. Seiten des Würfel № IV.

Kreise. Drei schwarze, ein weißer drücken das Verhältniß von 3. 1. aus.

IV.	IV.	IV.	IV.	IV.	IV.
G.	R.	T.			R.
25/10	15/6	30/12			15/6

6. Seiten des Würfel № V.

Zahlen. Würkung der unvortheilhaft placirten Geschütze.
Kreise. Vier schwarze gegen einen weißen, drücken das Verhältniß von 4. 1. aus.

Links: Treffende Points einer Batterie von 6 Kanonen und 2 Haubitzen. Rechts: Treffende Points einer Batterie von 8 Haubitzen.

	V. (G.)		V (G.)		V. (T.)		V. (R.)		V. (R.)		V. ( )		
Kleine Kartätschen	25	20	25	20	30	25	20	15	12	8	12	8	Kartätschen auf 400 Schritt
Große Kartätschen	16	12	16	12	20	15	15	8	8	6	8	6	do auf 800 do
Bogenschuß	6	8	6	8	8	10	5	7	4	5	4	5	Würfe mit kleinen Ladungen
Rollschuß	3	5	5	3	4	5	2	4	1	2	1	2	Grösste Wurfweite
Kreis	25/10		25/10		30/12		15/6		15/6				

Bei dem Würfel № I hat der Angreifende das Recht zu entscheiden, ob den Gegner die schwarzen oder weissen Kreise schlagen sollen. – Bei den Würfeln № II. III. IV. V. schlagen die schwarzen Kreise den Benachtheiligten. – Fällt eine leere Stelle, so wird der Wurf wiederholt. – Die erste Zahl unter den Kreisen giebt den Verlust eines geschlagenen halben Bataillons Infanterie, und die zweite Zahl, den Verlust einer geschlagenen Escadron Cavallerie an.

Lith. von F. Tornow.

thus a rehearsal of the kind of midlevel staff work that was necessary for the effective management of a modernized military campaign. Actual warfighting, in other words, would involve the same maps, the same study of lines of march and approach, the same analysis of terrain, the same plotting of contingencies and writing of orders as occurred in the Kriegsspiel. The corresponding sensory referents for the game were less the smell of gunpowder and the screams of horses and men than the scratching of the pen and the scent of a freshly blotted page. So it was that when the junior Reisswitz arranged a demonstration for Karl von Muef-fling, the chief of the Prussian General Staff, the latter is recorded to have exclaimed, "It's not a game at all! It's training for war! I shall recommend it enthusiastically to the whole army" (qtd. in Perla 1990, 26).

Unlike his father's game, the younger Reisswitz's Kriegsspiel lent itself to mass production. Complete sets, consisting of a copy of the rules, the lead playing pieces, the specialized dice, and suitable maps were distributed throughout the Prussian military, and play of the game proved popular. Johann himself, however, was to meet a sadder fate: professional enemies eventually saw him transferred to a remote outpost, and, believing his career at an impasse, he killed himself with his sidearm. In the absence of a dedicated steward, the inevitable happened as players tinkered with the game, evolving ever more baroque rules and procedures in the elusive pursuit of realism. Advances in tactics and weaponry likewise meant additional layers of detail to keep pace with the times. Gamers will be gamers, and Kriegsspiel's adherents thus gradually splintered into factions favoring either "rigid" Kriegsspiel or "free" Kriegsspiel, the former emphasizing an elaborate rules-based approach and the latter the umpire's decision making. The best-known proponent of free Kriegsspiel was Julius von Verdy de Vernois, who in 1876 published *Beitrag zum Kriegsspiel* (Contribution to the wargame). Wilhelm von Tschischwitz's "rigid" version developed in 1862, *Anleitung zum Kriegsspiel* (Instructions for the war game), became the basis for both the British and the American adaptations of Kriegsspiel that appeared, including Charles A. L. Totten's *Strategos* (1880), which was in turn to become an influence on *Dungeons & Dragons* (Gygax and Arneson/Tactical Studies Rules, 1974).

Here is not the place to sort the tangled genealogy of *Dungeons & Dragons* (Peterson 2012 is the definitive account), but we can at least observe that the distinction of so-called free and rigid Kriegsspiel seemingly anticipated much of what was to follow for the tradition of tabletop history and adventure gaming that proved so popular in the second half of the twentieth century. Rigid Kriegsspiel saw its direct descendants in hobby wargame publishers such as Avalon Hill, Simulations Publications, Inc., and Game Designers' Workshop, and it retains its disciples today, not only in the continued play of the original Reisswitz rules (translated by William Leeson, they are available in an attractive electronic edition[2]) but also in games such as *Advanced Squad Leader* (Avalon Hill, 1985), whose manual is hundreds of pages long to cover every conceivable contingency of World War II small-unit combat. Free Kriegsspiel, meanwhile, can be seen as the forerunner to the role-playing tradition over which the gamemaster reigned supreme—aided by dice and charts, to be sure, but ultimately arbitrating the fates of players with equal measures of whit and whim. Professional policy gaming likewise frequently takes the form of so-called BOGSAT (Bunch of Guys [and Girls] Sitting around a Table) games—essentially free-form role-playing exercises. Wargaming in both of these guises has even infiltrated the business world: BOGSAT sessions are a mainstay of corporate strategy retreats, and a consulting firm in the United Kingdom has seized upon the original Reisswitz Kriegsspiel as just the instrument to instruct future business leaders in timeless principles of command, control, and communication—participants in their all-day seminars are promised the opportunity to "plan and conduct a military campaign against the other team using the blocks and maps of the Prussian strategic leadership development tool in exactly the same way as it was used almost 150 years ago."[3]

We habitually associate the Prussian state with a legacy of martial prowess, dating from at least the Brandenburg army of Frederick the Great. In fact, however, the Kriegsspiel made its debut during a period of crisis and malaise in the Prussian military establishment less than a generation after its decisive defeat by Napoleon at the Battle of Jena in 1806. Kriegsspiel subsequently found its application as part of a totalizing scheme of reform and modernization, one that incorporated a wide variety of training regimens for the officer corps as well as Europe's most advanced body of theoretical writing on warfare, including Clausewitz's contributions. Clearly it would be a mistake to attribute the resurgence of Prussian arms in the European wars of the later nineteenth century solely to the influence of the Reisswitz game; nonetheless, Kriegsspiel did play a documented role in the planning of several of that century's pivotal military campaigns, and the tradition of instructional practices it inaugurated remained widely used through World War II. These institutionalized practices in fact are the "sand table" exercises of the General Staff that Friedrich Kittler invoked to gnomically frame his analysis of modern media on the opening page of *Gramophone, Film, Typewriter* ([1986] 1999); the book's famous first line, usually translated as "media determine our situation," has specific resonance in the construct of the wargaming situation or scenario—in German, *Lage*. Both the German way of war (adopting the phrase introduced by military historian Robert M. Citino [2005]) and the German way of *wargaming* thus emerged from a unique set of geopolitical, intellectual, and technoscientific contexts whose inheritance to this day is Kriegsspiel's uniquely ludic episteme, arguably the very *situation* of modernity.

Notes

I am deeply grateful to Jon Peterson for his careful reading of this essay as well as to Henry Lowood for his help with the translations. Any errors, of course, are my own.

1. My own account of the Prussian Kriegsspiel is synthesized from these various sources.

2. Available at http://toofatlardies.co.uk/Kriegsspiel.html.

3. The advertisement for Poppyfish's seminar "Business Kriegsspiel" is available at http://www.poppyfish.co.uk/Poppyfish_Business_Kriegsspiel_Flyer_3pager.pdf.

Works Cited

Citino, Robert M. 2005. *The German Way of War: From the Thirty Years' War to the Third Reich*. Lawrence: University of Kansas Press.

Kittler, Friedrich. [1986] 1999. *Gramophone, Film, Typewriter*. Trans. G. W. Young and M. Wutz. Stanford, CA: Stanford University Press.

Kostlbauer, Jösef. 2013. "The Strange Attraction of Simulation: Realism, Authenticity, Virtuality." In *Playing with the Past: Digital Games and the Simulation of History*, ed. Matthew Wilhelm Kapell and Andrew B. R. Elliott, 169–183. New York: Bloomsbury.

McHugh, Francis J. 1966. *Fundamentals of War Gaming*. 3rd ed. Newport, RI: US Naval War College.

Perla, Peter P. 1990. *The Art of Wargaming*. Annapolis, MD: Naval Institute Press.

Peterson, Jon. 2012. *Playing at the World: A History of Simulating Wars, People, and Fantastic Adventures from Chess to Role Playing Games*. San Diego: Unreason Press.

RSG. n.d. Radical Software Group—Kriegspiel. http://r-s-g.org/kriegspiel/about.php.

Vego, Milan. 2012. "German War Gaming." *Naval War College Review* 65 (4): 106–147.

Von Hilgers, Philipp. 2012. *War Games: A History of War on Paper*. Cambridge, MA: MIT Press.

34 MACHINIMA

Jenna Ng

Like so many discoveries, the term *machinima* emanated from accident. Its origins are attributed to Anthony Bailey (of *Quake Done Quick* fame), who in 1998 had initially proposed the word *machinema*, inventing it as a portmanteau of *machine* and *cinema*. However, it was subsequently misspelled and disseminated as *machinima*, and the new spelling stuck. Happily, Bailey (2007) later approved because he also liked in the new term what he called the "derivation" from the word *anima* (meaning "life"); other scholars have also commented on the relevance of animation (as implied in the term) for machinima in terms of their similar aesthetic (Quaranta 2009).

Among this host of connotations in the neologism, the machine remains the singular element for understanding machinima, which are commonly defined as films made by real-time three-dimensional (3D) computer graphics–rendering engines (or, as Bailey put it, "pieces of cinema that are made using 3D engines" [2007]). The term *machine*, then, points to the complex amalgam of computing hardware, accessories, and software as well as the mechanical nature of the "engine" that facilitates machinima's creation. The key to making machinima is the targeted use of rendering software; in that sense, machinima is unlike both animation, which builds its imagery with geometrical shapes, and live-action cinema, whose images are recorded primarily by a camera in front of a live event or actor. In comparison, machinima redeploys preexisting rendering engines to generate its imagery, sound software to record and synchronize its soundtrack, and editing software to piece it together into a desired narrative. It is born-digital media, produced entirely on a computer. To that extent, the sense of the "machine" runs through the entirety of its creative conception, development, and ontology.

It should be no surprise, for several reasons, that video games were the first engines used to make machinima. First, the development of 3D game engines in the early 1990s, heralded first by *Wolfenstein 3D* (id Software, 1992) and then by *DOOM* (id Software, 1993), had achieved unprecedented sophistication; with such advances, the ground was technologically laid for using game engines to create 3D imagery. Second, the practice of modification inherent in redeploying video game engines to make films lies within a tradition of subversion that runs deep in gameplay, which can be traced to themes of subversive play and entertainment in Dadaist and surrealist movements. Such subversion can be seen in methods of surrealism such as automatism: for example, the language game of "automatic writing," as a play against literary codes, instructs you to "sit at a table with pen and paper, put yourself in a 'receptive' frame of mind, and start writing. Continue writing without thinking about what is appearing beneath your pen. Write as fast as you can" (Gooding and Brotchie 1995, 17). As Mel Gooding and Alastair Brotchie write, "It was through games, play, techniques of surprise and methodologies of the fantastic that [the surrealists] subverted academic modes of enquiry, and undermined the complacent certainties of the reasonable and respectable" (1995, 10). Third, game design and software also facilitated such modification: *Quake*, released by id Software in 1996, quickly became subject to modification as "it became clear that id would provide more information and tools for modifying *Quake* than they had for *Doom*" (Lowood 2006, 31). Finally, added to this mix is the development of a competitive multiplayer game culture in the mid-1990s, particularly where players teamed up to battle out death matches. Indeed, as Henry Lowood writes, "machinima comes out of—in fact would have been impossible to imagine outside of—the context of competitive multiplayer games" of the 1990s (2009). Although multiplayer gaming first started with *DOOM*, which allowed up to four players to play, such gaming rapidly grew in popularity with id's subsequent titles, such as *DOOM II* (1994) and in particular *Quake*.

It is also worth noting that before the term *machinima* was coined, the recording of gameplay was called "*Quake* movies" specifically because of the connection between *Quake* and the phenomenon of recorded gameplay within this game's community and playing teams. By the end of the decade, by far the most popular games on the scene were team-based games such as the *Quake* sequels (*Quake II* [1997]; *Quake III: Arena* [1999]), *Unreal* [Epic MegaGames/Digital Extremes, 1998], *Half-Life* [Valve, 1999] and its sequels, and *Battlefield 1942* [Digital Illusions, 2002]). In this fusion of modification, recording technology, team-based play, and the desire to show off skilled gameplay, players began to document game techniques as recorded imagery for exhibition and circulation, initially made on demos (recorded films playable only with the engine installed in the computer) and later generated or "recammed" as moving-image files.

Machinima thus developed out of these initial recordings of virtuoso gameplay, beginning in the mid- to late 1990s, with titles such as *Quake Done Quick* and *Quake Done Quicker*, which recorded the game *Quake* being completed as quickly as possible, including the uses of any shortcut or trick but without cheats. Yet by virtue of their having been generated by a game engine rather than through animation or recorded live action, these moving-image recordings are unprecedented in terms of ontology. In terms of genre, these machinima are visual documentation of a new reality—namely, the dynamic situations that emerge and unfold in game worlds. In that respect, they can be compared to *La sortie des usines* (1895) by the Lumière brothers. Arguably the first motion picture ever made, *La sortie* was a recording of workers leaving the Lumière factory and as such a recording of a reality in the Lyon of 1895. A new medium began to take shape then, and so with the recording of gameplay the new medium of machinima likewise emerged.

Besides gameplay recording, machinima during this time was also marked by other developments. In 1996, a group of players called the Rangers hacked *Quake* to make a short film called *Diary of a Camper*. This film differs from the recordings that preceded it in two significant ways. First, it does not record gameplay. Rather, it consists of an independent narrative: two Ranger members from a team of five are sent ahead to scout a room but are ambushed by a waiting camper and killed. As revenge, the remaining three Rangers return fire and kill the camper. Although gaming elements are present in this machinima (it takes place within the *Quake* game space, and there are gaming references, such as the camper), the players were not playing the game while making this machinima. Second, the machinima was recorded from a third-person visual perspective, clearly departing from the first-person shooter outlook of *Quake*. Consisting of a single uncut shot, *Diary* begins with a static high-angle shot showing all the player-characters before the camera turns and tracks through a connecting corridor to the room where the ambush takes place, recording the attack as yet another high-angle shot; it makes this journey twice more as the rest of the Rangers return to avenge their fallen comrades. The unbroken camera movements underscore the real-time nature of the game, yet its independent angles present the action as one clearly separate from gameplay, which takes place from a first-person perspective.

Machinima as presented by *Diary* thus profoundly changed the nature of video games, from being a media object with a specific goal (i.e., to play the game and to achieve its stated purpose according to its rules) to being used to make *Diary*. The video game essentially became an open-ended tool kit for creative use with purposes removed from its original goal of gameplay and unrealized by the game developers themselves. In view of this subversion, Lowood, paying homage to Marcel Duchamp's "found object," calls machinima "found technology": like Duchamp's objects, the meaning, definition, and significance of the video game is transformed when it is transferred from the realm of gaming to the realm of filmmaking, a shift that creates a deliberate slippage between gameplay (part of game reality) and fictional cinema and places a paradoxical confusion at the heart of what each now means and how each is defined. For example, *Leeroy Jenkins* (2006), one of the most popular machinima with more than 44 million hits to date and counting, shows what appears to be genuine gameplay in *World of Warcraft* (Blizzard Entertainment, 2004). Members of the guild PALS FOR LIFE are discussing their strategy for defeating their enemies in the next room when Leeroy, who is absent during the discussion and thus ignorant of the battle plan, leaps ahead into the attack with his now-famous war cry ("Leeerooy Jenkiiiins"); his fellow horrified guild members quickly follow him and are promptly wiped out (as the machinima ends with one the of the players commenting, "Leeroy, you're stupid as hell").

However, *Leeroy Jenkins* was not actually played in a game but choreographed, acted, and scripted with some improvisation. The "playing of the game" to make the film creates an ambiguity in the definitions of both gameplay and filmmaking: Were the guild members playing the game, or were they making a film? Or were they only pretending to play the game in order to make a film? The machinima *Leeroy Jenkins* shades into all three questions. Another well-known machinima, *Serenity Now—Crash a Funeral in Winterspring* (2008), which has more than 6 million hits so far, is a film of how a *World of Warcraft* guild called Serenity Now raided the funeral of a player who had died in real life and whose funeral was now held not only in the game world but in a contested zone of a PvP server. The guild members of Serenity Now subsequently decimated every player who had turned up for the funeral. Although the attack serves as an anthropological textbook case for behavior in the game world versus behavior in the real world, it also demonstrates the ambiguity between game and cinema. The attack was part of the guild's gameplay, but it was at the same time specifically recorded as film footage to be made into a video, which shows cinematic sophistication in the use of editing, crosscutting, and contrasting pieces of nondiegetic music. If gameplay had previously been relatively straightforward—the controlled fulfilling of the game's prescribed objectives—machinima, with its facilitation of filmmaking via the game engine, now creates a definitive ambiguity between different media forms.

If anything, these two media may even be made to overlap, whereby fictional cinema creation is specifically made a part of gameplay (i.e., playing with a game). For example, Valve included Faceposer software with its game *Half-Life 2* (2004) "so that machinima creators could tweak the facial expressions of characters" (Thompson 2005). Such capabilities are now not limited to the games themselves; gaming consoles have followed suit. The Xbox 360, for instance, is not just a machine for playing games but comes with Game DVR to record gameplay video clips, so any previous 30 seconds of gameplay can be recorded by using the app or simply by saying, "Xbox, record that!" Where companies are directly bringing machinima practice into games' play environment, the message is that the nature of the game is open to recording and filmmaking, thus not only diffusing the focus of the medium on gameplay but also diversifying the medium's uses, purposes, and ends. In this sense, *machinima* is more than a term that captures the technical aspects of the various media involved; it is also a social and cultural concept relating emerging technologies and evolving practices.

The recording and creating of films in game worlds also expands those worlds. Machinima not only introduces games to nongaming communities (including both machinimators and machinima viewers) but also stretches the possibilities of that world and of being in it. Michael Nitsche calls this development "outside-in," or "the use of game engines as tools for traditional animation and story-telling independently from games," as opposed to "inside-out," which refers to "game players using machinima as expression and recording of their play" (2007). This distinction also outlines the two broad creative paradigms of machinima—namely, films that are independent of the game and its reality, including gameplay ("outside-in"), and films that record gameplay, and that refer to the game even if they are not explicit recordings of gameplay ("inside-out").

"Outside-in" machinima cover a wide aesthetic and technological range. For example, Friedrich Kirschner's machinima *The Journey* (2003), made with the game *Unreal Tournament* (Epic Games/DigitalExtremes, 1999), was so heavily modified that it in no way resembles the game. The aesthetic is stark and minimalist with largely black-and-white, simple shapes (for example, people are represented by stick figures and TV screens by coarsely drawn squares), in contrast to the visual intricacies and detail of the game itself. However, *The Snow Witch* (2006), made with the game *Sims 2* (Maxis, 2004) to retell a Japanese ghost story adapted from Lafcadio Hearn's tale *Yuki-Onna* (*Kwaidan*), featured few changes, and even those were made merely to lighting and coloring. Nevertheless, such machinima do not refer to the games from which they originate and instead use the game engine to generate images to express nongaming narratives. Other examples include reenactments, such as *A Few Good G-Men* (2005) by Randall Glass, which reproduces the climactic courtroom scene from the film *A Few Good Men* using the *Half-Life* game engine, and original stories, such as *The Awakening* (2005), a self-reflective machinima made with *The Sims* (EA Maxis, 2000) about a Truman-esque character who investigates his suspicion that he lives in an unreal world controlled by someone else (which of course he is).

"Inside-out" machinima, in contrast, rely largely on gaming references and inside jokes to underpin their narratives; they unsurprisingly tend to be made by gaming fans. For example, *Halo Warthog Jump* (2002), also by Randall Glass and called a "Halo Physics Experiment," features a series of Warthogs (light reconnaissance vehicles used in the game *Halo* [Bungie, 2001]) being launched into the air using dropped grenades. Edited into a tightly timed series and accompanied by the music of the Frank Sinatra tune "Fly Me to the Moon," this machinima achieves an almost balletic grace from the visual spectacle of armored vehicles flying high and falling fast. The machinima is premised on *Halo* not only for how it is created (by *Halo's* game engine) but for its very meaning. The visual extravaganza of flying vehicles makes sense only in the context of understanding (and marveling) how these Warthogs are launched—namely, by exploding dropped grenades: the grenades are the specific game realities that cause the Warthogs to "behave" in such a way. These details matter to any *Halo* gamer: the grenades get dropped, as Glass explains, only when a player shoots his or her own soldiers: "shoot 5 of your guys (dropping about 8 grenades total). They usually drop 0–2 grenades when they die" (interviewed in Halo Nation n.d.). Other methods for making a Warthog jump include "the Sputnik

Skull" and putting a Warthog "on top of a pile of Fusion coils." It is only in this context of the game that flying Warthogs make sense and the humor of that action is carried across.

However, "outside-in" and "inside-out" point to extended ends of a spectrum, and many machinima occupy places within that range. One example is the *Red vs Blue* (*RvB*) series by Rooster Teeth, possibly the most successful machinima to date. Created in various versions of the *Halo* game, including *Halo 2* (2004), *Halo 3* (2007), and *Halo: Reach* (2010), the series spans 11 full seasons, 5 mini-series, and more than 170 episodes, each ranging from 5 to 10 minutes in length, and is available both over the Internet and on DVD. More significantly, perhaps, following Rooster Teeth's machinima success, Microsoft hired it to produce more *RvB* videos to advertise *Halo* in game stores (Thompson 2005), even commissioning it to "produce an episode to run during a big software-developer conference; the clip recounts a fictionalized falling-out between Microsoft CEO Steve Ballmer and one of the game's soldiers" (Delaney 2004). These endorsements are telling, if anything, of machinima's commercial (and lawsuit-free) acceptance in the gaming and software industry—and across other entertainment industries as well: *RvB* videos are even featured in concerts by the rock group Barenaked Ladies. *RvB* relates the fictional adventures of the Red and Blue Teams as they engage with each other in war. Although there are a few nods to the game, such as indirect references to Smart AI, the Human–Covenant War, and Forerunners, as well as to gameplay, such as the introduction of a few characters as "rookies" and using "Capturing the Flag" as the basis of one plotline, *RvB*'s storyline is largely independent of any narrative extant in *Halo*, a decision made, as its makers explain, so as not to alienate nongamers (Delaney 2004). Yet in another sense it may be argued that the *entire* premise of the *RvB* series is the game, so that, even without explicit reference, *RvB* never departs very far from *Halo*. Much, if not all, of *RvB*'s humor, irony, and meaning is derived from the gameplay, which centers on two teams of faceless fighting machines whose goal is to decimate each other. In *RvB*, however, the team members are instead infused with vivid personalities, back-life stories, zany adventures, and Beckett-esque moments of wondering about their very existence as faceless fighting machines whose goal is to decimate each other. Much of these meanings would be lost if the game structure did not underpin the series.

A significant consequence of *RvB*'s popularity and success is its heralding a shift in thinking about copyright from the industry. Alongside practices such as sampling and remix, machinima similarly appropriates intellectual property (such as unlicensed footage and game assets) from the game company and technically faces cease-and-desist legal threats. However, the game developers at Bungie, the Microsoft subsidiary that developed *Halo*, were fans of the series and offered support, going as far as to add a button in future *Halo* releases that lowered a player's gun, an act that has no value in the game but eases the making of machinima (Delaney 2004). Today, most game developers and virtual-world owners generally recognize the legality of making machinima with their game engines, albeit with the significant caveat that their ownership extends to the machinima created.

Not all machinima is made with subverted preexisting game engines; many have also been made with game engines that were themselves developed specifically *to* make machinima. An early example of such an engine is Machinimation, developed by Katherine Anna Kang using *Quake III*'s engine, which allows one to set up and control all elements of a scene by oneself. With Machinimation, Kang made a film titled *Anna*, a visual depiction of a flower's life cycle from a seed to a withered husk, which won a Best Technical Achievement award at the Machinima Film Festival in 2003. In this sense, programs such as Machinimation move away from the original "machine" premise of machinima, in which the machinima maker relies on skills not originally part of filmmaking—hacking and subverting a game—to make films. All that is needed with machinima engines is the mastering of a program that *is* specifically purposed to make films. Possibly the most famous example of a machinima created in this way is *The French Democracy* (2005), which Alex Chan made with *The Movies* (2005), a business-simulation game from Lionhead Studios. Although *The Movies* is not expressly a filmmaking program, the goal of the game is to manage a film studio, which includes creating movies and thus in that sense has a dedicated engine for machinima. *The French Democracy* relates the anger and frustration of three Moroccan men in France who participated in civil unrest after facing various forms of discrimination. In the wake of race riots in Paris in 2005, *The French Democracy* had a huge impact, attracting media attention and controversy from publications such as *Libération* and the *Washington Post* for its suggested connection between racism and the Parisian riots (Varney 2007). It also raised the profile of video games as tools for making political statements, with many commentators remarking on the video's emotional strength and authenticity in its take on the issue of the disenfranchisement of foreigners in France, as in Mathieu Kassovitz's film *La Haine* (1995) (Brown and Holtmeier 2013; Lowood 2008). They focused on *The French Democracy*'s merits, specifically its simple aesthetic and amateurish technique—the editing was rough, the subtitles were ungrammatical, the sound quality was poor—because it was made out of a video game rather than with professional filmmaking technologies. A wider discussion thus ensued on the possibilities for citizenship and empowerment from technological innovations such as machinima and on the potential for video games to open up further channels for such expressiveness (Russell 2007).

From the mid-2000s, machinima expanded well beyond games, to be made not only in various 3D animation programs on the market by then, such as Moviestorm and iClone, but also in social virtual worlds, most notably *Second Life* (Linden Lab, 2003). High-profile examples include *Molotov Alva and His Search for the Creator: A Second Life Odyssey*, an American documentary machinima made by Douglas Gayeton in 2007. Filmed entirely in *Second Life*, *Molotov Alva* tells the story of how a man disappears from his "real" life in Petaluma, California, only to continue existing inside *Second Life*. It was first commissioned in 2006 by a Dutch television broadcaster, and in 2007 its North American broadcast rights were purchased by HBO for a reported six-digit figure, also representing the first acquisition of its kind by a premium cable channel. In *Second Life*, machinima has also expanded internationally, with machinimators from different corners of the globe, such as Cao Fei, a widely exhibited artist based in Beijing and a Hugo Boss Prize finalist in 2010, who makes machinima in her *Second Life* project, RMB City. One such work is *i.Mirror* (2007), "a sad, dreamy, but ultimately optimistic thirty minute epic in three parts which first aired at the Venice Bienniale" and was "later acquired for the collection of a renowned Italian fashion designer" (Au 2007). Another *Second Life* machinima curated by Cristina García-Lasuén (a.k.a. Aino Baer, her *Second Life* avatar name) was exhibited in the Madrid pavilion at the Shanghai World Expo in 2010, another first of its kind. The University of Western Australia organizes the annual Machinima Challenge, whose judges include well-known filmmakers such as Peter Greenaway as well as other academics, new-media bloggers, and art historians. Use of machinima has also expanded in diverse directions; for example, it has become a teaching tool in the university classroom, where students make short films, typically in *Second Life*, about topics discussed in class (Barwell and Moore 2013; Ng and Barrett 2013). Even as more machinima is being made with games—immensely popular machinima, for example, is being created in *Minecraft* (Mojang/4J Studios, 2009)—it is also being applied in diverse contexts and made with different software. To that extent, machinima has arguably broken loose from its original ties with games, heralding much fruitful creativity ahead and demonstrating its ever-changing nature.

Works Cited

Au, Wagner James. 2007. "The Second Life of Cao Fei." http://www.caofei.com/texts.aspx?id=21&year=2007&aitid=1. Accessed September 5, 2014.

Bailey, Anthony. 2007. "Origins of the Word 'Machinima.'" http://anthonybailey.livejournal.com/33236.html. Accessed September 5, 2014.

Barwell, Graham, and Christopher Moore. 2013. "World of Chaucer: Adaptation, Pedagogy, and Interdisciplinarity." In *Understanding Machinima: Essays on Filmmaking in Virtual Worlds*, ed. Jenna Ng, 207–226. New York: Bloomsbury.

Brown, William, and Matthew Holtmeier. 2013. "Machinima: Cinema in a Minor or Multitudinous Key?" In *Understanding Machinima: Essays on Filmmaking in Virtual Worlds*, ed. Jenna Ng, 3–21. New York: Bloomsbury.

Delaney, Kevin J. 2004. "When Art Imitates Videogames, You Have 'Red vs Blue.'" *Wall Street Journal*, April 9. http://www.wsj.com/news/articles/SB108145721789778243. Accessed August 25, 2015.

Gooding, Mel, and Alastair Brotchie. 1995. *A Book of Surrealist Games*. Boston: Shambhala.

Halo Nation. n.d. "Warthog Jumping." http://halo.wikia.com/wiki/Warthog_jumping. Accessed September 8, 2014.

Lowood, Henry. 2009. "Game Capture: The Machinima Archive and the History of Digital Games." *Mediascape*, Fall. http://www.tft.ucla.edu/mediascape/Spring08_GameCapture.html. Accessed September 8, 2014.

Lowood, Henry. 2008. "Found Technology: Players as Innovators in the Making of Machinima." In *Digital Youth, Innovation, and the Unexpected*, ed. Tara McPherson, 165–196. Cambridge, MA: MIT Press.

Lowood, Henry. 2006. "High-Performance Play: The Making of Machinima." *Journal of Media Practice* 7 (1): 25–42.

Ng, Jenna, and James Barrett. 2013. "A Pedagogy of Craft: Teaching Culture Analysis with Machinima." In *Understanding Machinima: Essays on Filmmaking in Virtual Worlds*, ed. Jenna Ng, 227–244. New York: Bloomsbury.

Nitsche, Michael. 2007. "Claiming Its Space: Machinima." *Dichtung-Digital*. http://www.dichtung-digital.de/2007/nitsche.htm. Accessed September 8, 2014.

Quaranta, Domenico. 2009. "Eddo Stern—Machinima Animation & Animated Machines." Rhizome, March 21. http://rhizome.org/discuss/42048/. Accessed September 5, 2014.

Russell, Adrienne. 2007. "Digital Communication Networks and the Journalistic Field: The 2005 French Riots." *Critical Studies in Media Communication* 24 (4): 285–302.

Thompson, Clive. 2005. "The Xbox Auteurs." *The New York Times Magazine*, August 7. www.nytimes.com/2005/08/07/magazine/the-xbox-auteurs.html?_r=0a. Accessed August 25, 2015.

Varney, Allen. 2007. "*The French Democracy*." March 13. http://www.escapistmagazine.com/tag/view/the%20french%20democracy. Accessed September 8, 2014.

35 MECHANICS

Miguel Sicart

Few concepts seem as integral to the formal structure of computer games as mechanics (Järvinen 2008; Sicart 2008; Rollings and Adams 2007; Adams and Rollings 2007, 2003; Bateman and Boon 2006; Cook 2006; Rollings and Morris 2004; Hunicke, LeBlanc, and Zubek 2004; Lundgren and Björk 2003; Church 1999). Much like rules, all games have mechanics, and these mechanics are often used both colloquially to describe what happens in a game and technically as the elements that engage users in satisfactory gameplay. In fact, the relation between mechanics and gameplay seems to be so close that games such as *Dear Esther* (The Chinese Room, 2012) and *The Graveyard* (Tale of Tales, 2009) challenge what games are and can express by challenging the very relation between mechanics and games.

Both of these games provide users with a limited set of actions to be performed in an interactive environment. Few of those actions have direct impact on the environment, and none of them has a direct relation with goals, achievements, or victory. *Dear Esther* and *The Graveyard* appropriate the rhetoric of games to explore their expressive potential when they cease to be conventional, agonistic games, and they do so by decoupling the mechanics of interaction from a rule-based competitive framework. The creative excellence of these games, then, is derived from their rhetorical manipulation of game mechanics. To deeply understand the importance of these works, however, we first need to understand what game mechanics are.

In this chapter, I propose an expanded definition of game mechanics that builds on my previous work. I do not add anything substantially new, but I expect this definition to be more friendly to developers and design researchers and to pose some new questions regarding concepts in game ontology that I argue are deeply related to mechanics, such as game loops and the "space of possibility." The definition of game mechanics I propose here is still a formalist approach to one of the conceptual instruments that game ontology, from both a design perspective and a critical perspective, uses to describe, analyze, and create objects to play with. Even though I try to keep the player in mind at all moments (in the style of Björk and Holopainen 2005; Koster 2005; Avedon 1971), this definition is concerned with understanding the role of the concept of mechanics in the formal description of game designs, both from an analytical and a creative perspective. Similarly, the reflections on game loops and the space of possibility are meant to be conceptual applications of the theoretical notion of game mechanics, illustrations of how to use this concept not only for critical or design purposes but also for the exploration of interesting technical and definitional elements that articulate games.

Defining (Game) Mechanics

I can start here by expanding my previous work. Starting here allows me to advance from a particular case of mechanics connected to the structure of a game, a point from which I expect to define the broader concept of mechanics in games. I define *game mechanics* as the rule-based methods for agency in the game world, designed for overcoming challenges in nontrivial ways. Let's unpack this definition. The phrase "rule-based methods" connects game mechanics with the concept of rules. Game mechanics are rule based because they are conceptually and experientially connected to the rules that structure any game. All games have rules or frames that act both as creators of the game as an object and as a cultural activity (Juul 2005). Rules define the boundaries of the activity, the goals of the activity, as well as other formal elements that constrain a particular set of actions to give them the meaning of being a game. Rules here should be understood as the formal structures that articulate a game rather than as the cultural rules that emerge from player communities. A typical rule would be that given a variable—say, "health"—when the value of that variable reaches 0, the player is not allowed to play anymore, and the game loop is stopped.

Game mechanics are instruments for player agency within the boundary of formally defined rules. Player agency is here defined as a "method," following object-oriented terminology (Weisfeld 2000). In very basic terms, methods allow for the exchange of data in a computer program (Abelson and Sussman 1985). In games, methods can be seen as the actions that agents can trigger to interact directly with the game rules in order to alter the game state. Shooting to goal in a soccer simulation is triggering an action that evaluates the state of the game (Does the player have legal control of the ball? Can the player perform the action of shooting?) and that, given some conditions, has an effect in the state of the game (the ball will move from the player in a particular direction with a particular velocity). In more layman terms, a game mechanic is a "verb" that can be used within the bounds of a rule system. Games are ontologically and "designerly" defined both by rules, as creators and frames of the activity, and by mechanics, as the modes and types of actions that a particular game affords to players.

However, we should not think of players exclusively as "the human players"—at least not from a formal, abstract perspective. Mechanics are available to any agent within the game world—hence, the use of the concept of agency. We traditionally tend to think that mechanics are available only to human players. However, this perspective is fairly restrictive if we think of those computer games in which artificial agents can play a role in the progress of the experience. An illustrative example can be taken from any "emergent sim" game, such as *Dishonored* (Arkane, 2012), in which artificial agents have a large degree of autonomy (understood as the capacity to make decisions in their environments free from prescripted constraints), an autonomy that is implemented by allowing these agents to have access to game mechanics similar to those available to human players. In other words, mechanics are independent of agency, just as agency is independent of humanity: any agent, human or not, can have access to game mechanics.

Game mechanics are contextualized in a "game world." Even though we tend to think of game worlds as the complex simulated environments in which video game actions take place, game mechanics occur in all kinds of sociotechnically defined game worlds. For instance, the mechanic of betting is contextualized by the poker game world, which is a construct of humans and technological elements defined by a mutual agreement on rules and contexts. Similarly, a mechanic such as typing in the text game *Blackbar* (Neven Mrgan & James Moore, 2013) makes sense in a textual world rather than in a three-dimensional world. And in a more conventional game-world world, the mechanic of "inspecting" articulates many of our interactions with the world of the game *Papers, Please* (Lucas Pope, 2013).

The purpose of mechanics in games is to help players "overcome challenges." That is, any game is designed to be a series of challenges that players need to overcome in order to complete the specifications demanded by the rules either to reach an end state or to keep on playing. Mechanics are designed as actions related to these challenges: limited by those challenges so that they are engaging but at the same time created to overcome them. For instance, in the climbing game *GIRP* (Bennett Foddy, 2011) the mechanics of holding to a stone in order to climb are coupled both to the rules of physics that determine the behaviors of the avatar on screen as well as to the actual keyboard layout. The challenge is created by the physics system as well as by the careful disregard for ergonomics in the mapping of the keys: it is difficult to find the next key to press and to time that input with the avatar's physics-based swing. However, that mechanic is also the only tool we have to try to win the game. The rules create the challenges by constraining the mechanics in engaging ways. And remaining within that design perspective, game mechanics are designed to overcome challenges in "nontrivial ways"—that is, mechanics need to be sufficiently complex to require a type of investment from players, be it an investment of their skills or their emotions, so the actions are perceived as meaningful within the context of the game.

Game mechanics are thus the actions afforded to any agent that allow the agent to overcome challenges in a game. But game mechanics are only a subset of the broader concept of mechanics, which can be defined as any methods afforded to agents within a game world. Any action that is allowed to an agent in a game is a mechanic, and those actions that are explicitly related to completing the goals of a game as defined by the rules are game mechanics.

Works such as *Dear Esther* operate within the expressive and rhetorical boundaries of games, deflating the mechanics of their "gameness" to explore agency and being in virtual environments. If these games are interesting, it is because, among other reasons, they force us to play with mechanics that are not mechanics and are not agonistic evaluations of performance against rules. These actions are designed in the twilight between game mechanics and mechanics: resonant of "gameness," engaging us in a playful mood, yet afar from the agonistic evaluations of conventional games. Game mechanics are the formal building blocks of games and as such can be used to understand more complex structures that appear in games. In the next section, I analyze game loops and the concept of the space of possibility using the previous definition of (game) mechanics.

Understanding Game Loops

From a programming point of view and on a certain level of abstraction, all games are loops: once input is given, they process the data, actualize the state of the game, provide feedback, and wait for the next input. These loops are broken only when the game reaches a state in which a rule dictates the loop needs to break—for example, when a player reaches the goals established by the game or when a player fails.

From a design point of view, the concept of loops—phrased in different ways but with the same essential idea behind all the formulations—has become common in describing the design of gameplay, particularly in free-to-play games and in gamification projects (Deterding, Dixon, Khaled, et al. 2011; Deterding, Sicart, Nacke, et al. 2011; Zichermann and Cunningham 2011; Fullerton 2008; Schell 2008; Rouse 2005; Pedersen 2003). Designers discuss the creation of core loops and how they can be used to engage players, and they tie that engagement with monetization strategies. These strategies are often described as the "metagame." Game loops are thus essential formal elements that articulate the flow of interaction of a game, but how are they related to mechanics?

Game loops can be defined as dynamic links between rules and mechanics designed to structure a game's input, computation, and feedback processes. If mechanics are bound by rules, loops are the structures in which a mechanic is coupled with a rule to create identifiable processes of interaction. Whereas mechanics are the actions afforded to players, what we perceive when playing are the game loops: mechanics coupled with rules that change the state of the game and give us feedback on the process of playing.

For example, a basic game loop in a resource-management game consists of harvesting resources, relocating them, processing them, and turning them into materials required to achieve the game's goals. In more detail, a game can ask players to mine metals and transport them to a furnace to turn them into weapons so they can build an army. In any of these links between mechanics and rules, designers can add challenges (resource scarcity, extended production ties), which can be meaningful both from a play experience perspective and from a monetization angle.

The most interesting concept to expand through the concept of mechanics is that of core game loops. A core game loop defines the main actions that a player has to perform to play the game and that identify the game's structure and genre. For example, resource-management games are defined by loops based on time challenges and resource scarcity: harvest, build, and expand faster than your opponents. Competitive shooters such as *Counter Strike* (Valve, 2000), in contrast, are defined by loops structured around the player's quick movements in three-dimensional space to occupy privileged locations while scouting the environment and shooting at other players.

Understanding game loops as linkages between mechanics and rules opens up the possibility of thinking formally about issues such as game balance, which can be defined as a consequence of the formal relation between rules and mechanics, focused in particular on the potential actions that a player can perform given his or her skills at a particular time. Or, if looking at the monetization of games, a designer might argue that to be successful at being engaging but also profitable, a game loop should be constructed to be engaging on its own yet carefully modular so the addition of paid mechanics enhances the essential enjoyment of the game.

Game loops are therefore the consequence of the combinations of games' formal building elements. If we increase the abstraction level a bit more, we might argue that what game loops generate is a particular set of actions afforded to players that enable them to interact with the game—or what designers have called a "space of possibility." In the following section, I look at the concept of space of possibility from the perspective of game mechanics.

Space(s) of Possibility and Mechanics

In this formal definition of game mechanics, I have related the rules of the game, understood as frames and evaluators of the game situation, with mechanics, understood as the actions afforded to the player. Those actions as game mechanics are directly tied to overcoming the challenges proposed by the rules of the game. I have presented these two concepts as interrelated using a metaphor of space: rules frame a context in which players act by using mechanics. Let's extend that spatial metaphor a bit more.

If the notion of game loops can be used to explain not only the connections between rules and game mechanics but also the structures of games, we should consider how these loops are interrelated and specifically what they tell us about the structure of a game as a device designed to make people play. That is, game loops are formal concepts useful for design and analysis, but what players experience might be different from the specifications of the formal system. In design research, the concepts of system image, user image, and designer image as well as the notions of gulf of execution and gulf of evaluation (Norman 2002) have been used to explain this process. In games, a similarly productive approach can be reached by applying the concept of space of possibility (Salen and Zimmerman 2004). This concept has been used to describe the potential actions available to a player at any given time in the game. Phrased this way, the concept is clear, illustrative, and useful in solving specific problems of design, such as the amount of information available to a player for solving a specific challenge. But the concept can be even more useful from a game design and analysis perspective if we examine it from the perspective of game loops and game mechanics. Let's complicate things.

Given our definition of mechanics, we can define at least four different types of spaces of possibility in games:

- An abstract game space of possibility comprises all the possible actions that a player can take at any time in order to complete the game using game mechanics. For instance, the game Tic Tac Toe, given its simplicity, can have its abstract space of possibility totally defined, with every movement available at any state defined. This abstract game space of possibility might be absolute, comprising all possible actions in the game, or relative to a particular state from which we analyze it.
- An abstract space of possibility comprises all possible actions in the game situation, both game related and performative, or not directed to the completion of the game. For instance, some actions can take place in a game that are not directly related to winning the game but form part of players' rituals or habits. Or there can be modes of interaction that are decoupled from game rules and just present in the game for aesthetic or simulation pleasures.
- A perceived game space of possibility comprises all the actions a player perceives at a particular point in time as available to her in order to play the game.
- A perceived space of possibility comprises all the actions that a player perceives as possible in the context of a game, be they related to the game or not.

The concept of space of possibility, then, becomes productive when approached through a formal definition of game mechanics. What designers construct is two things: an abstract (game) space of possibilities and a perceived (game) space of possibility, which is what they are presenting to players. However, if we take a step away from game design theory and we look into player behavior, we can see how playing a game is a process of understanding and engaging with game loops to create perceived spaces of possibility, with the task of the designer being to ensure that the actual and the perceived spaces of possibility are at all times sufficiently closed so that the experience of the game by players is close to the experience envisioned and authored by the designers.

Works Cited

Abelson, Harold, and Gerald Jay Sussman, with Julie Susman. 1985. *Structure and Interpretation of Computer Programs*. Cambridge, MA: MIT Press.

Adams, Ernest, and Andrew Rollings. 2007. *Fundamentals of Game Design*. Englewood Cliffs, NJ: Pearson Prentice-Hall.

Adams, Ernest, and Andrew Rollings. 2003. *On Game Design*. Indianapolis, IN: New Riders.

Avedon, E. M. 1971. "The Structural Elements of Games." In *The Study of Games*, edited by E. M. Avedon and Brian Sutton-Smith, 419–427. New York: Wiley.

Bateman, Chris, and Richard Boon. 2006. *21st Century Game Design*. Hingham, MA: Charles River Media.

Björk, Staffan, and Jussi Holopainen. 2005. *Patterns in Game Design*. Hingham, MA: Charles River Media.

Church, Doug. 1999. "Formal Abstract Design Tools." www.gamasutra.com, July 16.

Cook, Daniel. 2006. "What Are Game Mechanics?" LostGarden, October 23. Available at: http://lostgarden.com/2006/10/what-are-game-mechanics.html. Accessed March 26, 2008.

Deterding, Sebastian, Dan Dixon, Rilla Khaled, and Lennart Nacke. 2011. "From Game Design Elements to Gamefulness: Defining 'Gamification.'" In *Proceedings of the 15th International Academic MindTrek Conference: Envisioning Future Media Environments*, 9–15. New York: ACM.

Deterding, Sebastian, Miguel Sicart, Lennart Nacke, Kenton O'Hara, and Dan Dixon. 2011. "Gamification: Using Game-Design Elements in Non-gaming Contexts." In *Proceedings of the 2011 Annual Conference "Extended Abstracts on Human Factors in Computing Systems,"* 2425–2428. New York: ACM.

Fullerton, Tracy. 2008. *Game Design Workshop, Second Edition: A Playcentric Approach to Creating Innovative Games.* New York: Morgan Kaufmann.

Hunicke, Robin, Marc LeBlanc, and Robert Zubek. 2004. "MDA: A Formal Approach to Game Design and Game Research." http://www.cs.northwestern.edu/~hunicke/MDA.pdf. Accessed March 26, 2008.

Järvinen, Aki. 2008. *Games without Frontiers: Theories and Methods for Game Studies and Design.* Tampere, Finland: Tampere University Press. https://tampub.uta.fi/handle/10024/67820. Accessed August 23, 2015.

Juul, Jesper. 2005. *Half-Real: Videogames between Real Rules and Fictional Worlds.* Cambridge, MA: MIT Press.

Koster, Raph. 2005. *A Theory of Fun for Game Design.* Scottsdale, AZ: Paraglyph Press.

Lundgren, Sus, and Staffan Björk. 2003. "Game Mechanics: Describing Computer-Augmented Games in Terms of Interaction." In *Terms of Interaction: Proceedings of TIDSE 2003*, 5–56. Darmstadt: Germany. http://citeseerx.ist.psu.edu/viewdoc/summary?doi=10.1.1.13.5147. Accessed September 1, 2008.

Norman, Donald A. 2002. *The Design of Everyday Things.* New York: Basic Books.

Pedersen, Roger E. 2003. *Wordware Game and Graphics Library: Game Design Foundations.* Plano, TX: Wordware.

Rollings, Andrew, and Ernest Adams. 2007. *Fundamentals of Game Design.* Englewood Cliffs, NJ: Pearson Prentice-Hall.

Rollings, Andrew, and Dave Morris. 2004. *Game Architecture and Design: A New Edition.* Indianapolis, IN: New Riders Press.

Rouse, Richard, III. 2005. *Game Design Theory and Practice.* Plano, TX: Wordware.

Salen, Katie, and Eric Zimmerman. 2004. *Rules of Play: Game Design Fundamentals.* Cambridge, MA: MIT Press.

Schell, Jesse. 2008. *The Art of Game Design: A Book of Lenses.* Amsterdam: Morgan Kaufmann.

Sicart, Miguel. 2008. "Defining Game Mechanics." *Game Studies* 8 (2). http://gamestudies.org/0802/articles/sicart. Accessed August 23 2015.

Weisfeld, Matt. 2000. *The Object Oriented Thought Process.* Indianapolis, IN: Sams Publishing.

Zichermann, Gabe, and Christopher Cunningham. 2011. *Gamification by Design: Implementing Game Mechanics in Web and Mobile Apps.* Sebastopol, CA: O'Reilly.

36 MENU

Laine Nooney

In "The Culture Industry: Enlightenment as Mass Deception," Max Horkheimer and Theodor W. Adorno comment that "the culture industry perpetually cheats its customers of what it perpetually promises. ... [T]he promise, which is actually all the spectacle consists of, is illusory; all it actually confirms is that the real point will never be reached, that the diner must be satisfied with the menu" ([1944] 1993, 139). If Adorno and Horkheimer had hammered out their critical essay in some 1990s instantiation of Microsoft Word, they might have proposed that the *user* must be satisfied with the menu and suffered little loss of meaning. At its most basic, the term *menu* refers to some array of options supplied for our consideration, whether computational or gastronomic; its use in the former case clearly derives from the latter. For Adorno and Horkheimer, the diner with her menu was an analogue for the modern mid-twentieth-century citizen under mass capitalism: lulled by a rich variety of delicacies listed according to size, season, and expense, the diner mistakes the edges of the menu for the edges of the world. What these famed theorists understood about the base operation of a menu—that it gives the impression of totality through an iteration of carefully specified difference—is an observation with surprising relevance and depth for today, where the menu innocuously operates as both a structure and an aesthetic of contemporary digital culture.

Before Video Games, or How We Arrived at the Menu

The menu that Adorno and Horkheimer reference took several hundred years to develop, but our computational addition to this definition is fresh wrought, at least in etymological terms. The oldest, most arcane definition of the word *menu*, as listed in the *Oxford English Dictionary* (*OED Online*, 2001), betrays its origin in the Latin term *minūtus*, meaning "insignificant" or "minute." The word's earliest definition reflects this root: in Old French, the term *menu* once referred to "the common people," identifying those who were "small" or "unimportant."

The word transitioned in Middle French from the meaning "small" to the meaning "detailed," which allowed it to flower into its contemporary definition. Dining establishments through the eighteenth century offered only bills of fare, banquet-style listings of what an establishment's kitchen had to offer. If there were a list to speak of, it would have been descriptive rather than selective, a notice of what was being served rather than a list of options (Spang 2001, 7–9). It was not until late-eighteenth-century Paris that the menu proper emerged, an object from which one made choices. Here, notions of individual taste converged with an increased cultural awareness of bodily specificity and digestive sensitivity, creating an increased demand for individually tailored options. This new conception of the menu unlatched prepared food from the restraints of limited service hours and indicated that the establishment was paying attention to the varied constitutions of the body politic (Spang 2001, 66–68). A "weak-chested" Frenchman could slurp a restorative bouillon at any hour, as was his need, not just when the dinner bell sounded. For those with the wealth to invest in their sensibilities, the world was abruptly alive with notions of taste and preference—internal sensations that are essential to our contemporary experience of food (Spang 2001, 38).

In this notion of a menu as something from which choices are made, the restaurant concept of the menu and the computational concept of the menu find formal resonance. Open another classic piece of Microsoft software: *Minesweeper.* Click the game's icon, and it opens before you, likely in the standard beginner 9-by-9 grid. Play can begin immediately, or if you open the Options drop down menu (pre-Windows 10, anyway), you can choose preset difficulty levels; customize the width, height, number of mines; and possibly make other options, depending on platform and version. Both types of menus yield to three essential characteristics: first, listed options, presented simultaneously (or near simultaneously); second, required interaction, a demand that the user or diner provide input by making a choice; third, deferred output, leaving the execution of the choice to an exterior entity that is not the individual herself.

Yet, unlike a restauranteur's menu, the menus of *Minesweeper* are *configurative* rather than *selective.* This is their most important distinction: they allow you to arrange frameworks for the game, not explicitly choose its outcome. This is a shift as resonant in our language as it is in our play. Rarely do we discuss simply "choosing" from game menus. Rather, we "navigate" them, like captains gliding across digital surfaces. We might think of the video game menu, both functionally and aesthetically, as a mechanism for managing, arranging, and making accessible player opportunities in relationship to the limitations of a game's play experience; the more options integrated into gameplay, the more complex or tiered the menu systems become. The computational menu moves us from the diner's world of options to the digerati's realm of optimization. The menu is *of* the game but not *in* the game, a bracketed space set apart from the field of play. As Alexander Galloway puts it, menus permit "actions of configuration" which reside outside the authority of a game's diegesis. For Galloway, menu use is nondiegetic "precisely because nothing in the world of the game can explain or motivate it when it occurs" (2006, 13).[1] In video games, menus are the configurative boundaries of a game's user interface, as far behind the curtain as a player may reach without the assistance of modifications, hacks, cheats, or glitches. Yet selections made at the menu level may have a profound effect on the pace, quality, accessibility, and sensation of play.

The earliest menus emerged sporadically in computational systems, with little systematization, for the benefit of programmers, computer scientists, and other specialists. In cases where a computer system coupled command-line input with visual-display output, a menu made it easier to type command-line statements by shorthanding commands to one or two alphanumeric inputs. The menu was simply a bypass for someone who already knew how to input their commands in a more time-consuming fashion. In the 1970s, however, the increasing use of computer terminals by nonspecialists popularized the menu as a design technique. James Martin describes in his book *Design of Man-Computer Dialogues* (1973) an optimal menu as one that contains "a limited set of valid operator answers to a computer-initiated question," in which case "these [answers] may be listed and the operator asked to select one from the 'menu' of possible choices" (111). This notion of limited displayed choice is what separates the menu from any other input technique and is the identifying characteristic form that maintains the menu's definition even in the era of drop-downs, touchscreens, and other contemporary graphical menus. In this period, the menu was foremost a textual device, preceding the graphical user interfaces (GUIs) of Douglas Englebart's "Mother of all Demos" (1968), Xerox PARC's Alto (1973), Apple's Lisa (1983), or the first mainstream commercial computer with a GUI, the Apple Macintosh (1984). The menu was one design component in shifting conceptions of human–computer interaction and the development of the "user" as a practical category. In the introduction to *Design of Man-Computer Dialogues*, Martin details that the early decades of computing systems were "designed from the inside, out" (1973, 3). Emphasis was placed on efficient processing and storage, with human interaction as a peripheral concern. Martin's book accents the burgeoning transition to systems designed "from the outside, in," wherein "*man* must become the prime focus of system design" (3). In this regard, the video game is an idealized example of a "man–computer dialogue," a genre of software designed to produce pleasure through the delicate balance of affordances and limitations expressed through a computer's feedback-based graphical, auditory, and kinetic capacities.

The early history of games is short on menus because most pre-1980s games simply did not have gameplay mechanics, memory capacity, or screen resolution to make display-based game configuration necessary or feasible. Ludic computational experiments such as *Tennis for Two* (Higinbotham, 1958) and *Spacewar!* (Russell/MIT, 1962) had no menus because the hardware was not designed to provide user options via display terminal; optional features were managed using buttons or switches. Mainframe and minicomputer games of the 1960s and 1970s as well as microcomputer games of the late 1970s and early 1980s made frequent use of shorthand commands (assigning a number or letter to a specific action—what we might call a "hotkey" today) but would not necessarily list those options on screen more than once. An explanation of which keys did what was nested in an instructions section, which the player might transcribe or print out. Similarly, early consoles such as the Atari Video Computer System permitted users to cycle through different modes of a game such as *Breakout* (Atari, 1976) via its Game Select switch. Offloading such options to a toggle rather than to a selection screen and limiting opportunities for players to customize or optimize their play experience made sense in an era when programmers were attempting to cram a pleasurable play experience into comparatively small memory and graphical thresholds—and shows the material links between technological affordance and a culture of user choice. Arcade games of the 1970s and early 1980s also limited configurable choices to options such as one- versus two-player mode, typically selected with a joystick or button. Arcade games favored speed and coordination but rarely pensive strategy—the pause required for a menu is antithetical to arcade economics. In most cases, none of these configurations for console, arcade, or computer games qualify as a menu because multiple choices were not simultaneously displayed to a player. These details highlight the difference between a set of selections and a menu. Many games offer selective options, but a menu involves both a set of predetermined options and their visual anchor, the representation of those options, as a list: in other words, the menu itself.

The Menu in Video Games: Textual, Graphical, Interactive

When menus began to appear in video games of the 1970s and early 1980s, they were predominately textual devices in microcomputer games, composed of branching or individualized multiple-choice options chosen through alphanumeric or button input. Because the menu was first employed in computing as a way to move a user through a program or set of questions, it emerged only in the most straightforward, all-text digital quiz games, such as the sex scenario generator *Interlude* (1981), or in various entertainment software "party games," programs that could predict a partygoer's alcohol tolerance or life expectancy based on answers to multiple-choice questions (often with titles as self-evident as *Alcohol* or *Life Expectancy*). In such instances, the game *was* the menu, the only interface a player had with the software.

Traceable historic transitions in the implementation of menus correspond largely to technoaesthetic properties. Some of the first games to implement a textual menu within a graphical environment were role-playing games such as *Ultima* (California Pacific Computer Co., 1980), in which the player navigated both an overhead map view and a first-person dungeon view; alphanumeric menus popped up to permit interaction with nonplayer characters, assign weapons, and otherwise negotiate the avatar's stats and abilities. Similarly, *King's Quest* (Sierra On-Line, 1983) features a horizontal bar of drop-down menu categories such as Info, File, and Game. These categories provided a means for players not only to save and restore the game but also to adjust speed, examine inventory, and select specific avatar behaviors (such as "jump" or "swim"). These menus were both textual and graphical in the sense that the user could open, close, or move between them while still viewing the full screen of play. By the time *King's Quest V* (Sierra On-Line, 1990) was released, the erosion of text-based input and the transition to the (supposedly) more user-friendly point-and-click turned the horizontal, top-screen menu bar into a selection bay for different icon-based actions (a speech bubble for speaking, a hand for taking or moving, and so on), along with the inventory and a controls menu. The menu had entirely lost its textual character in favor of infographic icons, and the control panel used sliders to give players variable control of features such as sound. Such transitions moved in step with the mainstreaming of the mouse, the development of higher-resolution screens, and increased color palettes across the mid-1980s through the early 1990s; video game menus thus shifted in stride with the design standards of the GUI. Menus became increasingly graphical and integrated into gameplay and eventually, as we will see, morphed into objects of kinetic digital interaction that are themselves a site of experience and manipulation—something playful, if not gamerly, in their own right.

Probably the most ubiquitous, earliest, and longest-standing iteration of the menu is the Start Menu. The Start Menu serves as a title screen from which play may be initiated or other configurations of play may be selected for a game session or the entire game. The leanest form of the Start Menu is not a menu at all but simply allows the player to begin the game: in *The Legend of Zelda* (Nintendo, 1986), the game instructs players to "Push Start Button." In contrast, *Super Mario Bros.* (Nintendo, 1985) offers players a similar title screen but with a single choice: one player or two players. Adding a layer, *Tetris* (Spectrum Holobyte, 1987) displays a title screen followed by a menu that requires the player to select start level and starting height. Options for restoring, quitting, customizing the controller or audio settings, and viewing credits have become Start Menu standards because game companies have developed techniques allowing more configuration on the player's end (and thus more possibilities of play). For some genres, these preplay selection menus establish the sum total of a player's ability to configure the game, as is the case with many fighting, action, racing, and arcade games. A player cannot easily, if at all, change selections made to levels, character/vehicle, or difficulty in such games—for example, *Street Fighter* (Capcom, 1987) and *Spelunky* (Mossmouth, 2009)—once a match, round, or run has begun. As video game software produces an activity in which one is pleasurably challenged by the software itself (distinguishing it from an email client or photo-editing software), menus serve as one of the digital mechanisms for framing and shaping the player–machine relationship.

Attributes and Activities

The kinds of configurations video game menus make available can be divided into two types: there are menu options that govern the *attributes* of digital play and options that participate in the *activity* of digital play. Not all games have both, but if they do, these configurations do not always exist in equal proportion; however, this distinction is relevant to much of a game's information architecture and user-interface design. This taxonomy is not based on a spatial observation (Where do we find menus?), nor is it one of degree (To what extent does the menu dictate the game?). Rather, it is experiential: What in the gameplay is a menu permitting us to alter or select from?

Attribute menus give players the opportunity to define and customize the experience of the game. These menus are where one sets player preferences and adjusts play's accessibility, as in the one-player versus two-player choice in *Super Mario World* (Nintendo, 1990) or the difficulty-level menu in *DOOM* (id Software, 1993)—. But attribute menus also comprise the phenomenal proliferation of options embedded in many contemporary games, including options for adjusting audio and graphics, subtitles, the *y* axis, saving and restoring, in-app purchases, and fully reprogramming the controller to player specifications. These menus are carefully designed objects that anticipate frustration and desire as well as disability (for instance with subtitles) and remember that games happen in contexts (on bad TVs, in bright rooms, with players who can't finish a game in a single sitting). Barring attribute selections that initialize a game, these menus can typically be called up from any point of play but are often one step more removed than activity menus.

Menus governing the activity of play are menus in the game, however peripherally: the Pip Boy of *Fallout 3* (Bethesda Game Studios, 2009), the weapon selection wheel of *Grand Theft Auto V* (Rockstar North, 2013), the skill trees of The Elder Scrolls V: *Skyrim* (Bethesda Game Studios, 2012), the dialog boxes of *The Secret of Monkey Island* (Lucasfilm Games, 1990), the playbooks of *Madden* (Electronic Arts Tiburon), the character customization of *Mass Effect* (BioWare, 2007). As Galloway might put it, these menus exist within "an informatic layer once removed" (2006, 14). Whereas activity menus were once simply ways of gathering basic user input through branching selections or alphanumeric choices, menus in contemporary video games are frequently complex sites of player balance rather than straightforward selection. Activity menus are often sites of tweaking, shifting, adjusting, and switching, all of which produce cascading effects across increasingly granular subtleties of on-screen activity. Because activity menus are so frequently utilized, there tends to be tighter integration between these menus and the game; they are typically never more than a button-press away.[2]

In Adorno and Horkheimer's condemnation of capitalist plentitude, the menu is a template for the culture industry's ideological techniques. Martin's vision of the menu as an implement of humanist design strategies was one intended to broaden the computer's use to nonspecialists but not necessarily to enhance general human understanding of computer technology; this vision is Adorno and Horkheimer retrofit for the information economy, which heats and reduces all culture to "quantitative modulations and numerical valuations" (Galloway 2006, 17). In Galloway's terms, we are an economy of sampling and selection: "to live today is to know how to use menus" (17).

In and of itself, however, the menu is not just an allegory for the information age. The menu placed in the hands of the Parisian diner of the eighteenth century brokered a new realm of experience, and, as for any emergent medium, there was no guarantee the menu would stick. Before the menu became "a footnote to ... the algorithmic structure of today's informatic culture," it was open-source technology for a burgeoning model of modern subjectivity (Galloway 2006, 17). Its other futures may not have come to fruition, but they were, perhaps, at one point "options." We have had several hundred years to watch the diner (and the consumer) become templates of human experience. The history of the user—with her menu of whatever sort—has only just begun.

Notes

1. Galloway's innovative approach analyzes video games not as objects but as actions and categorizes these actions across a "diegetic vs. nondiegetic" *x* axis and an "operator vs. machine" *y* axis. For more information, see Galloway 2006, chapter 1. For Galloway, all menu activities constitute nondiegetic operator acts as well as other kinds of user interface inputs, such as hitting pause on a controller.

2. The distinction I make here is close to Galloway's own, in which he suggests that there are two basic types of nondiegetic operator acts: "setup" actions ("the interstitial acts of preference setting") and "gaming actions in which the act of configuration itself *is the very site of gameplay*" (2006, 13). In the case of the latter, Galloway is referring to real-time strategy, turn-based, and resource-management simulations such as *Warcraft III* (Blizzard Entertainment, 2002) or *Civilization III* (MicroProse, 2001), games that are so fully about their menus that they become the "very essence of the operator's experience of gameplay." In practice, however, many menus lie between these two types. Many games employ menus to facilitate interaction with the game environment but are not themselves reducible to a menu mechanic—which is why I suggest a more menu-specific distinction.

Works Cited

Galloway, Alexander R. 2006. *Gaming: Essays on Algorithmic Culture*. Minneapolis: University of Minnesota Press.

Horkheimer, Max, and Theodor W. Adorno. [1944] 1993. "The Culture Industry: Enlightenment as Mass Deception." In *Dialectic of Enlightenment*, 120–167. New York: Continuum.

Martin, James. 1973. *Design of Man-Computer Dialogues*. Upper Saddle River, NJ: Prentice-Hall.

Spang, Rebecca L. 2001. *The Invention of the Restaurant: Paris and Modern Gastronomic Culture*. Cambridge, MA: Harvard University Press.

37 METAGAME

Stephanie Boluk and Patrick LeMieux

For one thousand and one nights, Scheherazade delayed her execution at the hands of the Shahryar by telling him a never-ending story. Adapted for European audiences by French archaeologist and Orientalist Antoine Galland in 1704, textual compilations of Scheherazade's tale of tales were translated from Arabic, edited to remove some of the more erotic elements (along with most of the poems), and supplemented with oral folklore that had no literary precedent, such as "Aladdin, or the Wonderful Lamp" and "Ali Baba and the Forty Thieves." Although there are many versions of *One Thousand and One Nights*, Scheherazade's frame narrative is their common feature. Since the conclusion of Galland's twelve-volume publication in 1717, *One Thousand and One Nights* continues not only to fuel an entire genre of Orientalist fantasy but also serves as an archetypical example of metanarrative: stories about stories. From oral storytelling in West and South Asia to literary fairy tales in Europe to concert halls, ballet stages, playhouses, and silver screens around the world, Scheherezade eventually found herself depicted within the collectible card game *Magic: The Gathering* (Wizards of the Coast, 1993) (see figure 37.1).

After the initial release of *Magic* on August 5, 1993, the game's creator and lead designer, Richard Garfield, worked on a strict deadline to finish its first expansion by Christmas that year (Garfield 2002). Authored entirely by Garfield and based explicitly on *One Thousand and One Nights*, the *Arabian Nights* expansion set included new cards based on myths and legends, such as Flying Carpet, Mijae Djinn, and Ydwen Efreet. Other cards in the collection reflected the cultural imaginary of post–Gulf War America, such as Army of Allah, Bazaar of Baghdad, and Jihad.[1] Finally, *Arabian Nights* portrayed classic characters from Gallard's *One Thousand and One Nights*, such as Aladdin, Ali Baba, and, of course, Scheherazade. Illustrated by Kaja Foglio and printed for a limited time between December 1993 and January 1994, the Shahrazad[2] *Magic* card matched the character's myth. When the card is played, "Players must leave game in progress as it is and use the cards left in their libraries as decks with which to play a subgame of *Magic*. When subgame is over, players shuffle these cards, return them to libraries, and

Figure 37.1
Shahrazad, a *Magic* card designed by Richard Garfield and illustrated by Kaja Foglio. Originally included in the *Arabian Nights* expansion of *Magic: The Gathering* in December 1993. Wizards of the Coast.

resume game in progress, with any loser of subgame halving his or her remaining life points, rounding down" (Garfield 1993).

With the Shahrazad card, stories within stories become games within games. Thematically and mechanically, the figure of the storyteller stands in for *Magic* itself, a card game Garfield designed both to cultivate and to capitalize on previously ancillary aspects of gaming such as collecting, competition, and community not explicitly included within the rules of the game. After all, the Golden Rule of *Magic* is that the rules printed on the cards take precedence over the rules printed in the manual—one thousand and one ways to play (Wizards of the Coast 2014). One of the only cards banned in official tournaments (and one of the more valuable cards in the *Arabian Nights* set), Shahrazad is Richard Garfield's favorite *Magic* card and represents the game within the game that he calls the "metagame" (2002).

The word *metagame* does not appear in any dictionary. Although the term is used to denote a wide variety of activities related to games—from a specific subset of mathematical and economic game theory to the metaleptic slippage between in-game and out-of-game knowledge in role-playing games to the common strategies or passing fashions of competitive games—there is no unified definition of *metagame*. Whereas there have been numerous discussions surrounding the meaning of the word *game*, the etymology and meaning of the term's other constitutive element, *meta*, is not as heavily debated. Whether used to describe a story about stories, a film about films, a game about games, or any *x* about *x*, the adjective *meta* in English (and mainly in the United States) typically suggests "a consciously sophisticated, self-referential, and often self-parodying style, whereby something reflects or represents the very characteristics it alludes to or depicts" (*Oxford English Dictionary* [*OED Online*, 2014]). There has also been some slippage between this general meaning of *meta* and the more specific concept of recursion.[3] These adjectival uses of the term are derived from the more universal prefix *meta-*, which signifies an abstraction from, a second-order beyond, or a higher level above the term or concept that it precedes (*OED*).[4] For example, when prepended to a field of study, *meta-* "denote[s] another [subject] which deals with ulterior issues in the same field, or which raises questions about the nature of the original discipline" such as metaeconomics, metaphilosophy, and even metalexicography (*OED*). Based on the ancient Greek preposition *μετά*, meaning "with," "after," "between," or "beyond," the prepositional origin of the prefix *meta-* continues to characterize its modern use even if *μετα-* was also combined with verbs to express "change (of place, order, condition, or nature)" (*OED*). Etymologically, the term *metagame* does not simply signify the general category of games that reference themselves or other games but is also characterized by the deeply specific, relational quality of prepositions as parts of speech. In the same way a preposition situates the noun that it precedes, the meaning of *metagame* emerges within the context of specific practices and historical communities of a given game. A signifier for everything occurring before, after, between, and during games as well as everything located in, on, around, and beyond games, the metagame anchors the game in time and space.

One of the earliest concatenations of the terms *meta* and *game* occurred within the branch of mathematics known as game theory (as distinct from game studies). First formulated by John von Neumann in 1928 and later expanded with the help of Oskar Morgenstern in their book *Theory of Games and Economic Behavior* (1944), game theory is "the study of mathematical models of conflict and cooperation between intelligent, rational decision-makers" (Myerson 1991, 1). During the Cold War, von Neumann and Morgenstern's quantitative "science of decision making" influenced both US and USSR policies, including strategies of deterrence based on mutually assured destruction between two global superpowers. The canonical thought experiment that simultaneously popularized and challenged the underlying premises of game theory was the "prisoner's dilemma." First named by Albert Tucker in 1950 as a way to thematize the ideas of RAND researchers Melvin Dresher and Merrill Flood, the dilemma describes how two prisoners are arrested for the same offense, held separately, and given a choice to betray one another, with three possible outcomes:

(1) If one confesses and the other does not, the former will be given a reward ... and the latter will be fined. ...
(2) If both confess, each will be fined. ... At the same time, each has good reason to believe that
(3) If neither confesses, both will go clear. (Poundstone 1992, 118)

For game theorists, the Cold War represented a global prisoner's dilemma with potentially apocalyptic consequences. Considering the disturbing fact that the "rational" decision (i.e., confessing) would result in mutually assured destruction, the only way to win this logical paradox is to not play.

Rather than choosing infinite deferral or rational suicide, Nigel Howard's book *Paradoxes of Rationality: Theory of Metagames and Political Behavior* (1971) solves the prisoner's dilemma based on a "nonquantitative" and "nonrational" approach to game theory that he calls "the metagame," or "the game that would exist if one of the players chose his strategy after the others, in knowledge of their choices" (1, 2, 23). This is the earliest substantive use of the term *metagame* we have found. Whereas in the original dilemma there are only two options—to confess (*C*) or not to confess (*D*)—according to Howard's metagame theory, Player 1 can also make additional, "extensive" choices based on Player 2's *possible* actions (1971, 11). By projecting an opponent's potential behavior, the simple decision to confess or not to confess exponentially multiplies into four new "metachoices": confess if they confess (*C*/*C*), don't confess if they don't confess (*D*/*D*), confess if they don't confess (*C*/*D*), and don't confess if they confess (*D*/*C*)—a game theory within a game theory. Considering the implications of this metagame from Player 2's perspective, the projected possibilities exponentially branch again from a metagame with four choices to a meta-metagame with *sixteen* choices. This infinitely branching tree of possible choices "is the mathematical object studied by the theory of metagames" (Howard 1971, 55). When Player 1's "metachoices" are cross-referenced with Player 2's "meta-metachoices," additional points of "metarational metaequilibrium" appear and offer alternative, favorable outcomes (i.e., not confessing)—a mathematical solution to the problem of mutually assured destruction based on mutually assured metagaming (Howard 1971, 59) (see figure 37.2).

Famous for saying, "If you say why not bomb them tomorrow, I say, why not today" (qtd. in Blair 1957, 96), von Neumann mathematically reinforced his militant belief in deterrence (if not in preemptive nuclear strike) through game theory. The logical consequences of Howard's metagame, in contrast, lead to a different point of equilibrium: mutual disarmament. Although von Neumann's game theory and Howard's metagame analysis may seem far afield

	C	D
C	3,3	1,4
D	4,1	2,2

	C/C	D/D	C/D	D/C
C	3,3	1,4	3,3	1,4
D	4,1	2,2	2,2	4,1

	C/C	D/D	C/D	D/C
C/C/C/C	3,3	1,4	3,3	1,4
D/D/D/D	4,1	2,2	2,2	4,1
D/D/D/C	4,1	2,2	2,2	1,4
D/D/C/D	4,1	2,2	3,3	4,1
D/D/C/C	4,1	2,2	3,3	1,4
D/C/D/D	4,1	1,4	2,2	4,1
D/C/D/C	4,1	1,4	2,2	1,4
D/C/C/D	4,1	1,4	3,3	4,1
D/C/C/C	4,1	1,4	3,3	1,4
C/D/D/D	3,3	2,2	2,2	4,1
C/D/D/C	3,3	2,2	2,2	1,4
C/D/C/D	3,3	2,2	3,3	4,1
C/D/C/C	3,3	2,2	3,3	1,4
C/C/D/D	3,3	1,4	2,2	4,1
C/C/D/C	3,3	1,4	2,2	1,4
C/C/C/D	3,3	1,4	3,3	4,1

Figure 37.2

Nigel Howard's solution to the original prisoner's dilemma (*left*) involves drafting a metagame (*middle*) and a second-order metagame (*right*) that reveals two new points of metaequilibrium (*shaded cells*). According to Howard, "The symbol 'W/X/Y/Z' represents the policy 'W against C/C, X against D/D, Y against C/D, Z against D/C" (1971, 59).

from game studies, the distinction between a game defined by an individualistic, selfish form of abstract rationality and a metagame that acknowledges the collective, historical conditions of decision making parallels the self-referential and prepositional metagame—the game to, from, during, and between the game—deployed by Richard Garfield when he designed *Magic: The Gathering.*

Two decades after Howard's analysis and the dissolution of the Soviet Union, Garfield incorporated the term *metagame* into his design vocabulary shortly following the publication of the Shahrazad card. In the spring 1995 issue of *The Duelist* (1994–1999), an official Wizards of the Coast magazine, Garfield offered a preliminary discussion of metagames in his column "Lost in the Shuffle." Although the stakes are far less consequential than nuclear apocalypse, Garfield's article begins with an anecdote about "backstabbing [his] allies in *Diplomacy*," Allan B. Calhamer's wargame from 1954 in which play takes the form of discussion, negotiation, and other social dynamics (1995, 87). After treating each game as an individual, autonomous conflict (and losing more and more), Garfield realized that his relationships to other players, to the larger social structure in which games are embedded, and even to the physical or economic constraints of certain rules functioned "not as ends unto themselves but as parts of a larger game" or "metagame" (1995, 87). Following Howard, this is the second major theorization of the term *metagame.* Five years after publishing his initial definition, Garfield delivered a presentation on metagaming at the Game Developers Conference in San Francisco before revising and printing the talk in *Horsemen of the Apocalypse* (2000b). Whereas metagaming in role-playing games such as *Dungeons & Dragons* (Gygax and Arneson/Tactical Studies Rules, 1974) usually refers to the use of out-of-character knowledge to make in-character decisions, Garfield expanded the definition to encompass "how a game interfaces beyond itself" (2000b, 16). In the same way Howard applied the term *metagame* to describe a more pragmatic and contextual form of decision making or "policy" (1971, 59), Garfield argued "There is of course no game without a metagame ... a game without a metagame is like an idealized object in physics. It may be a useful construct but it doesn't really exist" (2000b, 16).

Based on the metagame's actual rather than ideal relation to games, Garfield divided it into four prepositional categories: "what a player brings to the game" (e.g., *Magic* decks, tennis rackets, personal abilities); "what a player takes away from a game" (e.g., prize pool, tournament rankings, social status); "what happens between games" (e.g., preparation, strategizing, storytelling); and "what happens during a game other than the game itself" (e.g., trash talking, time-outs, and the environmental conditions of play) (2000b, 17–8, 18, 20). To, from, during, and between: hearkening back to the prepositional etymology of the term *meta* as well as to Howard's practical solution to the prisoner's dilemma, Garfield's metagame describes players' lived experience and historical contexts in which games are played. Whereas Garfield's metagame has become synonymous with the current strategies and changing trends in the culture surrounding competitive games such as *Magic* and even video games such as *StarCraft* (Blizzard Entertainment, 1998) and has been taken up by scholars and designers such as Katie Salen Tekinbaş and Eric Zimmerman (2004),[5] there is an even more radical interpretation of the metagame.

Although most of Garfield's talk from the Game Developers Conference in 2000 was published later that year, a transcript from the event reveals a slightly different definition of the metagame: not "how a game interfaces beyond itself" (2000b, 16) but "how a game interfaces with *life*" (2000a, 1, emphasis added). This small but significant departure from the printed definition reveals Garfield's original commitment to the metagame as the only kind of game we play.[6] The metagame is not just how games interface with life; they are the environment within which games "live" in the first place. Like Mark Hansen's definition of media as "an environment for life" (2006, 297), metagames are an environment for games. Metagames are where and when games happen, not a "magic circle" within which unnecessary obstacles and voluntary pursuits play out, but a "messy circle"[7] that both constrains and makes games possible in the first place. Inside this second circle, the ideological desire to distance leisure from labor, play from production, or games from life breaks down—we don't play games; we constantly and unconsciously make metagames.

Nowhere is this ideological conflict more clear than in the relationship between metagames and video games. Although metagames have always existed (albeit subtly) alongside video games, over the past decade and with the rise of social media and sharing services such as Steam (2003), YouTube (2005), and Twitch (2011), the term *metagame* has become more commonly used to describe the community practices within, around, outside, and about video games. For example, when Narcissa Wright speedruns *The Legend of Zelda: Ocarina of Time* (Nintendo, 1998) and *The Legend of Zelda: The Wind Waker* (Nintendo, 2003), the addition of optional constraints, like a simple timer, radically change the way the game is played. Speedrunning is a metagame that encourages the discovery and manipulation of mechanical exploits not immediately evident to the player or accepted as legitimate forms of play. As Wright remarks, when new techniques such as "Dry storage got discovered [in *The Wind Waker*] ... I was reinvigorated to stay on top of the current metagame" (Wright 2013). Speedrunning is not only a metagame contingent on the virtuosic performance of real-time play but is also a community practice based on discovering exploits such as geometry clipping, cutscene skipping, sequence breaking, and memory manipulation—games *within* the game. More closely aligned with the spirit of Garfield's definition, other players recover the histories of shifting strategies and tournament trends around specific video games as a playful and productive form of spectatorship—a game *around* the game. For example, Richard "KirbyKid" Terrell encodes VHS tapes to track the history of competitive *Super Smash Bros. Melee* (HAL Laboratory, 2001) because "Every game with a metagame worth understanding deserves a devoted videogame historian" (Terrell 2011), whereas Daniel "Artosis" Stemkoski studies and commentates competitive *StarCraft II* tournaments in Korea (Blizzard Entertainment, 2010) because "The Metagame is not the same everywhere. ... Korea, China, NA [North America], and Europe ... each have their own Metagame" (Stemkoski 2012). The metagame can even take the form of a game *outside* the game, as is the case when Alex "The Mittani" Gianturco expands his *EVE Online* (CCP Games, 2003) empire via cyberwarfare and offline espionage—what he identifies as a "metagame which doesn't require booting the program up at all" (qtd. in Goldman and Vogt 2014). Finally, reminiscent of Garfield's favorite *Magic* card, Shahrazad, Andy Baio observes that the term *metagame* can also refer to "playable games about videogames" (2011)—a definition that recalls the use of the term *meta* in other media genres, as in *metafiction*, but departs from the way in which Garfield and much of the gaming community apply the term. Thus, when Jonathan Blow appropriates image, text, level designs, and thematic tropes from *Super Mario Bros.* (Nintendo, 1985) and *Donkey Kong* (Nintendo, 1981) in his independently developed game *Braid* (Number None Inc., 2008), he is making a game *about* games. *Braid* not only revisits the history of

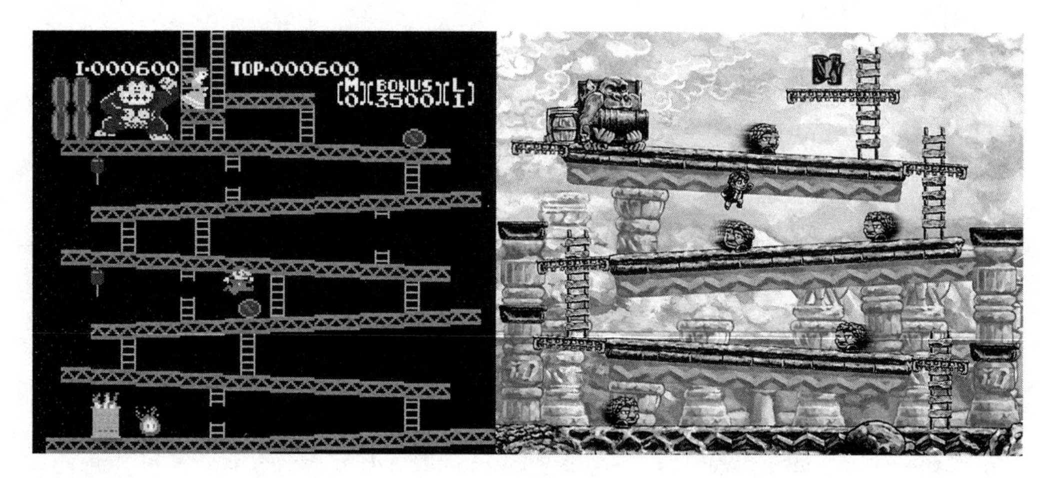

Figure 37.3
In search of lost time: in *Braid* (*right*), Jonathan Blow references the early history of platform games such as Nintendo's *Donkey Kong* (*left*) not only for nostalgic effect but also to make a metagame about the conventions, culture, and psychogeography of video games. Nintendo. Number None, Inc.

platform games as a genre but also makes a metagame about platforming itself through the addition of time manipulation mechanics (see figure 37.3). From speedrunning *The Legend of Zelda* to the professionalization of competitive *Smash Brothers* and *StarCraft* to the elaborate espionage surrounding *EVE Online* and the hyperreferentiality of the indie game boom in the late 2000s exemplified by *Braid*, the metagame includes these player-produced games *within*, *around*, *outside*, and *about* video games.

In their recent essay "Metagames, Paragames, and Orthogames: A New Vocabulary," Marcus Carter, Martin Gibbs, and Mitchell Harrop argue that for players "there is a broad, conceptually muddled use of the term [*metagame*] that encompasses a wide variety of different play types and styles for which a single term is not useful" (2012, 11). Rather than attempting to redefine the term or privilege a more narrow definition, we believe that the prepositional character of *meta* and the genealogy of both Howard's and Garfield's definitions of *metagame* linguistically and conceptually perform the term's common use: to locate the specific community practices, material configurations, and historical transformations of twenty-first-century play. Beyond the term's etymology and various definitions, metagames are not simply self-referential games about games or recursive games inside games. They are not just games before and after, to and from, or during and between games. They are not just games in, on, around, or through games. Instead, the metagame expands to include the contextual, site-specific, and historical attributes of human (and nonhuman) play. Attitude, affinity, experience, achievement, status, community, opponent, strategy, series, spectatorship, statistics, economics, politics, practice:

the metagame explodes the logic of the game, escaping the formal autonomy of both ideal rules and utopian play via those phenomenal, psychological, material, and historical factors not immediately enclosed within the game. What the metagame identifies is not the history of the *game*, but the histories of *play.*

Notes

1. Connotations of the Gulf War in *Magic: The Gathering* are as much a reference to Neil Gaiman's comic book series *The Sandman* (1989–1996) as to US geopolitics in the 1990s. Cards such as City in a Bottle explicitly cite issue 50 of the comic "Distant Mirrors: Ramadan," in which mythical Baghdad is framed as a story within the story of a beggar in war-torn Iraq (Garfield 2002; Gaiman 1993, 31).

2. Garfield titled the card based on the Persian transliteration "Shahrázád" rather than on the French transliteration "Scheherazade."

3. Douglas Hofstadter's book *Gödel, Escher, Bach: An Eternal Golden Braid* ([1979] 1999) popularized the slippage between an understanding of the term *meta* as "self-reflexive" and an understanding of the term *meta* as "recursive." To build his "metabook" around the concept of the "strange loop," Hofstadter braids descriptions of Bach's endlessly rising canon and other "metamusical offerings," Escher's recursive artworks such as *Metamorphosis* (1939–1940), and Gödel's paradoxical incompleteness theorem based on "metamathematics" or "metalogic" (viii, 15, 23, 10). A "strange loop" is a recursive "phenomenon [that] occurs whenever, by moving upwards (or downwards) through the levels of some hierarchical system, we unexpectedly find ourselves right back where we started" (10). Recursion, however, is not entirely commensurate with self-reflexivity. A self-reflexive film about film, such as *Man with a Movie Camera* (Dziga Vertov, 1929), may include references to itself and its medium, the feedback generated when a camera points at a screen produces recursive instances ad infinitum.

4. The conflation of *meta-* with second-order concepts and a notion of the beyond can be attributed to a misconception surrounding the title of Aristotle's *Metaphysics.* Although the content of *Metaphysics* may describe an ontology "beyond" the physical, the title *Metaphysics* originally signified the order of a series of books. As Peter van Inwagen notes, "An editor of [Aristotle's] works ... entitled those fourteen books '*Ta meta ta phusika*'—'the after the physicals' or 'the ones after the physical ones'—the 'physical ones' being the books contained in what we now call Aristotle's *Physics*" (2013).

5. Salen Tekinbaş and Zimmerman recognize the importance of the metagame in *Rules of Play* (2004). Examining the metagame from a design perspective, they argue that "game design is

a second-order design problem. ... [M]ost of any given game's metagame is beyond the reach of the game designer, for it emerges from play communities and their larger social worlds" (484). Considering the ways in which the metagame impacts play, Stewart Woods also deploys Garfield's concept throughout his study of tabletop gaming culture in *Eurogames* (2012), including lengthy descriptions and responses to surveys he conducted regarding the metagame. Finally, Garfield's own coauthored book *Characteristics of Games* (2012) includes a chapter that schematizes and compares various metagames according to six general categories: "status, money, socialization, achievement, knowledge, and fantasy" (Elias, Garfield, and Gutschera 2012, 209–10).

Although Salen Tekinbaş has continued to explore the metagame through studies on the video game culture surrounding e-sports such as *StarCraft* (Kow, Young, and Salen Tekinbaş 2014), Zimmerman has approached the concept less as a descriptive tool and more as part of his game design philosophy. In 2006, Zimmerman and Frank Lantz developed a prototype of *The Metagame,* a board-based trivia game about video games originally planned for Will Wright's "game issue" of *Wired* magazine (Zimmerman 2011). Eventually redesigned with the help of Colleen Macklin and John Sharp in 2011, *The Metagame* evolved into a card game about video games, a clever reference to both *Magic* and Richard Garfield's theories while still functioning literally as a game about games.

6. Garfield writes that the metagame "breathes life into the experience of game-playing" and "[inducts players] into a larger game community" (1995, 87). He stresses that "there is something magical, even infectious, that happens in the metagame" and that "Metagames tend to have application and meaning beyond the game itself; often, they seep into real life" (1995, 87).

7. As Ian Bogost contends in "Videogames Are a Mess," whether defined as "kilobytes [of data]," "a flow of RF modulations," "a mask ROM," "a molded plastic cartridge," "a consumer good," "a system of rules," "an experience," "a unit of intellectual property," "a collectible," or "a sign," video games and in this case metagames are a mess (2009).

Works Cited

Baio, Andy. 2011. "Metagames: Games about Games." *Waxy.org*, February 1. http://waxy.org/2011/02/metagames_games_about_games/.

Blair, Clay, Jr. 1957. "Passing of a Great Mind." *LIFE* magazine, February 25.

Bogost, Ian. 2009. "Videogames Are a Mess." *Bogost.com*, September 3. http://bogost.com/writing/videogames_are_a_mess/.

Carter, Marcus, Martin Gibbs, and Mitchell Harrop. 2012. “Metagames, Paragames, and Orthogames: A New Vocabulary.” In *Proceedings of the International Conference on the Foundations of Digital Games (FDG ’12)*, 11–17. New York: ACM.

Elias, George Skaff, Richard Garfield, and K. Robert Gutschera. 2012. *Characteristics of Games*. Cambridge, MA: MIT Press.

Gaiman, Neil. 1993. “Distant Mirrors: Ramadan.” *The Sandman*. June. New York, NY: DC Comics.

Garfield, Richard. 2002. “The Making of *Arabian Nights*.” *Wizards of the Coast*, August 5. http://archive.wizards.com/Magic/magazine/article.aspx?x=mtgcom/feature/78.

Garfield, Richard. 2000a. “Metagames.” *GDC 2000 Proceedings Archive*, March, 8–12. https://web.archive.org/web/20081221121908/http://www.gamasutra.com/features/gdcarchive/2000/garfield.doc.

Garfield, Richard. 2000b. “Metagames.” In *Horsemen of the Apocalypse: Essays on Roleplaying*, ed. Jim Dietz, 14–21. Charleston, IL: Jolly Roger Games.

Garfield, Richard. 1995. “Lost in the Shuffle: Games within Games.” *The Duelist: The Official Deckmaster Magazine*, Spring, 86–88.

Garfield, Richard. 1993. “Shahrazad” (game card). In the card game *Magic: The Gathering*. Seattle: Wizards of the Coast.

Goldman, Alex, and P. J. Vogt. 2014. “#11—RIP Vile Rat.” *TLDR*, January 22. http://www.onthemedia.org/story/11-rip-vile-rat/.

Hansen, Mark B. N. 2006. “Media Theory.” *Theory, Culture & Society* 23 (2–3): 297–306.

Hofstadter, Douglas. 1999. *Gödel, Escher, Bach: An Eternal Golden Braid*. New York: Basic Books.

Howard, Nigel. 1971. *Paradoxes of Rationality: Theory of Metagames and Political Behavior*. Cambridge, MA: MIT Press.

Kow, Yong Ming, Timothy Young, and Katie Salen Tekinbaş. 2014. *Crafting the Metagame: Connected Learning in the* StarCraft II *Community*. Irvine, CA: Digital Media and Learning Research Hub.

Myerson, Roger B. 1991. *Game Theory: Analysis of Conflict*. Cambridge, MA: Harvard University Press.

Poundstone, William. 1992. *Prisoner’s Dilemma: John von Neumann, Game Theory, and the Puzzle of the Bomb*. New York: Doubleday.

Salen, Katie, and Eric Zimmerman. 2004. *Rules of Play: Game Design Fundamentals*. Cambridge, MA: MIT Press.

Stemkoski, Daniel "Artosis." 2012. "Mapping Out the Metagame." *SC Dojo*, October 25. http://scdojo.tumblr.com/post/34233737871/mapping-out-the-metagame.

Terrell, Richard. 2011. "Metagame Meditations Pt. 4." *Critical-Gaming Network*, February 1. http://critical-gaming.com/blog/2011/2/1/metagame-meditations-pt4.html.

Van Inwagen, Peter. 2013. "Metaphysics." In *The Stanford Encyclopedia of Philosophy*. Stanford, CA: Stanford University Press. http://plato.stanford.edu/archives/win2013/entries/metaphysics/.

Von Neumann, John, and Oskar Morgenstern. 1953. *Theory of Games and Economic Behavior*. Princeton, NJ: Princeton University Press.

Wizards of the Coast. 2014. "*Magic: The Gathering*, Comprehensive Rules." June 1. http://media.wizards.com/images/magic/tcg/resources/rules/MagicCompRules_20140601.pdf.

Woods, Stewart. 2012. *Eurogames: The Design, Culture, and Play of Modern European Board Games*. Jefferson, NC: McFarland.

Wright, Narcissa. 2013. "Cosmo's List of Speedrun Progress." *Pastebin*, September 19. http://pastebin.com/FsUEBVyh.

Zimmerman, Eric. 2011. "Secret History of the Metagame." *Being Playful*, February 18. http://ericzimmerman.wordpress.com/2011/02/18/secret-history-of-the-metagame/.

38 MODIFICATION

Hector Postigo

Modification is a long-standing practice in video game culture. Its history goes hand in hand with the emergence of video gaming as a prevalent, persistent, and increasingly ubiquitous media-consumption practice. Its origins and continued practice lie squarely within computer game culture, and modification has not yet penetrated video game culture in the console and mobile platform markets.[1] In video game culture, modification is what can be called an *entry-point practice*. This practice allows users savvy enough to know their way around code and design a means to retell the narratives of their favorite games, appropriate them, and tweak them. It also serves as a testing ground for burgeoning designers, who may have little or no institutional training in design or computer programming but who, through their communities of practice, learn the craft of their possible profession. For the purposes of this chapter, I confine my engagement with the term *modification* to the practice involving game design at the code and game design levels. Tangential and related practices in hobby culture involve consumers modifying their automobiles, their writing implements, and their video game consoles and computers with "mod-chips" or stylized hardware cases. These elements of hobby/tinkerer culture are purposefully not given cursory exposition herein but rather left for other opportunities that allow deeper discussion. Because modification of code has had the most significant impact in blurring the boundaries between producers and consumers of video game content, I focus on it here.

To modify, to change, perchance to own: this riff on a line from Hamlet's famous soliloquy is admittedly an apt point of departure to understand the meaning of the term *modification* in game culture. Modification, in the context of productive participatory game culture, is commonly referred to as "modding" by its practitioners. It is a concerted and well-thought-out intervention into almost every element of a commercial game's design. Modders, practitioners of modding, may seek to change every element of a video game: its code, sound, textures, physics, architecture, communication tools, and rules of play. They will also, either by design or as a consequence of their productivity, add dimensions to the structure of the industry. The term *modding* itself, as it may apply to video games, has an indeterminate origin and can at best be traced back to the first modifications (mods) to video games in 1980 (Postigo 2003). When the practice first took hold, modification may have included substantive changes to the game engine source code, an admittedly far more technically difficult practice than modifying video game assets such as texture and sound elements. One can distinguish those modifications by thinking of them as hacks to the engine source, which may not have necessarily resulted in a discernable impact on the player experience.

Modifying Video Game Software and Design

The number of modifications and users modifying a game has grown historically. Today thousands of users are modifying games and distributing their modifications on the web. Mods can range from relatively simple rearrangements of in-game elements to total conversions. Total conversions are the most ambitious modifications because they attempt to convert gameplay completely. For example, a game originally designed as a first-person shooter, such as *Call of Duty* (Infinity Ward, 2003), might be modified to be a role-playing game set in feudal Japan. Modders not only produce changes to the games but also post on the Internet tutorials for various elements of the modification enterprise to encourage the novice hobbyist to contribute and learn the craft (Postigo 2003).

If any one video game title can be credited with being the first to afford modification to video game code, it is Silas Warner's classic *Castle Wolfenstein* (Muse Software, 1981). Developed in the 1980s for early Apple computers, *Castle Wolfenstein* was a success across the personal computer (PC) gaming market, and the user-generated modification—called *Castle Smurfenstein*—redesigned elements of this game away from its World War II context to include instead elements from the then-popular Saturday morning animated show *The Smurfs*. *Castle Smurfenstein* stands as arguably the first modification to commercial video game content (Au 2006).

We must remember that modification of video games was in its infancy at this time and those who engaged in it were still squarely situated within the emerging computer programming profession. What made modification at the user level possible was not so much any changes in coding languages or development tools but rather the availability of game source code among programming enthusiasts and initial penetration of PCs into the workplace and hobby culture. Hobbyist and tinkering culture and its adoption in video game culture are an extension of the hobbyist culture that moved the PC into the home. Computer hobbyists such as Michael Dell and Steve Jobs made an industry from their home-garage hobbies much the same way the radio and television transmitter tinkerers played a role in the emergence of the electronics industry in Japan (Takahashi 2000).

In the case of modifications to game code and a game's visual presentation, we see the evolution of some of the most technically and artistically talented user intervention into media production and media industries. As of this writing, the most modded games currently in the computer games market are *Skyrim* and *Minecraft*. *Skyrim*, released in 2011 by the well-established development company Bethesda Softworks, is a single-player role-playing game that is expansive and rich in artwork, world design, and narrative. Yet one of the most attractive elements of the game is that the user can create modifications to textures, character models, and gameplay rules via Bethesda's Creation Tool Kit and distribute them though Valve's Steam Workshop. As of September 2012, more than 10,000 modifications of *Skyrim* were available on the Steam Workshop alone. That figure does not include other modifications made available through other distribution venues. *Minecraft,* an "indie" development sensation developed by Marcus Persson in 2009 and released by Mojang in 2011, is premised largely on making modification the essence of play. *Minecraft* is categorized as a "sandbox"-style game,[2] so a significant portion of play is premised on breaking and building structures and environments and on exploring. Players are very creative and have built environments that are replications of the Notre Dame Cathedral in Paris (figure 38.1), entire cities (figure 38.2), the USS *Enterprise* (figure 38.3), and other architectural marvels. They have also built more modest structures such as huts and small homes. The scale of the total built content is vast, and the labor at play is staggering. With *Skyrim* and *Minecraft*, modification in the PC market is firmly part of the game design paradigm. In the console market, one of the few games to take this paradigm seriously

Figure 38.1
Replication of Notre Dame Cathedral made with *Minecraft* (Mojang, 2011). Image taken from trendminecraft.blogspot.com.

Figure 38.2

High Rossferry City. A city that was entirely constructed on *Minecraft* by players. Image taken from www.planetminecraft.com.

Figure 38.3

Replication of *Star Trek*'s USS *Enterprise* made with *Minecraft*. Image taken from http://www.pcgamer.com/2010/09/28/somebody-built-the-starship-enterprise-in-minecraft/.

was Sony's *Little Big Planet* (2008), but given the dearth of content in the game it did not see the success that *Skyrim* or *Minecraft* have seen.[3]

Modifying the Industry

User practices that modify game software and design remain an element in the video game industry's business model with respect to the PC market. Mods can play a role in extending the sales of the original game or in developing a devoted fan base. Previous study of modification in video games has attempted to approximate the value of the labor that goes into modifying a game title as a matter of hobby or to build a design portfolio (Postigo 2007). Top-selling games in the PC market typically allow users to generate modifications that in sum can approximate millions of dollars of labor costs. Sometimes modifications can do the work of "patching" bugs or design flaws before the game company has had a chance to release an official patch. The 2013 title *Total War: Rome II* for PC and Mac was highly anticipated, but media outlets reviewing the game noted it had significant design flaws. Within weeks of its release, however, user-generated modifications were addressing those issues long before the designer, Creative Assembly, began releasing its official patches.

Free labor and hobby culture notwithstanding, modification remains a practice that can afford users with an affinity for code and design a way of developing a portfolio that can be attractive to a game-development company. Although not the norm, modders have found employment in the industry, sometimes without any formal training. Modification as part of a team can be understood as apprenticeship. For those who have not undertaken formal education in coding or design, learning to modify from modification veterans in the video game fan community is a way of entering the world of game design, a world where they get to decide how a game looks, what its narratives will be, and how it will represent the world. Modding is empowering (Postigo 2010). The apprenticeship model puts pressure on the institutional validation that programming and design professions place on university degrees. When a relative novice, after years of apprenticeship in a modding group, can generate content on par with that produced by professionally trained programmers, it becomes difficult to justify the expense in time and resources needed to acquire a university degree.

On occasion, a modification can become a stand-alone game. The most well-documented cases of this transition are *Counter Strike* (1999), originally a modification of Valve's game *Half-Life*, and the *Team Fortress* franchise, originally a modification of id Software's game *Quake* (1996). These happy instances of success are few and far between, but they provide inspiration for narratives that drive modification as an *entry-point practice*. Hobby culture is not without precedent in this regard, so modding, as a technosocial practice, has a lineage that crosses technological platforms from automobiles to programming, ham radio, and YouTube stardom.

Modifying Hardware

In video game history, what began as software and design modification has spread to hardware as well. That users tinkered with and modified early PCs is an indisputable fact of the early PC history. Modification to video game hardware followed as a logical extension of that practice when users modified their Nintendo Entertainment Systems to work around the regional restrictions that prevented North American consumers from playing games sold exclusively in the Japanese market. Over time, consoles have been the site of some of the most elaborate modification attempts, made so difficult because of hardware design restrictions.

Over the years, consoles have been modified with "mod chips," processors soldered onto console motherboards to allow gamers to install emulators or break the copy-protection restrictions hard-coded on endogenous chips for PlayStation, Xbox, and Nintendo platforms. Beyond internal modification, mods also exist for controllers. Users will open Xbox controller cases, for example, and solder in place LED repeater lights over the trigger circuitry, which causes a series of trigger inputs to register at rates faster than a human finger can generate. This type of mod is considered a cheat by the gaming community, especially in games where quick twitch reflex is a competitive advantage, but it remains part of the controller mod repertoire. Another controller mod is the lag switch, also considered a cheat. With the switch spliced into the Ethernet cord connecting a console to the Internet, the lag switch can create an artificially strong latency in networked play for the players not hosting the game, so the host not only benefits from low latency but also creates a severe latency for other players. In the game, this modification is rendered as a player-character skipping across the visual field and experienced by other players as a freeze in frame rate.

Other controller modifications, such as those from Scuf Gaming for Xbox controllers, remap console controller buttons to installed paddles on the bottom of the controller (see figure 38.4). In the case of controller modifications, a third-party industry has taken hold, providing users with a redesigned controller that is more suitable to a particular genre of game (such as first-person shooters) or that may serve the needs of users living with disabilities that render use of their hands difficult.

Despite possible patent and copyright disputes, modification of hardware and software is generally accepted, and the game industry does not as a general practice pursue modification with punitive action in the form of injunctions or lawsuits. Regardless, there are some cases where modification has run aground of legal protections afforded to game and console

Figure 38.4
Scuf Gaming's Xbox controller Classic. Image taken from http://scufgaming.com/scuf-shop/.

manufacturers. In those instances, legal measures are enough to dissuade modification (Postigo 2008). What impact these measures may have on a video game's allure remains unknown, and third-party modification of hardware or software certainly remains a possible area of expansion for the game industry.

Critically Speaking

Modification is a rare user hobby that is clearly an example of free labor present in many business strategies that afford user-generated content. Free labor on the Internet is an extension of cultural production as a business venture with the added intervention and value of the appropriation and commodification of user-generated content, which Tiziana Terranova initially pointed out when in 2000 she considered the sort of maintenance work needed to sustain the sociality of the early web. Valuation of user-generated products has become a central concern for platform owners and users alike. As the ecology of social media platforms intersecting on the web grows (video game forums with modification-distribution capacity as well as Facebook and Twitter to inform users), who owns the content and who has legal rights to extract profit from derivative works that border on transformative are becoming pressing questions for the video game industry.

The boundaries between producers and consumers are effectively blurred in the case of game software modification and hardware tinkering. Because the results of software and game design modification can be so easily distributed over online venues, modification's impact as a massively distributed good is not to be ignored. Critical scholars have for some time concerned themselves with invisible labor. Beginning with feminist scholars who situated women's household labor as part of the supportive unwaged labor that capital is dependent on, critical researchers have been quick to point out other manifestations of invisible labor that remain unaccounted for in corporate ledgers but are clearly responsible for profit (Banks and Potts 2010; Kucklich 2010; Postigo 2009; Terranova 2004; Downey 2001). Because digital media afford digital goods a permanence and a persistence that other sites of invisible labor do not (the home, for example), it is fairly easy to catalog and account for the material goods generated via modification in the game industry.

Video game modification work is not hidden by patriarchy or the power of discourses that have been guilty of devaluing invisible work such as women's labor. Critical scholars have not found it hard to argue for a more critical approach to the concept of ownership and innovation in digital media goods such as video games. If digital media such as video games afford users the ability to intervene not only as part of participatory culture but also as part of a business enterprise, how can arguments premised on intellectual property rights conveyed solely to the original creator hold? In the case of modifications, that issue has already been contested. In moments of contention, as modders agree to refrain from infringing on third-party copyrights, they question the legitimacy of copyright over derivative works that can potentially be lucrative for game companies and third parties as well as recruit more players (Postigo 2008).

Conclusion

Modification in video games is a continuously evolving moving target. With the advent of mobile games and the increasing potential for augmented-reality interfaces such as Google Glass, it's difficult to claim any closure for the term *modification* or for its history because it is still in the making. That being said, the idea of modification has changed in some important regards. It is now integral to play and business in some genres; it involves hardware that has found a market niche; and for the very young, it serves as an alternative to formal training or as an entry point to design and programming. Despite these historical shifts, the ever-present regulatory force of patent and copyright continues to serve as a break on what has the potential to be a creative burst of Cambrian proportions. In that sense, the history of what has been accomplished by modification in game culture needs to account for what has not been accomplished because of regulatory inhibition—work that still remains to be done.

Notes

1. Modification herein is conceptualized as changes to code or design made by users to the original game content and distributed online. It does not include manufacturer patches or add-in version updates, nor does it include player glitch or bug exploits during gameplay.

2. In video game design, a game that has an open structure with only limited rule sets for victory or completion are considered sandbox games, where play is the central activity, and victory and reward conditions are secondary. In a sandbox game, good play is its own reward.

3. *Skyrim* has seen success on console platforms yet not to the degree it enjoys on the PC. *Minecraft* was ported to console, but, although content creation is possible via consoles, its distribution is limited.

Works Cited

Au, Wagner James. 2006. "Triumph of the Mod." *Salon,* June. http://dir.salon.com/story/tech/feature/2002/04/16/modding/index.html

Banks, John, and Jason Potts. 2010. "Co-creating Games: a Co-evolutionary Analysis." *New Media & Society* 12:253–270. doi:.10.1177/1461444809343563

Downey, Greg. 2001. "Virtual Webs, Physical Technologies, and Hidden Workers." *Technology and Culture* 42 (2): 209–235.

Kucklich, Julian. 2010. "Precarious Playbour: Modders and the Digital Games Industry." *Fibreculture*, February 5. http://journal.fibreculture.org/issue5/kucklich_print.html.

Postigo, Hector. 2010. "Modding to the Big Leagues: Exploring the Space between Modders and the Game Industry." *First Monday* 15 (5). http://firstmonday.org/htbin/cgiwrap/bin/ojs/index.php/fm/article/view/2972/2530.

Postigo, Hector. 2009. "America Online Volunteers: Lessons from an Early Co-production Community." *International Journal of Cultural Studies* 12 (5): 451–469.

Postigo, Hector. 2008. "Video Game Appropriation through Modifications: Attitudes Concerning Intellectual Property among Modders and Fans." *Convergence* (London) 14 (1): 59–74.

Postigo, Hector. 2007. "Of Mods and Modders: Chasing Down the Value of Fan-Based Digital Game Modifications." *Games and Culture* 2 (4): 300–312.

Postigo, Hector. 2003. "From *Pong* to Planet Quake: Post-industrial Transitions from Leisure to Work." *Information Communication and Society* 6 (4): 593–607.

Takahashi, Yuzo. 2000. "A Network of Tinkerers: The Advent of the Radio and Television Receiver Industry in Japan." *Technology and Culture* 41 (3): 460–484.

Terranova, Tiziana. 2004. *Network Culture: Politics for the Information Age*. London: Pluto Press. http://www.loc.gov/catdir/toc/ecip0415/2004004513.html.

Terranova, Tiziana. 2000. "Free Labor: Producing Culture for the Digital Economy." *Social Text* 18 (2): 63.

39 NARRATIVE

Marie-Laure Ryan

Playing games and exchanging stories are possibly the oldest and certainly the most widespread forms of cultural entertainment. Evolutionary psychologists attribute them to a cognitive ability that presents great adaptive advantages (Boyd 2009), the ability to explore the realm of possibilities: possible moves for games, possible worlds and events for stories.

The temptation is strong to try to combine the strategic pleasure of games with the imaginative pleasure of stories for an even more fulfilling experience. An early attempt to achieve this combination was to decorate game boards—especially the boards of dice games played on monocursal labyrinths, such as *Chutes and Ladders* or what is known in French as *Le jeu de l'oie*—according to narrative themes. From *The Path to Good Life and Heaven*, the ancient Indian version of this game, to the *Labyrinth of Ariosto* in the seventeenth-century, which represented episodes of the poem *Orlando Furioso*, and the *Game of Life* in the nineteenth century, decorated games boards injected narrative interest into a set of rules that allows few or no choices.

Whereas decorated game boards can be considered an attempt to narrativize an abstract game, the converse project of "gamifying" or playing concrete stories is represented by the strategic wargames that developed from the nineteenth century on. Based on either imaginary situations or real historical events (Napoleon's battles, the American Civil War, and World War II are favorites), wargames simulate military operations by means of props, dice, and a rulebook. Considered the precursors of *Dungeons & Dragons* (Gygax and Arneson/Tactical Studies Rules, 1974), those table-top role-playing games focus on the enactment of variations on a loosely scripted narrative scenario through a combination of dice throws that determine events and players' dramatic improvisations.

The convergence of narrative and play gained considerable momentum with the development of digital technology. Though the historical approach to video games is less productive with respect to narrative than with respect to other topics because of the difficulty of identifying the first appearance of story-enabling technological features, because games (like films and novels) can present variable degrees of narrativity, and, last but not least, because of the difficulty of defining what a narrative is, we can safely claim that the development of narrative video games is closely connected to the growth of computing power. In part owing to the low-resolution graphics of early personal computers (PCs), computer games of the 1980s, such as *Pac-Man* (Namco, 1980) and *Tetris* (ELORG, 1984) were played on abstract playfields that looked more like a chessboard or a soccer field than a concrete world. On a soccer field, the lines correspond to rules, which means that they are strategically significant, but they do not represent anything that exists independently of the game, such as people or cars. The playfields of early PC, arcade, and console games were usually shown in map or elevation view, projections in which no object hides any other, but the objects are reduced to barely recognizable shapes. Low-resolution graphics did not, however, prevent producers from courting narrative interest as they packaged games in boxes decorated with realistic images of fantastic worlds. An extreme example of this discrepancy between the art on the box and the appearance of the display is *Ultima I: The First Age of Darkness*, developed by Richard Garriott and released in 1981 for the Apple II (Demaria and Wilson 2002, 120). It took a great deal of imagination to see the shapes on the screen as dragons or princesses, but experiments conducted in the 1940s with films featuring squares and circles had already suggested that people have a spontaneous tendency to narrativize the movements of abstract shapes (Heider and Simmel 1944). *Pac-Man*, for instance, is easily interpreted as the attempt of the "good guy" (the shape controlled by the player) to escape monsters intent on killing him (or her, as in *Ms. Pac-Man* [Namco, 1982]). The fact that the monsters have names is a strong incentive to narrative interpretation.

With the improvement of graphic resolution and processing speed, the space of computer games evolved from abstract playfields to concrete worlds seen from the perspective of a human body, and the representation of agents evolved from simple shapes to realistic images of persons, animals, and supernatural creatures. Once game space became a world, it was not difficult to imagine this world as the setting of a story, the characters as intelligent, autonomous agents engaged in the pursuit of their personal interests, and the player's actions as the events that help the characters reach their goals. Setting, characters, and actions: these are the basic ingredients of narrativity. The development of games in the first decades of the twenty-first century has only accelerated this trend toward narrativization. Although the best-seller charts remain dominated by shooters with a rather stereotyped plot (save the earth or your country or civilization by killing the bad guys), many games devote increasingly sophisticated resources to both the story and its presentation. Characters are now voiced by highly trained actors, and their movements are made more realistic through body-capture techniques. Moreover, rather than treating nonplaying characters (NPCs) as one-dimensional enemies or helpers, games may now present them as mentally complex creatures who inspire emotions: players may be faced with tough moral decisions when they have to eliminate NPCs to progress in the game (Nitsche 2008). Meanwhile, narrative discourse (i.e., the dynamic presentation of the story) has adopted many of the techniques of film and literary narratives, such as flashbacks, stream of consciousness (*Heavy Rain* [Quantic Dream, 2010]), and alternation between reality and dream sequences (*Max Payne* [Remedy Entertainment, 2001]). Cinematic cutscenes, during which the player cannot interact, are increasingly used to build up the game's narrative scaffolding. In many independent art games and so-called serious games, the story being presented takes precedence over gameplay. The purpose of games such as *Darfur is Dying* (Susana Ruiz, 2006), *Escape from Woomera* (never released),[1] and *1979: The Game* (about the Iranian Revolution, forthcoming at the time of this writing) is to make players aware of real-world events and get them to empathize with the victims rather than to challenge their problem-solving skills or eye–hand coordination. But it would be simplistic to regard game narrativity as totally dependent on the realism of representation. As Jesper Juul points out, a game such as *The Marriage* by Rod Humble (2006) reverts to telling stories by means of abstract shapes: the partners are represented by squares and the influences on their relationship by circles (2014, 181).

To regard games as capable of narrativity, it is necessary to expand both the notion of game and the notion of narrative beyond their standard manifestations. Juul has defined the prototypical game situation through a set of conditions inspired by a game such as chess: based on rules; leading to different, quantifiable outcomes to which players attach value (they want to win and hate to lose); requiring player effort; and presenting negotiable consequences (you can play for money, for the title of world champion, for your life, or just for fun) (2004, 30). Gerald Prince has defined the prototypical narrative situation as the recounting of events by a narrator to narratee (1987, 58), though he later modified it to make it compatible with drama. The phenomena of both games and narratives are infinitely more diversified than these prototypes. Roger Caillois proposed a game typology that includes not only *ludus*, or games based on rules and competition, but also *paidia*, or games involving free play and characterized by "diversion, turbulence, free improvisation, and careful gaiety" ([1958] 2001, 13). *Paidia* games are much more narrative-friendly than pure *ludus* games; one needs only think of the fictional scenarios invented and enacted by children during games of make-believe, which evolutionary psychologists regard as the wellspring of narrative fiction (Boyd 2009). Far from being limited to competition, computer games run the full spectrum from *ludus* (exemplified by first-person shooters) to *paidia* (represented by *The Sims* [EA Maxis, 2000], *Dance Dance Revolution* [Konami, 1998], and *Second Life* [Linden Lab, 2003]), and they have developed many original ways to combine these two types. As for the definition of narrative given earlier, it works well for oral storytelling and literary fiction, but it cannot account for dramatic modes of presentation. (Or, to borrow Plato's terminology, it works for diegetic but not mimetic narrative modes.) This limitation can be circumvented by defining narrative as a use of signs that activates in the receiver a mental representation that fulfills a certain number of semantic requirements, such as being about characters who take actions and change the state of the world. If we call this mental representation a story, it is obvious that many games, especially computer games, have storytelling abilities.

Regarding play as compatible with story does not mean, however, that games can match the narrative diversity of literature and film. The problem here is that game stories must be interactive; in other words, their plot must integrate the user's input in a nontrivial way, whereas the plots of standard narratives are controlled by the author. Interactive storytelling is the object of intense activity among specialists in artificial intelligence and scholars of digital media (for instance, in the annual International Conference for Interactive Digital Storytelling), but the magic formula that would solve what Ruth Aylett and Sandy Louchart have called the "narrative paradox" (2003, 244)—reconciling the author's top-down design with the user's bottom-up input in a way that gives the user a sense of freedom while satisfying her desire for a well-formed story—is still eluding researchers. Part of the problem resides in the kind of activity that game controls afford. Whereas in life as well as in literature and film people (or characters) interact with each other through both language and physical actions, in computer systems interaction is limited largely to physical gestures because it would take very advanced language-understanding and language-generating systems to sustain a coherent verbal exchange between player-characters and NPCs. The repertory of actions that can be easily simulated through game controls consists of moving around the game world, picking up objects, using them (i.e., activating the behaviors attached to these objects), and selecting items from a menu. The menu can be used to simulate a variety of actions that cannot be directly performed through game controls, especially conversations with NPCs, but it breaks the flow of the game and hinders immersion. The limited repertory of controls-compatible actions explains in part the predominance of violent physical action in video games.

From a structural point of view, game narratives can be classified into four broad, occasionally overlapping categories: the journey or epic narrative; the mystery or epistemic narrative; the world or spatial narrative; and the system or emergent narrative. (The second and fourth categories are also part of the system proposed in Jenkins 2004; for another taxonomy based on different criteria, see Ryan 2006, 107–122.)

The journey narrative focuses on the adventures of a solitary hero in a world full of danger. Its prototype is the archetypal narrative pattern described by Vladimir Propp (1968) and Joseph Campbell ([1968] 1973): a hero receives a mission, fulfills it by performing various tasks, and gets rewarded if he succeeds. In the journey pattern, the story can be endlessly expanded by giving new tasks to the hero and by adding new episodes, which correspond to distinct levels or sequels. The player may have the impression of moving freely through the game world, but he or she progresses along a storyline that has been fully plotted by the designers, even when the plot presents multiple branches. After each level is completed, the branches come together so as to place the game world in the same state for all players at the beginning of the next level. Juul calls this structural type "games of progression," and he observes that because of the designers' strong top-down control, the game of progression is the most amenable to narrativization (2004, 5).

The literary forbear of our second structural type, the mystery or epistemic plot, is the detective story, a genre that arose in the nineteenth century. As Henry Jenkins observes, the mystery story structure is characterized by the embedding of a fixed story, the events that took place in the past, in a variable story constituted by the investigation (2004, 126). The epistemic plot runs a close second to the epic plot in its compatibility with user interaction. It casts the player in the well-defined role of detective, combines an authorially defined story with a variable story created in real time by the actions of the player, takes advantage of the visual resources of digital systems by sending the player on a search for clues disseminated throughout the game world, and is fully compatible with the types of action that can be easily performed by game controls: moving across the world, finding documents, examining them for clues, and interrogating NPCs through a menu of canned questions.

The third kind of narrative configuration, the world design, is found in single-player games such as *Grand Theft Auto* (Rockstar Games, 1997) and in multiplayer online worlds such as *World of Warcraft* (Blizzard Entertainment, 2004). I call it *spatial narrative*, a term suggested by Jenkins (2004) because it does not consist of a linear progression along a predetermined plot but rather of the exploration of a two-dimensional (or maybe three-dimensional) space that contains not just one but multiple stories. These stories are attached to specific locations within the game world. For instance, when the player reaches a certain place, she may encounter an NPC who will tell her about the legends, gossip, and history relating to this place. Further in her wandering, she may meet another NPC who will give her a task to fulfill—a quest—that will take her to various other locations. Quests consist typically of a backstory told by the NPC that explains the need for action and of a sequence of steps that, when performed by the player, will complete the story. The structure of the individual quests is therefore similar to the structure of the journey narrative, but the world design does not present an overarching story, and the world rarely evolves as a whole. In multiplayer games, the player's actions have an impact on the avatar's life because they may lead to his death or to his advancement, but because the same task can be performed by many different players, its performance has no lasting impact for the game world. If a player slays a dragon as part of a quest, the dragon will "respawn" when another player reaches the same spot.

In the fourth kind of narrative, the emergent, there is no prescripted story but rather a system of existents—characters and objects—capable of various behaviors. The player creates a story by manipulating these entities from an external or "god" perspective. When the system contains many existents capable of a variety of actions, and when these actions have side effects for other existents, the system becomes too complex to be predicable, and the stories that it produces acquire a quality of emergence. The prototype of this type of structure is *The Sims*, a game in which the player creates a family and selects from a menu the actions performed by its various members (Ryan 2011). These actions have an impact on the mental state of their participants and affect how they feel about each other. For instance, if a character constantly irritates another and then asks her to marry him, it is likely—but not certain—that he will be rejected; if he is, he may sink into depression, lose his job, and have to sell his house. When the player selects an action, the system computes its consequences and updates the current state of the game world, opening up a new set of possible actions. The computer can also play the role of blind fate by occasionally throwing in random events, such as neighbors dropping by unexpectedly or death claiming a character. Yet even when the computer takes a turn at implementing events, it does not operate on the basis of predetermined narrative templates. The game simulates the randomness of life rather than the teleology of most narrative genres, but we can nevertheless call its output a story because it consists of a causally connected sequence of events affecting individual characters. Investigating what stories the system is capable of producing is the point of the game—a point that captures in its purest essence the spirit of *paidia*.

Will this taxonomy still be valid ten years from now, or will radically new hybrids of game and narrative develop? In other words, is the taxonomy based on logical considerations reflecting permanent properties of narrative and inherent limitations of digital media so that future development will be variations on the same basic categories, or are games an "emergent form undergoing rapid technological change and aesthetic experimentation" (Jenkins 2005, 80) whose narrative future cannot be predicted? It is too early to answer this question.

Note

1. An incomplete version was released in 2014.

Works Cited

Boyd, Brian. 2009. *On the Origin of Stories: Evolution, Cognition, and Fiction*. Cambridge, MA: Belknap Press of Harvard University Press.

Caillois, Roger. [1958] 2001. *Man, Play, and Games*. Trans. M. Burach. Urbana: University of Illinois Press.

Campbell, Joseph. [1968] 1973. *The Hero with a Thousand Faces*. 2nd ed. Princeton, NJ: Princeton University Press.

Demaria, Rusel, and Johnny L. Wilson. 2002. *High Score! The Illustrated History of Electronic Games*. Berkeley, CA: McGraw Hill and Osborne.

Heider, Fritz, and Marianne Simmel. 1944. "An Experimental Study of Apparent Behavior." *American Journal of Psychology* 57:243–259.

Jenkins, Henry. 2005. "Computer Games and Narrative." In *Routledge Encyclopedia of Narrative Theory*, ed. David Herman, Manfred Jahn, and Marie-Laure Ryan, 80–82. London: Routledge.

Jenkins, Henry. 2004. "Game Design as Narrative Architecture." In *First Person: New Media as Story, Performance, and Game*, ed. Pat Harrigan and Noah Wardrip-Fruin, 118–130. Cambridge, MA: MIT Press.

Juul, Jesper. 2014. "On Absent Carrot Sticks: The Level of Abstraction in Video Games." In *Storyworlds across Media: Toward a Media-Conscious Narratology*, ed. Marie-Laure Ryan and Jan-Noël Thon, 173–192. Lincoln: University of Nebraska Press.

Juul, Jesper. 2004. *Half-Real: Video Games between Real Rules and Fictional Worlds*. Cambridge, MA: MIT Press.

Louchart, Sandy, and Ruth Aylett. 2003. "Solving the Narrative Paradox in VEs—Lessons from RPGs." In *Intelligent Virtual Agents*, ed. Thomas Rist, Ruth Aylett, Daniel Ballin, and Jeff Rickel, 244–248. Berlin: Springer.

Nitsche, Michael. 2008. *Video Game Space*. Cambridge, MA: MIT Press.

Prince, Gerald. 1987. *The Dictionary of Narratology*. Lincoln: University of Nebraska Press.

Propp, Vladimir. 1968. *Morphology of the Folk Tale*. Trans. L. Scott. Revised by A. Wagner. Austin: University of Texas Press.

Ryan, Marie-Laure. 2011. "Peeling the Interactive Onion: Layers of User Participation in Digital Narrative Texts." In *New Narratives: Stories and Storytelling in the Digital Age*, ed. Ruth Page and Bronwen Thomas, 35–62. Lincoln: University of Nebraska Press.

Ryan, Marie-Laure. 2006. *Avatars of Story*. Minneapolis: University of Minnesota Press.

40 PLATFORM

Caetlin Benson-Allott

Scholars writing about computational and game platforms exhibit a peculiar tendency to insist on (and then allegedly resolve) a quintessential confusion about their key term. Whatever a *platform* is, it is neither self-evident nor easily defined; we do not know it when we see it or even when we read about it. And yet, these authors insist, platforms are the fundamental and necessary bases of game design, game development, gameplay, and, by extension, game theory. In their attempt to inaugurate a field of *platform studies*, Nick Montfort and Ian Bogost describe a platform as "an abstraction," "an operating system," "whatever the programmer takes for granted when developing, and whatever, from another side, the user is required to have working in order to use a particular software" (2009, 2). In that article, they insist that "a platform must take a physical form" (2009, 2), but, less than a year later, they would acknowledge that "the question of whether something is or isn't a platform may not ever have a useful answer[;] the real question should be whether a particular system is influential and important *as a platform*" (Bogost and Montfort 2009). Yet if a platform can be defined as whatever is influential and important as a platform, then how is one to articulate the unique role that platforms play in video games? Is a platform a concept, a material thing, or—as Bogost and Montfort's final definition suggests—a tautology? The answer, of course, is that it is all three because platforms represent the inextricable combination of meaning and matter. They are where the rubber meets the road (or the silicon meets the code). As that metaphor suggests, platforms are necessarily both concepts and things. They teach us that things have the same power and agency as concepts and as subjects. Inasmuch as platforms are the basis of computer and video games, they point to the fundamental imbrication and interdependence of the material, the cultural, and the political.

Thus, the term *platform* is no more or less confusing than its twenty-six-page definition in the *Oxford English Dictionary* (*OED Online*, 2014]) would suggest. When platform theorist Tarleton Gillespie notes that "a term like 'platform' does not drop from the sky," he could hardly be more correct; its etymology goes all the way back to the sixteenth century (Gillespie 2010,

359). It comes from the Middle French *platte forme*, meaning raised or level ground "on which something may stand" (*OED*). As Peter Keating and Alberto Cambrosio suggest, the original geological definition of the term then spread by metonymy until it denoted all kinds of supporting structures, including railway platforms, artillery platforms, oil-drilling platforms, diving platforms, stage platforms, and even platform shoes.[1] *Platform games* continue to employ the term in this architectural sense, as their gameplay comprises avatars jumping, climbing, swinging, and falling between different stages and structures (e.g., *Donkey Kong* [Nintendo, 1981], *Super Mario Bros.* [Nintendo, 1985]).

Over the course of the sixteenth and seventeenth centuries, the term *platform* also took on metaphorical meaning as the starting point or basis for action. From "a raised level surface on which people or things can stand," it quickly came to connote "the ground, foundation, or basis *of* an action, event, calculation," or plot (*OED*). The English philologist Reverend Walter W. Skeat notes that *platform* is used synonymously with *plot* in several sixteenth-century plays and suggests that *plot* may in fact be a derivation of *platform* (1901, 219–220). To this day, the *OED* continues to define *platform* as "a plan, a strategy" as well as "a design, a concept, an idea" and "a plan or scheme of government or administration; a plan of political action." Thus, *platform* often refers to "the major policy or set of policies on which a political party (or later also an individual politician) proposes to stand," a usage that could derive from either the geological or the tactical etymologies of the term or from both.

All these discursive lineages come together in the term *computational* or *game platform*, although they do not do so cleanly or harmoniously. The *OED* defines a computer platform as "a standard system architecture; a (type of) machine and/or operating system, regarded as the base on which software applications can run." It sounds so simple: the platform is the computer's infrastructure, the physical and organizational resources necessary for its functioning. But that "and/or" gestures to greater confusion, even controversy, over what counts as a computer platform. Philosophically minded game scholars such as Bobby Schweizer suggest that, "broadly speaking, a platform is anything built that makes it easier to build other things" (2012, 488). Schweizer's definition can include hardware, such as the custom-designed 8-bit Sharp LR35902 central processing unit (CPU) in Nintendo's first-generation Game Boy devices or software, such as Xbox 360's Dashboard operating system, or both. Such recursive logic suggests that platforms often—maybe even always—rely on and work with other platforms. Dashboard could not run without the Xbox 360's Xenon processor, for instance, nor could it play Xbox games without DVDs, which require additional hardware and decryption software. For most self-declared platform scholars, though, the term *platform* has to refer to computational infrastructure, to "the (so far neglected) specific basis for digital media work" (Bogost and Montfort 2009). Taking a stand against the recursivists, some even go so far as to claim, "If you can program it, then it's a platform. If you can't, then it's not" (Andreessen 2007). By that logic, the platform is the bedrock of computation. It cannot include hardware such as cathode ray monitors or DVDs because they cannot be programmed—even though one can hardly imagine the history of console-based video games without them.

In sum, the diverse etymological history of the term *platform* reflects ongoing confusion and conflict regarding what counts as a computer or game platform. Must it be material? Must it be programmable? As Montfort and Bogost suggest, such hair splitting is antithetical to both the intrinsic polyvalence of the term *platform* and the basic supposition of platform studies—namely, that we need to pay more attention to the ways that different infrastructures shape games, developers, and players.

Game scholars understand platform studies as a reaction against the ideological game analyses and narratology/ludology debates of the 1990s and early 2000s. However, one might equally consider our critical interest in platforms as part of a larger scholarly return to *materialism*—in other words, to asking how *things* and *matter* produce action and meaning in the world. By this logic, platforms matter because they are matter, as are we, as is this book, as is everything else in the universe. Platforms need to be studied because their limitations and affordances influence the creative works produced on them. An 8-bit processor cannot render photorealistic three-dimensional space, for instance, so game developers working with this processor will necessarily design their game space and aesthetics for two or two-and-a-half dimensions. As this example illustrates, platforms have consequences. By studying platforms as *things*, we uncover their political agency and significance. We also open ourselves up to a new ethical relation to the wider world.

Things, according to political philosopher Jane Bennett, are "vivid entities not entirely reducible to the contexts in which [human] subjects set them, never entirely exhausted by their semiotics" (2010, 5). To recognize an object as a thing, one must realize that it has an existence, a power in the world beyond its mere utility or associations for people. Things have "*Thing-Power*: the curious ability of inanimate things to animate, to act, to produce effects dramatic and subtle" (Bennett 2010, 6). Things can alter other things and courses of events. A fire can alter the shape of a house and the lives of the people, dogs, mice, and termites that live in that house; a television set can alter the activities, biorhythms, and even the body composition of the people who watch it. As things, platforms are part of an assemblage of human and nonhuman elements that together make something happen, such as a game. When I play *Candy Crush Saga* (King, 2012) on my iPhone, for instance, my hands, my iPhone, and the operating system iOS interact—or *intra-act*, as quantum physicist Karen Barad (2007, 33) puts it—to produce meaning, such as matching various candies to solve problems for the mascots of different game levels. My hands, the iPhone, and the electrical impulses of iOS together make different outcomes possible and different histories intelligible. As my right pointer finger begins to tire from the strain of flicking back and forth across the iPhone's touch screen, its material exhaustion may remind me of the material exhaustion of Foxconn factory workers who assemble the iPhone touch screen. When iOS glitches and *Candy Crush Saga* suddenly quits, I may also reflect on the planned obsolescence built into iOS's recurring updates, which render the system too complicated for my poor iPhone4 to manage and lead me to covet a brand new iPhone 5S. Thus do our encounters with things teach us to feel the world in new ways, both physically and emotionally.

Things, as physical matter, are "produced and productive, generated and generative" (Barad 2007, 137). All matter comes with history and produces more history; it is generated by natural and cultural forces and is itself generative of natural and cultural forces. For instance, the iPhone on which I play *Candy Crush Saga* requires rare-earth minerals, many of which are mined in Asia, South American, and Africa under deplorable environmental and political conditions. Before the Dodd–Frank Wall Street Reform and Consumer Protection Act went into effect in 2010, certain "conflict minerals" used to make consumer electronics for the United States came from militia-controlled mines in Central Africa (Bleasdale 2013). iPhones require tantalum capacitors, as do many other computational devices, and the United Nations has linked the capacitors used in iPhones to the tantalum mined in war-torn regions of eastern Congo (Gilson 2010). In short, the platform undergirding *Candy Crush Saga* is undergirded by a history of colonialism going back to King Leopold II of Belgium. That history is part of what makes an iPhone a thing, especially in that it is also productive of further violence because Apple's strategy of planned obsolescence produces an insatiable demand for conflict minerals. This is why Barad holds that "matter and meaning are not separate elements. They are inextricably fused together, and no event, no matter how energetic, can tear them asunder" (2007, 3).[2]

Thinking about platforms as things, then, can be a way of thinking about the history of video games as a political, cultural, and deeply material world history. Platforms show us how the material hardware that shapes game design and the experience of play is also shaped by larger social forces that determine who gets to play, who works to enable the play of others, and how our play is part of and affects the world in which it occurs. The lack of attention to materialism in game studies has been a lack of attention not just to chips and wires but to the relation of chips and wires to other things. Media archaeologist and cultural historian Laine Nooney (2013) therefore urges game scholars to consider not just the computing platform on which a game is designed but also the furniture underneath that computer. Nooney argues that we cannot just add Game Boy CPUs, Xbox Dashboards, and the experience of the Congolese to the way we think about video games. Rather, inasmuch as platforms are infrastructures, we need to account for how other sociomaterial systems condition the people and things that produce computational platforms.

One such system that Nooney suggests game scholars consider is gender. "Gender," Nooney writes, "is an *infrastructure* that profoundly affects who has access to what kinds of historical possibilities at a specific moment in time and space" (2013). She analyzes the work of Sierra On-Line game designer Roberta Williams to demonstrate how gender roles shaped the games Williams created. As a wife and mother in central California in the late 1970s, Williams designed her first game, *Mystery House* (On-Line Systems, 1980), on the platforms she had at hand—legal pads, unrolled wrapping paper, and the dining-room table: "The table would have been the *most obvious* surface in the Williams' home large enough to map on, and in most suburban tract houses, especially those of the California ranch style, layouts were such that kitchens gave the optimal domestic observation—a mother could 'work' in a kitchen and still observe the play of her children in another 'room' of the home" (Nooney 2013). Nooney argues that Williams's gender had a material and thus a socially meaningful impact on *Mystery House*, an adventure game for the Apple II. Some platform scholars might stop their investigation at the Apple II, with an analysis of how the five-and-a-quarter-inch floppy disc and Apple DOS influenced *Mystery House*'s structure and its player's experience. Nooney challenges them to look deeper, to consider the platform underneath the platform, the relations among gender, wrapping paper, tract houses, and game maps that make this intra-action of hardware and software, this game experience, possible.

Let's return to Marc Andreessen's claim that "if you can program it, then it's a platform. If you can't, then it's not" (2007). Andreessen makes this claim to restrict the possible meaning of the term *platform*, to limit it to computational platforms. The trouble with his definition is that the term *program* can mean both "to cause (a computer or other device) automatically to do a chosen task or perform in a chosen way" and "to train or condition (an animal, a person) to behave, act, or think in a predetermined way" (*OED*). By Andreessen's own definition, then, people are platforms too. That is the ethical insight that platforms can offer game developers, players, and scholars. Platforms are the material bases of games; they are the *thingness* of games that allows us to recognize how our own thingness works with that of a console or operating system and by extension the larger material and political world of which we all are part. Barad writes that "knowing is a matter of part of the world making itself intelligible to another part" (2007, 185). Platforms remind us that games are part of the world, as are we, and that we shape each other through our intra-action. Studying platforms, then, we study (or we *should* study) how games come to exist in the world, how they are produced, and what they produce. Thus, a platform is a concept, a thing, and a philosophy.

Notes

1. For even more thorough and typologized etymologies of the term *platform*, see Gillespie (2010, 349–350) and Keating and Cambrosio (2003, 26–30).

2. For models of such environmentally minded platform studies, see Parks (2014) and (2015).

Works Cited

Andreessen, Marc. 2007. "Analyzing the Facebook Platform, Three Weeks In." http://web.archive.org/web/20071021003047/blog.pmarca.com/2007/06/analyzing_the_f.html. Accessed April 18, 2014.

Barad, Karen. 2007. *Meeting the Universe Halfway: Quantum Physics and the Entanglement of Matter and Meaning*. Durham, NC: Duke University Press.

Bennett, Jane. 2010. *Vibrant Matter*. Durham, NC: Duke University Press.

Bleasdale, Marcus. 2013. "The Price of Precious." *National Geographic*, October. http://ngm.nationalgeographic.com/2013/10/conflict-minerals/gettleman-text. Accessed April 18, 2014.

Bogost, Ian, and Nick Montfort. 2009. "Platform Studies: Frequently Questioned Answers." *Digital Arts and Culture*, December 12–15. http://escholarship.org/uc/item/01r0k9br. Accessed April 18, 2014.

Gillespie, Tarleton. 2010. "The Politics of Platforms." *New Media & Society* 12 (3): 347–364.

Gilson, Dave. 2010. "The Scary Truth about Your iPhone." *Mother Jones* 31 (March). http://www.motherjones.com/environment/2010/03/scary-truth-about-your-iphone. Accessed April 18, 2014.

Keating, Peter, and Alberto Cambrosio. 2003. *Biomedical Platforms: Realigning the Normal and the Pathological in Late-Twentieth-Century Medicine*. Cambridge, MA: MIT Press.

Montfort, Nick, and Ian Bogost. 2009. *Racing the Beam: The Atari Video Computer System*. Cambridge, MA: MIT Press.

Nooney, Laine. 2013. "A Pedestal, a Table, a Love Letter: Archaeologies of Gender in Videogame History." *Game Studies* 13 (2). http://gamestudies.org/1302/articles/nooney. Accessed April 18, 2014.

Parks, Lisa. 2015. "Water/Access/Energy: An Ethnographic Analysis of ICT Infrastructure in Rural Zambia." In *Signal Traffic: Critical Studies of Media Infrastructures*, ed. Lisa Parks and Nicole Starosielski, 115–136. Champaign: University of Illinois Press.

Parks, Lisa. 2014. "Energy-Media Vignettes." *FLOWTV*, March. http://flowtv.org/2014/03/energy-media-vignettes. Accessed April 18, 2014.

Schweizer, Bobby. 2012. "Platforms." In *Encyclopedia of Video Games: The Culture, Technology, and Art of Gaming*, ed. Mark J. P. Wolf, 488–489. Westport, CT: Greenwood.

Skeat, Walter W. 1901. "Plot." In *Notes of English Etymology*, 219–220. Oxford: Clarendon Press. https://archive.org/details/notesonenglishet00skeauoft. Accessed April 18, 2014.

41 PLAYING

Jesper Juul

Many animals besides humans—notably mammals and birds—exhibit play behavior (Burghardt 2006), but here I discuss something more specific—namely *humans playing games.* We can think of *games* as rule-bound and mostly goal-oriented activities that differ from other goal-oriented activities in that the primary consequences of game playing are negotiable rather than obligatory (Juul 2005, chap. 2). In addition, we tend to discuss game playing as a voluntary act whose primary purpose is entertainment.

Game playing is therefore a subset of the larger set of play activities, which points directly to a juxtaposition: *play* is broadly associated with free-form and voluntary activities, yet games are also defined by rule structures that in part limit what players can do. This juxtaposition contains the fundamental question of game playing: Is game playing a free activity, or is it determined and controlled by the game rules?

Four Conceptions of Game Playing

There have historically been four central conceptions of the act of playing a game. They fall roughly on a scale from the assumption that the game dictates the playing of the game to the assumption that the player essentially creates a game by playing it. The four conceptions are:

1. Playing as **submission**, where the player is bound by the limits set forth by the game rules.
2. Playing as **constrained freedom**, where the game creates a space in which players acquire a certain amount of freedom and the opportunity to perform particular acts.
3. Playing as **subversion**, where the player works around both the designer's intentions and the game object's apparent limitations.
4. Playing as **creation**, where the game is ultimately irrelevant for (or at least secondary to) the actual playing.

Broadly stated, the first two conceptions are *game-centric* in that they focus on the game design's contribution to the game-playing activity, whereas the two latter conceptions are *player-centric* in that they emphasize the player's contribution. Generally speaking, the player-centric conceptions of game playing have appeared quite recently and often as a reaction to earlier game-centric theories.

These four conceptions are general claims about all game playing. I discuss hybrid and prescriptive conceptions later in the chapter and question whether it is at all correct to consider the issue of game playing as a conflict between games and players. In addition, each of the conceptions can be used to cast games in either a positive or a negative light.

Submission

In *Truth and Method*, philosopher Hans-Georg Gadamer argues that play is a type of submission to the game: "The real subject of the game (this is shown precisely in those experiences in which there is only a single player) is not the player but the game itself. What holds the player in its spell, draws him into play, is the game itself" (2004, 106). For Gadamer, this is not a bad thing; it is simply a particular trait of playfulness that Gadamer sees in both games and art. Many critical views of video games agree with Gadamer that game playing is a kind of submission but rate this relationship as profoundly *negative*, describing players as being controlled by the game to the detriment of their free ability to act or make decisions. For example, in *Mind at Play* (1983) Geoffrey Loftus and Elizabeth Loftus discuss elements of video game design in terms of Skinner boxes, referring explicitly to behaviorist experiments with rats.

Specifying the idea of game playing as submission, Scott Rettberg has argued that *World of Warcraft* (Blizzard Entertainment, 2004) involves players in a Protestant work ethic and that "the game is training a generation of good corporate citizens" (2008, 20). Game playing can thus be devalued on the assumption that submission is negative, but the same conception can alternatively be viewed through a positive lens because this very control trains players for future experiences of adversity. Benjamin Franklin thus extolled the lessons of perseverance that chess playing could teach the player: "We learn by chess the habit of not being discouraged by present bad appearances in the state of our affairs, the habit of hoping for a favorable change, and that of persevering in the search of resources" (1786, n.p.).

Constrained Freedom

Johan Huizinga famously declared play to be a voluntary activity, one that we can then use for "stepping out of 'real' life into a temporary sphere of activity with a disposition all of its own" (1950, 8). This definition does not posit game playing as a moment of absolute freedom but rather as a situation where players are given a freedom that is usually temporally and spatially delimited—that is, taking place within a magic circle (Salen and Zimmerman 2004; Huizinga 1950). Thus, conceiving game playing as a type of constrained freedom assumes that this freedom is *enabled* by the game design. For example, Katie Salen and Eric Zimmerman say that "play is free movement within a more rigid structure" (2004, 304). Bernard Suits (1978) similarly argues that although game rules on one level do limit player options—for example, by forbidding the golf player to carry the ball to the hole by hand or by disallowing chess players to introduce an extra king of their own—game rules *also* enable new meaningful actions that would not be possible without the game (see also Juul 2005). It is only when the rules of chess have been specified for a check or a checkmate that these actions become available and meaningful.

This conception of constrained freedom connects game playing with other art forms. Ragnhild Tronstad (2001) and Clara Fernández-Vara (2005) have in particular compared game playing to theater and to performances more generally. Fernández-Vara argues that "the performance of the player is a negotiation between scripted behaviours and improvisation based on the system" (2005, 7). Game playing, then, is like acting in a play, where actors are valued for their ability to express fixed material in interesting ways.

Subversion

Playing can also be construed as a type of subversion, where the player overcomes both the designer's intentions and the game object's apparent limitations. The difference between constrained freedom and subversion is that the former conceives the game as an enabler of player activity, but the latter, like submission, sees the game as fundamentally limiting. The subversion perspective argues that players actively overcome the limitations of the game structure as well as the game developers' possible intentions. For example, in *Cheating* (2007) Mia Consalvo stresses how players may act against designer intentions. Espen Aarseth (2007), following Gadamer, describes submission as the general law of game playing but argues that player subversions and transgressions are moments of "hope" when players see the possibility of escaping the submission that their game playing implies.

Placing subversion in a historical context, Jonas Heide Smith (2006) argues that for various historical reasons, notably the predispositions of some types of critical theory, the field of game studies has privileged the conception of the player as a subversive force, working against designer intentions. Joshua Tanenbaum (2013) similarly argues that both scholars and the game industry have set up a false conflict between designers and players when they view players as being primarily subversive.

Creation

Unlike subversion, which sees the game as an actual set of limitations that the player can overcome, the creation perspective sees the game rules as irrelevant or at least secondary to the player's contribution. Like subversion, creation comes in differing strengths. In the weaker variation, Mikael Jakobsson examines players of a *Super Smash Bros.* (HAL Laboratory, 2001) tournament and does not discount the existence of a game, but he argues "that the very nature of a game can change without changing the core rules" (2007, 293). Linda Hughes, studying Foursquare players, argues that "players can take the same game and collectively make of it strikingly different experiences" (1999, 94).

In stronger formulations, the game in effect becomes secondary to game playing. Laura Ermi and Franz Mäyrä say that "the essence of a game is rooted in its interactive nature, and there is no game without a player" (2005). In a yet stronger phrasing, Anne-Mette Thorhauge claims that game rules are in actuality created by players: "The player culture is not just something taking place 'on top' of the game, it rather defines the game as a product of the continuous communication and negotiation among players" (2013, 389).

Hybrid and Prescriptive Ideas

I have placed these four conceptions on a scale, and proponents of each conception also tend to present their own position as being placed on such a scale. Some theorists argue that insufficient emphasis has been put on game design, and some argue that players' contribution is undervalued. One agreement in the discussion, then, is that there is an underlying scale to be tipped, and that any emphasis on the game must be to the detriment of any emphasis on the player and vice versa. But what if this is wrong?

The alternative argument is that game playing operates differently depending on both the game and the player. A precursor of this argument can be seen in Roger Caillois's distinction between *paidia* (free-form play) and *ludus* (structured games) activities ([1958] 2001, chap. 2). However, Caillois still emphasizes that *paidia* is an *unstructured* activity, and he therefore does not consider the possibility that a strictly structured activity can give rise to freedom for the player. But the distinction between games of emergence and games of progression (Juul 2002) as well as the existence of "games" without goals (Juul 2009, chap. 7) suggest that a game design *can* be more or less open to different uses by players. In this case, structured game design can in and of itself enable player freedom, meaning that the perceived conflict between game-centric and player-centric viewpoints may be a misunderstanding.

A further option is to remember that arguments are often made for the benefits of particular types of design. For instance, calls for emergent gameplay (H. Smith 2003) argue for the value of games that leave room for the player. It is clear that such prescriptive ideas about game design must assume that design *matters* and that game playing will be different with different games. For example, the New Games movement of the early 1970s often exhibited a distaste for the playing-as-submission formula (New Games Foundation 1976) and therefore designed less-competitive and more loosely structured group activities.

Along similar lines but for different ends, the experimental game developer Tale of Tales argues that rules, goals in particular, "get in the way of playfulness" (Harvey and Samyn 2006). Here, the constraints that game rules and goals impose are seen as working against the expressive or experiential aspect of game playing. This question thereby connects to broader cultural conceptions of expressivity and *art*. In a controversial statement, late film critic Roger Ebert made the *opposite* argument—namely, that video games can never be art because "by their nature [they] require player choices, which is the opposite of the strategy of serious film and literature, which requires authorial control" (2005). This distinction illustrates how the different conceptions of game playing are invoked with references to broader cultural ideas, even though these ideas may not be as fixed as presupposed: in fact, there is cultural disagreement about whether art works, such as games, should control users or give them freedom.

The Value of Freedom and Constraint

Any discussion of freedom in games, such as this one, will automatically draw upon cultural assumptions about the value of freedom relative to duty, order, submission, and so on. In the present day and in the Western world, it may seem obvious that freedom and subversion are preferable to submission, but the truth is that this view has changed historically. For the New Games movement, working in California in the early 1970s, the value of freedom in the face of oppressive systems was a given. But beyond that, think only of how Christianity, Judaism, and Islam—the *Abrahamic* religions—have in different ways emphasized the story of Abraham's sacrifice of Isaac, holding forth unwavering submission to God's will as a something to be emulated. In other words, submission has often been understood as preferable to individual freedom, and approaching game playing with this belief would make it possible to reverse many of the arguments discussed earlier, seeing subversive gameplay as a problem rather than as a moment to be cherished.

The seeming opposition between submission, constrained freedom, subversion, and creation may then be an illusion: if a game can be designed such that players have more freedom to use it in the ways that they see fit, then the player contribution is not at the expense of the role of the game design or vice versa. Games may be more or less flexible, as players may be (Juul 2009), and these variations are characteristic of this particularly human phenomenon of game playing. A purely game-centric perspective will always have the flaw of denying the contribution by players, and a purely player-centric perspective will have to ignore that players express aesthetic preferences for particular games (Björk and Juul 2012). Game playing is therefore not a conflict between games and players but a moment where games and players fit together and mutually constitute each other.

Works Cited

Aarseth, Espen. 2007. "I Fought the Law: Transgressive Play and the Implied Player." In *Situated Play: Proceedings of the Third International Conference of the Digital Games Research Association (DiGRA)*, 24–28. Tokyo, Japan.

Björk, Staffan, and Jesper Juul. 2012. "Zero-Player Games, or What We Talk about When We Talk about Players." In *Proceedings of the Philosophy of Computer Games Conference*. http://www.jesperjuul.net/text/zeroplayergames/.

Burghardt, Gordon M. 2006. *The Genesis of Animal Play*. Cambridge, MA: MIT Press.

Caillois, Roger. [1958] 2001. *Man, Play, and Games*. Urbana: University of Illinois Press.

Consalvo, Mia. 2007. *Cheating: Gaining Advantage in Videogames*. Cambridge, MA: MIT Press.

Ebert, Roger. 2005. "Why Did the Chicken Cross the Genders?" *Answer Man*, November 27. http://rogerebert.suntimes.com/apps/pbcs.dll/section?category=answerman&date=20051127.

Ermi, Laura, and Frans Mäyrä. 2005. "Fundamental Components of the Gameplay Experience: Analysing Immersion." In *Changing Views: Worlds in Play: Proceedings of the 2005 DiGRA International Conference*. Vancouver, Canada. http://www.digra.org/dl/db/06276.41516.pdf.

Fernández-Vara, Clara. 2009. "Play's the Thing: A Framework to Study Videogames as Performance." In *Breaking New Ground: Innovation in Games, Play, Practice and Theory: Proceedings of the 2009 DiGRA International Conference*. Utrecht, The Netherlands.

Franklin, Benjamin. 1786. *The Morals of Chess*. Philadelphia: Columbian Magazine.

Gadamer, Hans-Georg. 2004. *Truth and Method*. London: Continuum.

Harvey, Auriea, and Michaël Samyn. 2006. "Realtime Art Manifesto." Athens. http://www.tale-of-tales.com/tales/RAM.html.

Hughes, L. A. 1999. "Children's Games and Gaming." In *Children's Folklore: A Source Book*, ed. Brian Sutton-Smith, 93–119. New York: Routledge.

Huizinga, Johan. 1950. *Homo Ludens: A Study of the Play Element in Culture*. Boston: Beacon Press.

Jakobsson, Mikael. 2007. "Playing with the Rules: Social and Cultural Aspects of Game Rules in a Console Game Club." In *Situated Play: Proceedings of the Third International Conference of the Digital Games Research Association (DiGRA)*, 386–92. Tokyo, Japan: Digital Games Research Association. http://www.digra.org/dl/db/07311.01363.pdf.

Juul, Jesper. 2009. *A Casual Revolution: Reinventing Video Games and Their Players*. Cambridge, MA: MIT Press.

Juul, Jesper. 2005. *Half-Real: Video Games between Real Rules and Fictional Worlds*. Cambridge, MA: MIT Press.

Juul, Jesper. 2002. "The Open and the Closed: Games of Emergence and Games of Progression." In *Computer Game and Digital Cultures Conference Proceedings*, ed. Frans Mäyrä, 323–329. Tampere, Finland: Tampere University Press. http://www.jesperjuul.net/text/openandtheclosed.html.

Loftus, Geoffrey, and Elizabeth R. Loftus. 1983. *Mind at Play: The Psychology of Video Games*. New York: Basic Books.

New Games Foundation. 1976. *The New Games Book*. Ed. Andrew Fluegelman. Garden City, NY: Dolphin Books.

Rettberg, Scott. 2008. "Corporate Ideology in World of Warcraft." In *Digital Culture, Play, and Identity: A* World of Warcraft *Reader*, ed. Hilde Corneliussen and Jill Walker Rettberg, 19–38. Cambridge, MA: MIT Press.

Salen, Katie, and Eric Zimmerman. 2004. *Rules of Play: Game Design Fundamentals*. Cambridge, MA: MIT Press.

Smith, Harvey. 2003. "Systemic Level Design for Emergent Gameplay." Paper presented at the Game Developers Conference, San José, CA, March 4–8. http://www.witchboy.net/wp-content/uploads/2009/08/systemic_level_design_harvey_smith_.ppt.

Smith, Jonas Heide. 2006. "Plans and Purposes: How Video Games Shape Player Behavior." PhD diss., IT University of Copenhagen.

Suits, Bernard Herbert. 1978. *The Grasshopper: Games, Life, and Utopia*. Toronto: University of Toronto Press.

Tanenbaum, Joshua. 2013. “How I Learned to Stop Worrying and Love the Gamer: Reframing Subversive Play in Story-Based Games.” In *DeFragging Game Studies: Proceedings of the 2013 DiGRA International Conference.* Atlanta, GA http://www.digra.org/wp-content/uploads/digital-library/paper_136.pdf.

Thorhauge, Anne Mette. 2013. “The Rules of the Game—the Rules of the Player.” *Games and Culture* 8 (6): 371–391.

Tronstad, Ragnhild. 2001. “Semiotic and Nonsemiotic MUD Performance.” In *Proceedings of Cosign 2001*, 79–82. Amsterdam, The Netherlands.

42 PERSPECTIVE

Jacob Gaboury

Perspective is the site of the legislation of seeing, but it has never operated under a single verdict or binding rule.
—James Elkins, *The Poetics of Perspective*

The simulation of depth by means of perspective projection is a broad cultural technique shared across a wide range of visual media. Indeed, it is the defining feature of much visual production from the Renaissance to the present, used for the simulation of embodied vision in drawing, painting, photography, film, and computer graphics. Most histories of perspective begin with its rediscovery by Filippo Brunelleschi in 1420s Renaissance Italy,[1] and subsequent formalization by Leon Battista Alberti in his treatise *De pictura* of 1435. Brunelleschi's linear perspective offered a formal method for the representation of a three-dimensional (3D) scene on a 2D surface. It accomplished this through the imposition of a geometric system consisting of a horizon line and a set of orthogonals that merge at a vanishing point on the horizon, which produces the illusion of 3D depth.

As with many origin stories, Brunelleschi's invention of linear perspective offers a compelling simplification of a complex historical milieu, but it is the site from which we can most clearly draw a line to the present because Brunelleschi's mathematical theory of perspective standardized the practice of 3D visual representation and was thus adopted and modified by a wide range of subsequent media forms. Arguably the most significant among these perspectival forms are those technologies modeled on the camera obscura—a tool for the reproduction of vision whose origins in art and engineering date back to ancient China and Greece. The camera obscura served as an important visual tool throughout the modern period, but it did not become a ubiquitous means of visual reproduction until the development of photography and film in the nineteenth century. Through the mechanization of vision in these "perspective machines," linear perspective became naturalized as a dominant representational form. What began as a technique for artistic representation was transformed through automation and mechanization into the de facto visual mode for the realistic and indexical reproduction of the visible world.

Of course, this description, too, is a gross simplification of the perspectival form and its evolution over the course of some 500 years, but the challenge of narrativizing this history points to the complexity of perspective as a deeply embedded way of seeing and representing the world.[2] Although perspective appears to be a unified system, over the course of its long history it has become overburdened as both technique and expression, method and metaphor. It is equated with mimesis and is demanded of all visual forms that purport to offer an accurate representation of the visible world. Thus, most narratives of computational perspective adopt this same history, equating the simulation of perspective projection with the rise of visual realism in computer graphics beginning in the early 1960s.

The earliest example of perspective in computer graphics comes from the graduate work of Lawrence Roberts at MIT in 1963. His dissertation, "Machine Perception of Three-Dimensional Solids" (1963), was the first program to produce a perspective projection that could be executed by a computer. Roberts's technique for producing linear perspective in many ways resembled traditional optical techniques. Indeed, Roberts referred to German textbooks on perspectival geometry from the early 1800s to properly identify the mathematics of perspective for his program (Manovich 1993, 10). This reference is unsurprising, however, given the mathematical basis for linear perspective and the computer's function as a mathematical device built for the simulation of machines or systems that may be described procedurally. Several scholars have identified Roberts's work as a critical moment in the history of the algorithmic image, some going so far as to suggest it was "as momentous, in its way, as Brunelleschi's perspective demonstration" (Mitchell 1992, 118). Yet, despite this comparison, the development of perspective by means of computation marked little more than the imposition of an existing visual model on a new media technology. In pulling this thread that connects the digital image to its Renaissance predecessors, we once again collapse perspective into a singular and unified form. Rather than adopt a narrative whereby perspective is carried from Brunelleschi and Alberti to the digital imaging technologies used in contemporary video games, we might take up Erwin Panofsky's claim in his field-defining book *Perspective as a Symbolic Form* (1991) and argue that perspective is but one culturally relative spatial system from which we might interpret those modes of knowledge, belief, and exchange that characterized the cultures and technologies from which they arose. Far from a singular practice, perspective is an enduring cultural technique whose perseverance is due precisely to its ability to transform and adapt to new media forms.

Nowhere is this multiplicity more apparent than in the history of video games. For decades, our understanding of perspective has been dominated by the camera obscura and its privileging of the single subject, but the computer is in no way tied to this perspectival form. A computer has no lens, no aperture—it merely simulates these devices in an effort to mimic that which has been naturalized as realistic and true. Indeed, if we look to the history of video games, we see a shattering of vision into an array of perspectival modes. These visual forms developed in part as a response to the technological limitations of the field's early history, but they came to constitute a set of cultural practices whereby the perspectival mode that each game or genre takes may be viewed as corresponding to the needs of its particular ludic form. By examining the history of video games, we begin to see perspective as a complex and divergent technique that coalesces across technical media.

2 to 3

The simulation of linear perspective through computation was first developed in the 1960s, but until the 1980s the realistic 3D perspective now common in many game genres was not possible on arcade cabinets, consoles, and personal computers (PCs), and even then such games were limited to extremely simple visual forms. Prior to this moment, the processing power needed to create real-time, 3D simulations was available only at massively funded research facilities with million-dollar machines used largely for military applications, such as flight and vehicle simulation (Gaboury 2015). Thus, in lieu of perspective projection by means of 3D object simulation, many early games adopted a flattened 2D system meant to simulate functional depth using a form of visualization rarely seen outside of architecture and engineering, a method known as isometric projection.[3] In isometric projection, all three coordinate axes are equally foreshortened and do not narrow with distance, as they do with perspective projection. In technical drawing, this equal foreshortening allows for more accurate measurement in construction and reproduction, but in games it allows for the presentation of a large plane of 3D objects with equal visibility (Ferguson 1994). Objects do not need to be scaled or transformed as they move across the game screen or recede into the distance, allowing for a flattened field of play that nonetheless evokes a 3D space. Isometric projection creates a detached 3D perspective untethered from any single viewing angle or presumed subject because the field or game space is presented as a broadly accessible plane on which objects may be placed or actions directed.

One of the earliest games to use this technique was the arcade title *Q*bert*, developed and published by the Chicago-based Gottlieb Corporation in 1982.[4] The game consisted of a set of isometric 3D cubes stacked in a triangular grid on which the player-character could ascend or descend. The goal of the game was to jump on each of the colored cubes while avoiding balls, snakes, and other objects. Using isometric projection, the play space presented the illusion of 3D depth, but that depth was an entirely visual effect akin to the M. C. Escher drawings that inspired the game's design. Since the 1980s, a number of game genres have utilized isometric projection, suggesting that the relational form afforded by such visual systems is in some way essential to particular modes of play. Popular "beat 'em up" titles such as *Double Dragon* (Technōs Japan, 1987) utilized an isometric perspective that allowed players access to limited depth along the *z* axis, increasing the game's play space and facilitating team play. City-building simulators such as *SimCity 2000* (Maxis, 1994) also utilized an isometric perspective that fused cartographic vision with the strategic placement of tabletop and board-game play. Over time, many game genres have evolved away from isometric perspective as the technical affordances of computers and game consoles have improved; however, others remain structurally dependent on the forms of vision isometric projection allows and continue to use modified isometric systems despite the availability of other perspectival modes.

By the 1990s, the field of computer graphics had developed more sophisticated hardware at a price accessible to the digital games market, and it is this period that saw the rise of games that more closely approximate cinematic forms of perspective projection. Two of the earliest titles to successfully adapt an embodied 3D perspective were *Wolfenstein 3D* (1992) and *DOOM* (1993), both of which were produced by the influential game developer id Software. To create the illusion of three dimensions, both games used a rudimentary form of ray tracing known as "ray casting," along with 2D sprites that could be scaled as they moved along the *z* axis. This technique for simulating 3D spaces or objects through technical workarounds and other methods is often referred to as "2.5D," and it allowed players to navigate a 3D space in real time when the use of fully interactive 3D characters was not possible on the PCs of the time.[5]

It was not until the mid-1990s and the development of hardware systems such as the Nintendo 64 and game engines such as the Quake engine that games using polygon-based 3D graphics became widely available and along with them the use of a linear perspective akin to Roberts's early model. This period marks an interstitial moment in the development of 3D gaming as multiple genres experimented with the affordances of 3D game space in a variety of ways. Perhaps the most emblematic genre to emerge from this moment was the first-person shooter (FPS), which grew out of games such as *Wolfenstein 3D* and *DOOM* but was also derived from existing visual schema in photography and the cinema. As Alexander Galloway (2006) has argued, cinema offered a kind of prototypical logic for the FPS's subjective view. Classical film style is replete with point-of-view shots that loosely take on the perspective of a character within the film's diegesis. Yet Galloway insists that, despite this visual norm, the visual form of the FPS more closely resembles the "subjective shot" wherein the viewer is made to feel that he or she directly occupies the body of a character. This technique is most often used to indicate some form of alienated, distorted, predatory, or machinic vision and is quite rare in most cinematic forms.[6] By contrast, in the FPS genre, this visual form becomes normalized and ubiquitous, informed by the visual tradition of the cinema but naturalized through a refusal of cinematic tropes such as montage, which would break its gamic vision.

Often left out of this discussion is a second game genre that readily adopts the perspective projection of 3D graphics: the flight simulator. Flight simulation was one of the earliest applications for 3D graphics at research sites such as the University of Utah, and the cockpit view of a flight simulator serves as a functional analog to the embodied vision of the subjective shot in narrative cinema. Significantly, the view from a cockpit may be read less as that of the pilot than as that of the plane itself, analogous to the machine vision that Galloway terms "gamic cinema" (2006, 62). This gamic view has a long history in both computer graphics and digital gaming, with early arcade games such as *Battlezone* (Atari, 1980) simulating basic vehicle movement and 3D perspective using vector graphics.[7] The genre was further developed through titles such as *Microsoft Flight Simulator* (1980) and has persisted as a game genre that relies almost exclusively on this form of perspective projection.

Nevertheless, although certain genres seem to lend themselves to this particular way of seeing, others had to train players in it. The release of the Nintendo 64 platform in 1996 came with two launch titles: the flight-simulation game *Pilotwings 64* (1996) and the breakout hit *Super Mario 64* (1996). Whereas *Pilotwings 64* followed an established genre whose convention was that of 3D perspective, *Super Mario 64* was the first title to attempt a complex 3D platformer, necessitating an external camera that could move and be moved by the player themselves. The game camera has become a fairly common convention in a number of contemporary genres, but *Super Mario 64* was the first to introduce this visual technique, and as such the game designers chose to integrate the camera system directly into the game's diegesis. The game therefore begins not with its principal character, Mario, but with a Lakitu character who floats on a cloud holding a fishing rod on which a film camera is suspended. The game then switches to the perspective of the suspended camera, as indicated by the appearance of a head-up display. It is only then that Mario emerges from a nearby warp pipe and the game's controls are explained, including the function of the Lakitu's external camera, which can be manipulated by the player at will. In this way, the game trained its players in new ways of seeing and interacting with a 3D world and attempted to transition the platformer genre from its 2D origins into a new 3D space, structured by the visual logic of perspective projection.

Perspectival Play

As the case of *Super Mario 64* demonstrates, the visual norms of a particular genre are deeply tied to the forms of play that structure it. This distinction is made explicit if we compare the relational affordances of two distinct game genres: the FPS and real-time strategy (RTS).[8]

First-person shooters, as I have argued, rely principally on embodied vision as a relational mode and as such utilize the visual form of perspective projection. It may be argued that this reliance is due to the nature of the simulation itself, which most often replicates the actions of an individual equipped with a weapon that must be aimed and fired within a limited field of vision. Therefore, the relational form that is most meaningful to this genre of play is the vector line (Galloway 2007). In an FPS, successful play is modeled on the ability to aim a weapon at a distant target and engage it. Certainly, the goal and function of a weapon or interaction may vary, but action is most often directed in a line across a visible distance to some effect. Like the vanishing point of Brunelleschi linear perspective, the FPS weapon reticle draws a line to infinity along which the player enacts various forms of action. Thus, the game's goals and the requirement for successful play are modeled through and supported by the particular perspective form the game deploys.

An RTS, by comparison, is more likely to utilize a disembodied isometric projection, such that a large section of the play space is visible at any given time. In a given game, a wide range of actions may be available to the player, but the vast majority of interaction between the player and the game world is predicated not on a system of vector lines but through a relational system of proximity. Objects must be in proximity in order to engage one another, and the range of attacks or abilities is of principal significance to a given topology. Galloway has argued that RTS games model the control structure of all digital media, whereby the player is asked to master and perform the algorithms that structure play in order to succeed (2006, 91–92). Such actions would be quite difficult in a game reliant on embodied perspective because we would lose all sense of the play space or system as a whole. In this particular genre Galloway's control allegory is made acutely visual, as the isomorphic game space of *Civilization III* (MicroProse, 2001) or *League of Legends* (Riot Games, 2009) foregoes individual perspective in favor of a cartographic vision that gives players an expansive view of the accessible field. To see and construct the field is to control the territory, to master the system. This control allegory is central to the gameplay of the RTS genre, and as such the perspectival form it deploys mirrors that function.

This is not to suggest that these qualities are essential to these genres or that all gameplay in these genres can be essentialized into these two geometric models of interaction. I would, however, suggest that these play conventions lend themselves to particular forms of visualization through perspective and that, despite the fact that improvements in computer graphics have made it possible to approximate more realistic forms of embodied play, we nonetheless see a wide range of visual models in digital gaming due to the ways in which these models facilitate unique forms of play. Unlike other media forms predicated on Renaissance or camera obscura models of visual realism, digital games play with perspective as a convention, laying bare its function as a means of structuring relations between objects through vision.

This final point is made most apparent in the recent return to retro-game aesthetics, a trend that is particularly visible in what is often referred to as the "art" and "indie" game scenes. Here there is a growing tendency to play with perspective as means of structuring play.[9] Notable games that have used perspective as a means of transforming play include the Xbox Arcade title *Fez* (Polytron, 2012) and the iOS mobile game *Monument Valley* (Ustwo, 2014). In *Fez*, the game appears at first glance to be a side-scrolling platformer lacking all depth. However, to play through each level, the player must rotate the game's perspective in a 90-degree circular motion, at once revealing the three dimensionality of the world and collapsing that dimensionality back into a 2D surface plane. In this way, the player can navigate otherwise impossible structures, which are traversed and connected by means of perspective transformation. *Monument Valley* offers a similar functionality, but in lieu of the flattened two dimensionality of a retro platformer, it mimics the isometric cubes of *Q*bert*'s M. C. Escher-inspired landscape. Like *Fez*, *Monument Valley*'s game space can be rotated and transformed, creating traversable optical illusions that move the player forward. By recognizing perspective as a ludic element that may be transformed within a given game, both *Fez* and *Monument Valley* create new and innovative forms of play.

Video games are unlike those visual media that came before them. Untethered by the material restriction of linear perspective and simultaneously hindered by the technical restriction of early computer graphics, video games evolved uniquely to support the multiple visual forms perspective may take. Certainly, the replication of realistic 3D graphics remains a goal for many games, and with each new generation video games move ever closer to the realistic visual simulation of objects and environments, but this realism is predicated on a visual form that has been culturally and historically determined, derived more from the grids and lenses of Renaissance Italy than from the circuits and memory of modern digital computers. Although this teleological push toward realism is important to the way we imagine and produce digital images, through video games we are able to apprehend other ways of seeing. Through the persistence of multiple perspectival systems and the unique forms of play they afford, we can observe the ways in which vision is mediated by the technologies that shape it, which makes video games truly unique.

Notes

1. Brunelleschi is often described as having "rediscovered" perspective because it is believed to have been invented in the ancient world and subsequently lost. Brunelleschi's original perspective experiments have also been lost but are described by his biographer, Antonio Manetti (1423–1497).

2. For a useful account of this long history, see Friedberg (2009).

3. Isometric projection is the most common form of axonometric projection, which is itself a form of orthographic projection. All three forms of axonometric projection are visually similar in that each foregoes the distortion seen in embodied perspective in lieu of a set of axes drawn along a consistent scale. In the case of isometric projection, that scale is equal for all axes, creating a symmetrical appearance in which multiple sides of an object are equally visible. Although games may use a variety of axonometric projections for visual effect, isometric was the earliest and most common form and remains the one most explicitly associated with digital games. Later games would adopt both dimetric and trimetric projection as alternate axonometric effects, but all three forms may be distinguished from perspective projection in that they do not distort the image as it approaches a vanishing point.

4. The first game to use isometric axonometric projection was *Zaxxon* (Sega, 1982), from which it derived its name.

5. The world in *Wolfenstein 3D* is built from a square-based grid of uniform-height walls meeting solid-colored floors and ceilings. To draw the world, a single ray is traced by the engine for every column of screen pixels, and a vertical slice of wall texture is selected and scaled according to where in the world the ray hits a wall and how far it travels before doing so.

6. Galloway's point is useful, but it should be noted that this defaulting to the cinematic as the primary relational mode for visual technology is extremely prevalent in early writing on digital games. Such comparisons often discount those alternate visual modes that inform and structure games and play—for example, the use of gun interfaces in analog amusement games in the early twentieth century.

7. The relationship between military technology, flight simulation, and the game industry is explored in Crogan (2011).

8. This relational comparison is drawn, in part, from an essay by Alexander Galloway titled "Starcraft, or, Balance" (2007), in which he compares the FPS *Counter-Strike* (Valve, 1999) to the massively multiplayer online role-playing game *World of Warcraft* (Blizzard Entertainment, 2004).

9. One of the earliest games to adopt this playful transformation was *Super Paper Mario* (Intelligent Systems, 2007), the third game in the *Paper Mario* series, which allowed players to navigate 3D game space as a 2D "paper" sprite. Unlike in earlier games in the series, where the characters' flattened dimensionality was little more than a diegetic conceit, in *Super Paper Mario* the player can control the dimensionality of the game space using nonplayer characters known as "Pixls," rotating the view to achieve 3D perspective.

Works Cited

Crogan, Patrick. 2011. *Gameplay Mode: War, Simulation, and Technoculture*. Minneapolis: University of Minnesota Press.

Elkins, James. 1996. *The Poetics of Perspective*. Ithaca, NY: Cornell University Press.

Ferguson, Eugene S. 1994. *Engineering and the Mind's Eye*. Cambridge, MA: MIT Press.

Friedberg, Anne. 2009. *The Virtual Window: From Alberti to Microsoft*. Cambridge, MA: MIT Press.

Gaboury, Jacob. 2015. "Hidden Surface Problems: On the Digital Image as Material Object." *Journal of Visual Culture* 14 (1): 40–60.

Galloway, Alexander. 2007. "Starcraft, or, Balance." *Grey Room* 28 (Summer): 88–107.

Galloway, Alexander. 2006. *Gaming: Essays on Algorithmic Culture*. Minneapolis: University of Minnesota Press.

Manovich, Lev. 1993. "The Mapping of Space: Perspective, Radar, and 3-D Computer Graphics." http://manovich.net/index.php/projects/article-1993.

Mitchell, William J. 1992. *The Reconfigured Eye: Visual Truth in the Post-photographic Era*. Cambridge, MA: MIT Press.

Panofsky, Erwin. 1991. *Perspective as Symbolic Form*. Brooklyn, NY: Zone Books.

Roberts, Lawrence. 1963. "Machine Perception of Three-Dimensional Solids." PhD diss., MIT.

43 PROCEDURALITY

Eric Kaltman

Procedural Language

Procedurality refers to breaking a problem down into more manageable subproblems, solving them, and then bringing the sub solutions together in the service of solving the original problem.
—Diane Poulin-Dubois, Catherine A. McGilly, and Thomas R. Shultz,
"Psychology of Computer Use"

The first notions of procedurality follow the creation of abstracted procedures for early computers. Computing machines originally had a single purpose and were designed for a specific mathematical or physical problem. They focused on calculating trajectories for ballistics and code breaking but lacked the ability to switch computational context without significant rewiring or reengineering. Computers as general computational devices began with the realization that the program itself, the set of instructions defining a computation, could also be stored electronically inside the machine. This stored-program architecture, still in use today, allowed for programmers to define commonly used procedures (such as calculating trigonometric functions) and store them on tape for later use (Ceruzzi 2003). These procedures consisted of numerous lower-level instructions abstracted into a reference of their collective function. Computing the sine of an angle now relied on calling the sine procedure and removed the need to rewrite the instructions explicitly every time. This freedom to define procedures separate from the main intent of a program (initially "subroutines" to the main "routine" list of commands) allowed for the creation of abstract, encapsulated computations (Wilkes 1982). Simpler references hid more complex underlying processes, allowing for more flexibility and reusability. Procedural programming and the creation of high-level computer languages followed this new abstraction technique and improved the organization and shareability of code.

Procedural languages represent a specific paradigm for computation in which the user specifies "the operations to be performed and the sequence of those operations" (Hansen and

Hansen 1991, 394). These languages are then distinct from declarative ones, such as HyperText Markup Language (HTML), that ask the user to define the result instead of the process. The initiation of *procedurality* as a term is based on the rise of procedural languages; its usage was tied to the level of procedural instruction afforded by a language or system. Something possessed "procedurality" if it was procedural, if it could receive and process a set of user instructions. By the 1970s, computer use started to shift from university and institutional mainframes to more quotidian settings, such as homes and public schools. New microcomputers and game consoles brought a wider range of people into contact with procedural concepts and computer programming. Many early systems, such as the Apple II, the TRS-80, and the Atari Video Computer System supported the Beginner's All-Purpose Symbolic Instruction Code (BASIC), a beginner-oriented programming language. Understanding the procedurality of a system and how to manipulate it through programming became an important skill for competent computer use. Researchers began to focus on interactive ways for new and young users to understand the procedural nature of computers.

Seymour Papert's Logo language, developed in 1967, allowed users to issue instructions to a cursor (the Logo "turtle") that drew geometric patterns. Logo is an early example of a computational system designed to teach procedurality to children. Not only is procedurality in this context a product of the language, but it is also linked to the ability to "think like a computer" (Papert 1980). To possess procedural thinking is to understand how to organize instructions to solve specific problems. Logo's goal was to imbue an understanding of computational process into the user; it was a hope for the embodiment of procedural thinking based on mastery of Logo's procedural model. Procedurality was present in the computational system (through its ability to act on a set of input instructions) and in the user's mind. Computational structures afforded procedurality to the user, who in turn engaged his or her procedurality to articulate their wishes to the machine. The epigraph at the head of this section is taken from a report on the success of using Logo to raise procedurality in students (Poulin-Dubois, McGilly, and Shultz 1989); "procedurality" is made an explicit trait of the learner and a quantifiable metric. The researchers' concept of the procedurality trait is adapted from George Pólya's (Pólya and Conway 2004) pedagogical method for solving mathematical problems. By engaging in decompositional analysis, students could break down more complex tasks into simpler procedures, a process analogous to the creation of computational sub-routines. According to the report published in 1989, an analysis of many other studies conducted with Logo throughout the 1970s and 1980s, this direct transfer met with limited success. It is important, however, as an illustration of procedurality's early meaning as the personal understanding of process-oriented creation.

Procedural Narrative

ELIZA [Joseph Weizenbaum's virtual psychiatrist] was brought to life by the procedural power of the computer, by its defining ability to execute a series of rules.
—Janet H. Murray, *Hamlet on the Holodeck*

The emergence of studies of narrative in digital media works aligned procedurality more closely with creative expression. Janet Murray's book *Hamlet on the Holodeck* (1997) provides an important example of this shift in meaning. She describes procedurality as one of the four properties of the computer as an expressive medium. This use of procedurality is distinct from yet related to the idea of user procedurality in the Logo-based studies described earlier. For Murray, a procedural machine is one that takes user input and adaptively responds based on its underlying programmed procedures. The idea of procedurality as a component of narrative method is tied to the grand vision of the holodeck on the Paramount television show *Star Trek: The Next Generation*, a machine that allows a user to be embedded in real-time into a fictional world whose narrative responds to the user's every action. Only computation can allow for this type of immersion and responsiveness, a property that older forms such as print and film lack. The expressive meaning of the computer is then based on how it interprets the user and how the feedback it provides is internalized and understood.

Computational media rely on procedures to generate meaning through authored processes that are then reified into graphical and narrative representations; it is important to Murray that designers understand how to control and influence procedural meaning.

Murray's procedurality is akin to Papert's procedural thinking. Her stress on the procedurality of a system is foremost an argument for the understanding of procedure by both creators and interactors (Murray's term for the user or player). By incorporating knowledge of a computer's procedural nature into their work, future authors can control and create interactive narrative systems that respond in rich, multicursal ways, which would allow for procedural authorship of narrative experiences in computational mediums. Murray stresses that procedurality is "the most important element the new medium adds to our repertoire of representational powers ... its ability to capture experience as systems of interrelated actions" (Murray 1997, 274). The ability to take ownership of the medium's expressive potential through the understanding and manipulation of its procedural nature is key. Not only is procedurality the foundational element differentiating new digital forms from old, but it also requires a new core competency in designers and creators, a literacy of procedural power and systemic representational schemes. This "procedural literacy" is then a basic component in the creation of any procedural representation (narrative or systemic) and any computer-based new media work or game.

Michael Mateas, a computer scientist and interactive narrative designer, advocates for procedural literacy as a way for new media practitioners to open the "black box" of computational process and regain control of the tools for creative expression. Like Murray, he is directly concerned with procedural narrative authorship, but he also references arguments for procedural competency from the 1960s (Mateas 2005). In this way, his procedural literacy is an amalgam of Murray's procedural authorship and Papert's procedural thinking, and it signals a new transition for the meaning of the term *procedurality*. "New media practitioners without procedural literacy are confined to producing those interactive systems that happen to be possible to produce within existing authoring tools" (Mateas and Stern 2005, 8). Creators then need to master the art of process, the art of programming, so as to design new tools and experiences without relying on others' conceptions of expressive potential. (Mateas's game *Façade* [with Andrew Stern, 2005] is in part a creative argument for this perspective.) Therefore, to embrace procedurality is to embrace programming as the method of authoring processes. However, procedurality is not only programming; it is the use of procedural knowledge to create semantic systems and configurations. Murray's book *Inventing the Medium* (2011), published 14 years after *Hamlet on the Holodeck*, reflects a move more in line with Mateas's position on the importance of procedural literacy to creative expressions. Large sections of Murray's work are devoted to foundational programming concepts. The work leads the reader through procedural and object-oriented programming schemes in hopes of helping designers understand procedural design and "the way information is represented and processed within the machine" (122). This is necessary because "the greater our power to design it ... the better we can communicate the basis of the machine's decision making to the interactor" (122).

The communication of meaning, which stems from an understanding of procedural literacy, is also important to the academic study of games. The preceding section focused on how people engage with procedurality to understand computational process and on how that understanding allows for new creative expression. Digital games are a major example of expressions constructed through processes; therefore, procedurality is central to debates about meaning in games. With respect to procedurality, the academic field of game studies has been concerned mainly with how players interpret the results of computational processes as representational. Do games such as *SimCity* (urban planning [Maxis, 1989]), *Civilization* (historical progress [MicroProse, 1991]), and *Balance of Power* (war [Chris Crawford, 1985]) form arguments for sociopolitical positions through play? If so, how do they do it? The current focus on procedurality and its location in this volume on game history arise from procedurality's ties to the literacy of the procedural systems found in games and to how that knowledge allows one to author specific rhetorical experiences. This connection leads to questions about what types of systems are possible to create and how their processes, their underlying systemic interactions, function in the creation of meaning.

Procedural Rhetoric

Procedurality refers to a way of creating, explaining, or understanding processes.
—Ian Bogost, *Persuasive Games*

Ian Bogost extends procedural literacy into the field of rhetorical analysis. Processes created by game designers are not neutral; the systems they create express a simulation with a particular viewpoint, a specific rhetorical position. Procedural rhetoric is a move away from the designer and user and back onto the system. Expression through process is a unique function of computational media and games, according to Bogost. While procedural rhetoric, as an analytical strategy, is fixed firmly on representations, Bogost's definition of procedurality as "a way of creating, explaining, or understanding processes" (2007, 2) is a bit more amorphous. Procedurality is now ontologically diffuse, occupying the expressive potential for a medium, its source of meaning, and the major property separating it from other forms. Even Bogost assigns procedurality as the source of expressive meaning and the creative process. Procedurality is "the fundamental notion of authoring processes" and "fundamental to computational expression" (2007, 12, 5). It incorporates both literacy as an enactive means of meaning production and rhetoric as the site of that production.

Procedural rhetorical claims fomented a debate about the persuasive power of games and about who was in control of their meaning. Bogost claims that if one could construct a procedural rhetorical argument through a functioning system, that system itself would embody a certain political or social position in its design (2007, 3). The player, in negotiating his or her interaction with the system, might then internalize the system's model and reflect on its procedural argument. Bogost's arguments for procedural rhetoric have taken hold as a legitimation for "serious" games aimed at advocating ideological positions (such as Molleindustria's *McDonald's Game* [2006] and Susana Ruiz's *Darfur is Dying* [2006]) as well as for advergames designed to promote certain products, lifestyles, and brands. Players engaged with a game might embody through play the persuasive messages of the underlying system. Some scholars, most notably Miguel Sicart (2011), have rejected the idea that a game's meaning is only a function of its rules. That a designer has complete control over a game's meaning to players is overreaching and mechanistic. The player in enacting, breaking, or ignoring the rules actually constructs a game's meaning through play. The title of Sicart's essay, "Against Procedurality," also illustrates a telling conflation of procedurality with procedural rhetoric. His procedurality "claims that it is the system of rules and mechanics that conveys the message of the game" (2011). Procedurality is again assigned as the site of meaning production and not as a property of the designer's literacy in authoring processes. Sicart is concerned with the procedurality of the system and its resultant rhetorical arguments. Procedurality is a strawman on which to hang a play-centric hat.

The tension between the so-called proceduralists and the play constructivists caused some in academic circles to try and reconcile the meaning of procedurality. These reconciliations hinge on trying to define new levels or valences for the procedurality of a given system. Michael Skolnik (2013) presents a weak/strong dichotomy for the amount of procedural influence in a given system, and Michael Treanor and Michael Mateas (2013) engage in a two-level semiotic analysis. They assign different meanings to both the internal processes of the machine and what those processes afforded for representational meaning to the player. The reconciliations further inflect the meaning of procedurality and reinforce Bogost's and Sicart's oppositional use of the term, which downplays the historical meanings of procedurality as a way to author, understand, and engage with computational representation instead of being dominated by it. Procedurality is affixed to the procedures of the system and removed from the embodied personal procedurality of the programmer, designer, or interactor. A focus on the games intended to persuade and enforce a specific rhetorical position moves procedurality into direct opposition with its earlier historical framings. Procedural thinking, authorship, and literacy function against the more technocratic notion of an affective procedural rhetoric. They make moves against the influential power of simulations because they try to engage the individual with the processes of the machine and in this way reveal the machine's procedural inner workings; they can be viewed as a defense against rhetorical claims in game systems through understanding. Historical procedurality is therefore a counterpoint to the current oppositional nature of the term.

Procedural Meaning

The situation [of critical interpretation of computation] ... discriminates among: (i) a program ... (ii) the process or computation to which that program gives rise ... and (iii) some (often external) domain or subject matter that the computation is about.
—Brian Cantwell Smith, *On the Origin of Objects*

Another goal of this collection is to try and affix a new or comprehensive meaning to the concept in question. The historical position of procedurality has oscillated between a focus on computational expression (both rhetoric and narratological) and a focus on process literacy. Procedurality has also been used as both a tool for better understanding procedural systems and a means of being controlled by them. As the meaning and creation of processes, it has been pigeonholed into the service of any discourse discussing the tricky nature of computational meaning, but it also remains a notably indistinct concept. It is not located anywhere specifically inside the computer but results from computational execution. One cannot simply point to the site of meaning making in the machine. Is the authoring of processes procedurality? How about the interplay of each level of game simulation? In the cross currents of various processes from the processor to the player that merge to create a playable experience, is procedurality all or only part of that system?

In the preface to a discussion about the site of semantic and syntactic meaning inside the computer in his book *On the Origin of Objects* (1996), Brian Cantwell Smith notes two levels of inscription, vaguely similar to those presented by Treanor and Mateas in their remedy to the modern procedural/constructivist divide. The relationship between program and process is the reification of program code into a series of logical instructions for a computer's processor. Computational processes are then assigned some externally mediated meaning or subject matter. Much of what has been said about procedurality focuses too intently on the concept as some distinct property of either people or machines stemming from the invention of computational architecture. Procedurality is historically either a person's internalized understanding of computational processes—of how to organize logical connections between the many subtasks needed to complete a goal—or the interaction of those processes in the generation of meaning through computation or play. My hope is that we will now look not only at what we think procedurality is and what it represents but also at how programming praxis and design have shaped our current understanding of the concept.

Works Cited

Bogost, Ian. 2007. *Persuasive Games: The Expressive Power of Videogames.* Cambridge, MA: MIT Press.

Ceruzzi, Paul E. 2003. *A History of Modern Computing.* 2nd ed. Cambridge, MA: MIT Press.

Hansen, Gary W., and James V. Hansen. 1991. "An Approximate User-Complexity Measure for Relational Query Language Systems." *Data & Knowledge Engineering* 6 (5): 393–407.

Mateas, Michael. 2005. "Procedural Literacy: Educating the New Media Practitioner." *On the Horizon* 13 (2): 101–111.

Mateas, Michael, and Andrew Stern. 2005. "Procedural Authorship: A Case-Study of the Interactive Drama." Paper presented at the Digital Arts and Culture Conference, December 1–3. Copenhagen, Denmark.

Murray, Janet H. 2011. *Inventing the Medium: Principles of Interaction Design as a Cultural Practice.* Cambridge, MA: MIT Press.

Murray, Janet H. 1997. *Hamlet on the Holodeck: The Future of Narrative in Cyberspace*. New York: Free Press.

Papert, Seymour. 1980. *Mindstorms: Children, Computers, and Powerful Ideas*. New York: Basic Books.

Pólya, George, and John Horton Conway. 2004. *How to Solve It: A New Aspect of Mathematical Method*. Princeton, NJ: Princeton University Press.

Poulin-Dubois, Diane, Catherine A. McGilly, and Thomas R. Shultz. 1989. "Psychology of Computer Use: X. Effect of Learning LOGO on Children's Problem-Solving Skills." *Psychological Reports* 64 (3c): 1327–1337.

Sicart, Miguel. 2011. "Against Procedurality." *Game Studies* 11, no. 3. http://gamestudies.org/1103/articles/sicart_ap.

Skolnik, Michael Ryan. 2013. "Strong and Weak Procedurality." *Journal of Gaming & Virtual Worlds* 5 (2): 147–163.

Smith, Brian Cantwell. 1996. *On the Origin of Objects*. Cambridge, MA: MIT Press.

Treanor, Mike, and Michael Mateas. 2013. "An Account of Proceduralist Meaning." Paper presented at the Digital Games Research Association Conference, August 26–29. Atlanta, Georgia.

Wilkes, M. V. 1982. *The Preparation of Programs for an Electronic Digital Computer: With Special Reference to the EDSAC and the Use of a Library of Subroutines*. Charles Babbage Institute Reprint Series for the History of Computing, vol. 1. Los Angeles: Tomash.

44 ROLE-PLAY

Esther MacCallum-Stewart

This chapter examines the development of the term *role-play* as well as the problematic nature of this evolution and briefly traces its progression via a gaming genres perspective.

The *Oxford English Dictionary* (2nd ed., 1989) indicates that the word *role-play* originated from the German *rollenspiel* (akin to the transition of *Kriegsspiel* to *wargame*). It was first used by psychotherapist Jakob L. Moreno to describe the performances given by participants during group-therapy exercises. After immigrating to the United States, Moreno included the terms *role-player* and *role-playing* in a 1943 paper.

In its broadest sense, role-play is an act of play, comprising mimicry and acting-out. In areas such as psychology, education, managerial studies, and performing arts, the term *role-play* refers to activities with a history and corresponding critical discourse of their own. The escapism that role-play engenders means that the term has informal associations elsewhere, including semiotic connections to bondage, sadomasochism, fetish play, and sexual dress-up. These multiple understandings have often led role-play to be regarded with suspicion and a degree of prurience. In addition, role-play as a playful activity has been marginalized because it is perceived to be a form of *paidia* without tangible outcomes, relying instead on an individuals' creative input. Early game scholars ascribe role-play to feminine activities; for example, Roger Caillois sees such play behavior as "the ultimate and humiliating metamorphosis of a serious activity" ([1958] 2001, 61) and uses the example of "playing with dolls, which everywhere allows a girl to imitate her mother or herself be a mother" (62).

Jon Peterson (2012) argues that the term *role-play* was first associated with games between 1956 and 1959, when it was written into reports detailing wargames carried out by RAND's Inter-Nation Simulation group. According to Harold Guetzkow, "the actors need to imagine many features of the military situation and respond to each other's moves in terms of these self-imposed role conceptions" (qtd. in Peterson 2012, 379).

Peterson finds that the term *role-playing game* was subsequently used beginning in the 1960s to describe certain activities in commercial wargames (2012, xix) and was common by the time that Gary Gygax and David Arneson published *Dungeons & Dragons* (*D&D*) in 1974.

The Role-Play "Genre"

Within gaming, the term *role-play* has several meanings. Instead of simply connoting a body of activity, it also implicates multiple genres and is a verb to describe play. It frequently becomes part of a mnemonic; for example, *massively multiplayer online role-play(ing) game* (MMORPG), *live-action role-play* (LRP or LARP), *tabletop role-play* (TRPG), and *computer role-playing game* (CRPG). Players sometimes simply refer to "role-play" without further qualifying what they mean, making the term at once a signifier of genre and indicative of enacted content.

It is also difficult to identify commonalities of role-play because the term has become a cross-genre mnemonic and signifier. Not only has it been attached to several different types of gaming, but these genres are subject to design evolution. Core aspects change, lose popularity, are revived, or develop. For example, in early TRPGs such as *D&D* (which was itself based on a miniatures-intensive wargame), miniature figures were an important element of play. As TRPGs became more focused on imaginative processes, miniatures became less important, although never entirely redundant. Now, however, board-game role-play games such as *Descent: Journeys in the Dark*, second edition (Fantasy Flight Games, 2012) use miniatures as a vital element of play. Thus, using miniatures has transitioned across role-play genres, not only demonstrating their importance as tropes but also signifying that they are not defining elements of either genre. Similarly, although Matt Barton claims that "the only common factor that stretches across the entire span of CRPGs is the statistical system that determines how characters fare in combat (or whatever tasks they are asked to perform)" (2008, 5), one CRPG, *To the Moon* (Freebird Games, 2011), avoids combat (apart from one satirical scene mimicking the conventions of this behavior).

Linearity and Cross-Pollination

Several texts try to locate role-play within a solid historical context, but the breadth of this task quickly becomes a barrier. These books often find that they must instead limit their discussion to more focused aspects of the genre (Barton 2008); follow one core title considered to epitomize role-play, such as *D&D* (Ewalt 2013; Peterson 2012; King and Borland 2003); or trace personal journeys in gaming to establish a chronology via experiential means (Gilsdorf 2009). Michael Tresca is perhaps the only author to trace the evolution of "fantasy role-playing games" as a cross-genre history; however, even he maps his own history of experience as moving from *D&D* to multiuser dungeons (MUDs) to video games and finally to MMORPGs, despite being an author of multiple TRPG supplements during this period (Tresca 2011, 1–3). This sequence unintentionally maps a rather false picture of role-play's evolution within gaming. These histories create a linear route whereby technologically superior games supplant their predecessors, for example, assuming that video games and MMORPGs replace earlier pen-and-paper incarnations. As a result, earlier role-play game genres are often cited as descendants rather than as contemporaries.

It is therefore important to regard all of the genres mentioned here as contiguous; no one genre ends with another taking its place. Most have instead developed alongside each other. For example, a new version of *D&D* was released in 2014, and a thriving indie scene of small, recently developed TRPG titles challenge or explore established ideas within the form. It is also possible to see a great deal of cross-pollination; for example, MMORPG terms such as *tank*, *dps*, *raid*, and *epics* have been seeded across role-play genres by exoteric gaming communities (Ensslin 2011)

This diffusion by players is also responsible for the generic title *role-play* gradually becoming synonymous with a number of very different genres. Although TRPGs may have been an origin text, 40 years later they are certainly not the point from which many players now first encounter the word *role-play*. This term may connote ludic practices (such as the system of physical and mental attributes that is measured by number) or more vaguely a genre in which players adopt the roles of one or more characters.

Historical Development

Dungeons & Dragons is universally recognized as the ludic origin of role-playing games, although the game box initially described the contents as guidelines for "fantastic medieval wargames campaigns." *D&D* is seen as a benchmark for both role-play games and role-play as a gaming activity.

Designed by Gary Gygax and Dave Arneson and originally entitled *The Fantasy Game* (1973), *D&D* was developed from a miniatures wargaming title called *Chainmail* (Gygax and Peren, 1970). *D&D* initially used *Chainmail*'s rules for resolving combat, but *D&D* introduced statistical attributes for each player-character based around personal traits, such as strength, stamina, charisma, intelligence, dexterity, and wisdom; a pool of points from which the player draws when casting spells; hit points to determine how much health the player has in combat; character classes that explain a player's career path; and abilities or spells that the player can choose to accompany these other characteristics. These player-characters act out events in the game on a personal level rather than players taking on the role of omnipotent generals moving pieces on a wargaming board, which gives *D&D* a more freeform, imaginative type of play. The fantasy setting eschews traditional historical battles and simulations, and the game is played as a team effort rather than as a competition, with each player having different attributes and abilities (although early versions of the game were often competitive, with players scoring via experience points and kills made).

The TRPG is a good example of history developing retroactively because although *D&D* ultimately became the dominant text, several other games that evolved at the same time initially outstripped it (*D&D* was not distributed very effectively at first, nor did it have an international presence until it was picked up by Games Workshop in 1975) (Ewalt 2013; Livingstone 2008). *D&D* rapidly spawned rival titles that developed from existing worlds or had precursor histories of their own. *Empire of the Petal Throne* (Barker, 1975) was based on the world of Tékumel, which scholar Muhammad Abd-al-Rahman Barker had created in the 1940s. The game therefore had a much larger narrative pool from which to draw, and although its rules were modeled on those of *D&D*, it used Tékumel's rich worldsphere to digress from the wargaming aspect of its predecessor. Sharron Appelcine's extensive review of TRPGs identifies several other companies with nascent products that almost immediately became popular TRPG franchises, including Chaosium (established in 1975 but predominantly TRPG related by 1977), Gamescience (established in 1965 as a wargaming company, dice based by 1975), and Fantasy Games Unlimited (established in 1975), which produced *Chivalry and Sorcery*, a TRPG intended to bring elements of historical realism to TRPGs, although Appelcine also admits that "D&D was the de facto standard" (2013, 355) for these evolutions.

These contemporaneous companies are rarely mentioned now because they were either amalgamated into Tactical Studies Rules (TSR, which owned *D&D* and aggressively bought out other titles and authors) or could not sustain the momentum of the *D&D* franchise (TSR quickly used its writers to produce modules and expansions for *D&D*, in this way flooding the market). Thus, Ken St. Andre says that he developed *Traveler* (1976) in response to his game *Tunnels and Trolls* (1975) (cited in Schick 1991, 314) rather than to *D&D*. *Tunnels and Trolls* also produced the first solo games (Schick 1991, 358), arguably making them a more comparable predecessor to early CRPG games. Steve Jackson and Ian Livingstone's *Fighting Fantasy* series of books (1982–present) targeted young, solo readers. The authors simplified play, making the role-play experience into a page-turning choice between two or three options, and simplified the combat and character statistics so that they required only two 6-sided dice.

Here, we can see how role-play games quickly started to diverge in format as different media and play ethos started to influence the contents and composition of each text. Even *D&D* modules comprised different styles of play. *The Temple of the Frog* was first run by Dave Arneson as a *Chainmail* game as early as 1972 (Appelcine 2013) and takes adventurers quickly from one encounter to another, whereas author Paul Jaquays is remembered more for creating modules containing rich imaginative worlds and vivid art (Maliszewski 2010). Although Jaquays' work might correlate most closely to my initial, imaginative interpretation of role-play, both are seen as an archetypal example of *D&D* modules.

Three main conclusions can be drawn from this history:

- First, the semiotics, ethos, and structural composition of *D&D* hugely influenced subsequent role-play games and genres.
- Second, the agency that *D&D* gave players—the potential to create their own stories or direct their characters independently—was key to developing the game beyond its origins. In many senses, *D&D* has always been a multiauthored text told by players, dungeon keepers and sanctioned authors through the acting out of role-play itself.
- Third, *D&D*'s constant evolution creates a flexible structure, enabling role playing to develop in multiple directions, to amend imperfect rules, or to tweak aspects of play to bring them more in line with each other, and this flexibility is a key element of role-play gaming today.

Downloadable content, expansions, and new versions of games or rules reflect this mutable development within role-play genres.

Beyond *D&D*

D&D not only provided the core rules of role-playing in a ludic context but also set in motion an ethos for its community, an ethos that is crucial to the way many video games are consumed (and sometimes developed) today. Role-players proved a dedicated fanbase, willing to share information initially through fanzines and newsletters such as *Owl and Weasel* (Jackson and Livingstone, 1975–1977) and *White Dwarf* (Games Workshop, 1977–present) and later through Usenet groups, forums, and websites. Role-play, in whatever form, is experimental—players are consistently encouraged to think creatively and try new solutions, which means they are willing to share information and ideas, often developing role-play texts as they do so. Both the Generic Universal Role-Playing System (GURPS; Jackson, 1986), and the Basic Roleplaying (BRP) system (Stafford and Willis, 1980) were written by fans who wanted to adapt *D&D*'s rules so that other players in turn could broaden their play experiences. Different companies created their own worlds and allowed players to experiment with new ideas. Some closely aped *D&D*—for example, the "grim and perilous world of adventure" depicted in *Warhammer Fantasy Roleplaying* (Games Workshop, 1986)—whereas others took the game into the worlds of Lovecraftian horror (*Call of Cthuhlu* [Chaosium, 1981]), steampunk (*Castle Falkenstein* [R. Talsorian Games, 1994]), vampires and other mythical races (*Vampire, the Masquerade* [White Wolf, 1991]), space opera (*Trinity* [White Wolf, 1997]), and a thousand others. Some allowed the players to develop their own systems around generic or basic role-playing rule sets—GURPS, BRP, Fudge (1992), and Fantastic Adventures in Tabletop Experiences (FATE, 2003) are systems that provide rules and structures, allowing the gamemaster and players to experiment with different settings and modes of play. Still other role-playing games, including James Wallis's *Adventures of Baron Munchausen* (1988), *Lady Blackbird* (2009), and *The Dresden Files* (2011), are small indie titles in which players experiment with a specific game dynamic or concept (the telling of outrageous fibs for entertainment, courtly love) or a novel dice system). Finally, *D&D* has continued to develop in the decades that have passed since its initial release, undergoing several revisions and upgrades.

All of these games, including versions unique to tabletop gaming groups or one-off events run for tiny groups of players, have had a huge influence on the subsequent development of role-play in games-related communities. Many fantasy and science fiction authors (Jim Butcher, Pat Rothfuss, Scott Lynch, Naomi Novik, Ernest Cline, Benedict Jacka) attribute their work to role-play games or mention them in their writing. The *Munchkin* games (Jackson, 2001) satirize the idiosyncrasies of tabletop gaming and have gone through multiple iterations. This relationship demonstrates the ongoing importance of role-play conventions instigated by *D&D* throughout multiple gaming/geek platforms, including video, card, board, and fantasy/science fiction. For example, open-world board and video games develop role-play conventions that allow players to develop the characters they play or experience what Geoff King and Tanya Krzywinska call a "rich text" (2005), a vibrantly built world that allows further role-play beyond the remits of the games' rules. Examples include the *Arkham* series by Fantasy Flight Games, the card game *Gloom* (Atlas Games, 2004), and the *Elder Scrolls* series of video games (Bethesda Game Studios, 1994–present).

Digital Role-Play

Within video games, the history of role-playing games suffers equally murky origins. Shay Addams claims that the progenitor of the CRPG, *Colossal Cave Adventure* (Crowther, 1975), predates *D&D* by several years; designer Will Crowther told him that he "wrote the original game in 1967 or '68" (Addams qtd. in Barton 2008, 26). Matt Barton argues that the earliest versions of digital role-play games existed almost simultaneously with *D&D*, approximations of Gygax and Arneson's game appearing illegally on many university campuses. Early text adventures such as Infocom's *The Hitchhiker's Guide to the Galaxy* (1984) aped Jackson and Livingstone's "choose your own adventure" style, with participants typing instructions to proceed. "Point and click" games matched these tasks to visual objects as graphics developed. Puzzle-solving elements were subsequently incorporated into more physical activities such as moving blocks or activating timed objects (*Tomb Raider* [Core Design, 1996]). Although these adventure games provided immersive worlds and situations as well as increasingly well-developed characters, they were not usually considered to epitomize role-play in video gaming. Video games instead appropriated "role-play" as a type of game—for example, the RPG or the MMORPG. The term *role-play* became a signifier of design elements, eschewing modes of free-form play. Titles such as *Baldur's Gate* (BioWare, 1998), *Planescape: Torment* (Black Isle Studios, 1999), *Neverwinter Nights* (BioWare, 2002), and *Dungeons & Dragons Online* (Turbine Inc., 2006) draw directly from the *D&D* license; however, RPG development also owes much to Japanese gaming, where a long-standing tradition has evolved synchronously. These games incorporate imaginative aspects of role playing by engaging with sustained narratives (e.g., the *Final Fantasy* games [Square Enix, 1997–present]), but the player's imaginative agency is largely removed. The emphasis is more on controlling a diverse party of characters, managing statistical advancements, killing monsters, and looting treasure. It is clear that although video gaming and role-play are closely linked, they suffer from semantic confusion, which makes charting a specific line of historical inquiry extremely problematic.

Conclusion

Although *role-play* is a floating term, defined best as a cross-genre behavior rather than as a specific artifact or genre, the development in complexity of video games is allowing role-play as an imaginative form of play to reemerge. The need to retain a player's interest means that many games involve a degree of visible or "quiet" role playing (imaginative role-play that the player undertakes alone, without needing to inform or demonstrate to others that it is taking place (MacCallum-Stewart 2011) because the player's experience is often enhanced by adding simple activities (for example, the emotes enacted by robots Peabody and Atlas in *Portal 2* [Valve, 2011] or the stylized illustrations of unit cards in *Total War: Rome II* [Creative Assembly, 2013]) that make gameplay a more expressive experience. In addition, game developers are concerned with introducing new generations of players to role-play. *D&D*'s fifth edition (2014) simplifies nearly 40 years of rules and revisions, intentionally allowing new players an easy entry point. Fantasy Flight Games has revisited popular tabletop titles such as *Call of Cthuhlu* through board and card games using the same worldsphere. These reimaginings blend gameplay with "flavor text" that allows players to immerse themselves in the game. This cyclical behavior is perhaps the best indication of how role-play continues to work as a cross-genre behavior that has a rich future ahead of it.

Works Cited

Appelcine, Sharron. 2013. *Designers and Dragons: The TSR Years*. Silver Spring, MD: Evil Hat Productions. http://www.evilhat.com/home/wp-content/uploads/2013/12/DD70s_PreRelease_TSR.pdf.

Barton, Matt. 2008. *Dungeons and Desktops: The History of Computer Role-Playing Games*. London: AK Peters, CRC Press.

Caillois, Roger. [1958] 2001. *Man, Play, and Games*. Trans. Meyer Barash. Urbana: University of Illinois Press.

Ensslin, Astrid. 2011. *The Language of Gaming*. London: Palgrave Macmillan.

Ewalt, David. 2013. *Of Dice and Men*. New York: Scribner.

Gilsdorf, Ethan. 2009. *Fantasy Freaks and Gaming Geeks*. New York: Globe Pequot Press.

King, Brad, and John Borland. 2003. *Dungeons and Dreamers: The Rise of Computer Game Culture from Geek to Chic*. New York: McGraw-Hill.

King, Geoff, and Tanya Krzywinska. 2005. *Tomb Raider and Space Invaders: Video Games in the 21st Century*. London: I. B. Tauris.

Livingstone, Ian. 2008. "Obituaries: Gary Gygax, Creator of *Dungeons & Dragons*." *Independent*, March 7. http://www.independent.co.uk/news/obituaries/gary-gygax-creator-of-dungeons--dragons-792724.html.

MacCallum-Stewart, Esther. 2011. "The Place of Role-Playing in Massively Multiplayer Online Role-Playing Games." In *Ringbearers:* Lord of the Rings Online *as Intertextual Narrative*, ed. Tanya Kryzwinska, Esther MacCallum-Stewart, and Justin Parsler, 70–92. Manchester, UK: Manchester University Press.

Maliszewski, James. 2010. "Interview with Paul Jaquays." *Grognardia*, July 21. http://grognardia.blogspot.co.uk/2010/07/interview-paul-jaquays-part-i.html.

Peterson, Jon. 2012. *Playing at the World*. San Diego: Unreason Press.

Schick, Lawrence. 1991. *Heroic Worlds: A History and Guide to Role-Playing Games*. New York: Prometheus Books.

Tresca, Michael. 2011. *The Evolution of Fantasy Role-Playing Games*. Jefferson, NC: McFarland.

45 SAVE

Samuel Tobin

Saving is hopeful. To save is both to play and to manage that play; it is at once inside and outside of play, at the same time instrumental and ludic. When players save games, they engage in a complicated series of negotiations between their needs, skills, and goals as players, between their everyday lives and the technical systems with which they play. As saving techniques and technology have changed, so too have possibilities for play and pleasure as well as a new range of problems and compromises. To address these issues we need an expanded understanding of saving that draws on these predigital forms and looks to the limit of what we might mean by the term *save*.

Games were being saved long before they went digital. Saving was accomplished through devices (for example, playing-card clips to hold hands and board-game covers), modes of recording and transmitting (scoring and note taking on pads, chess books with accounts of games, play-by-mail games, bridge problems presented in the newspaper), and the verbal recounting of anecdotes of past game playing, of losses and victories.

In the arcade, there were two key save practices: the entry of a high score into the system and the placing of a quarter (or token) on the bottom of the screen or marquee panel to "call next." The high score saved a record of player accomplishment, sometimes until the cabinet was switched off or in the case of battery and nonvolatile random-access memory for longer (forever?). The high score served as a challenge, a call to play. The quarter on the game saved a place in line and signaled intent to play and sometimes a challenge. The quarter on the cabinet, with its insurance of future play, encapsulates the vitality and promise of saving as a play practice. The high-score screen and the quarter on the cabinet represent twin impulses: to save to document and to save to ensure future play.

The high score as record of play reverberates across game culture, from Henry Lowood's How They Got Game project's *DOOM* videos to Twin Galaxies' endless lists of top scores to "Let's Play" YouTube videos. This kind of game saving, with its double relation to both archive and community, resonates with game preservation, museology, and the like. Lowood (2009) has shown how practices such as replay videos and machinima are inherently documentary in nature and save or capture play. In the communities that inform, surround, and support these postings of videos and lists of game-playing accomplishments, information of various kinds is exchanged. These forms of game saving, with their tips, tricks, and walkthroughs, constitute a huge repository of saved game knowledge online. Such forms of saving also had (and continue to have) a pre–World Wide Web print counterpart. The International Center for the History of Electronic Games in Rochester, New York, has an entire wall of game guidebooks, and every issue of each gaming magazine is filled with such guides.

Alongside such printed and web-based modes and more significant for a reconceptualizing of what it means to save, we need to consider a set of homemade, do-it-yourself, and personal modes of game saving. Home-console players of the 1980s and 1990s left themselves handwritten notations of game passwords. It was these passwords that would allow a player to come back to life, to start where he left off or even where his friend left off. Players wrote down these save codes in the backs of game manuals (which often had blank pages for "notes"), on scraps of paper, and even in ink on the backs of their hands. Such jotted-down save codes were part of the larger corpus of game notes maintained by players. Game writing of this sort has a history that precedes video game play, as in pen-and-paper role-playing games such as *Dungeons & Dragons* (Gygax and Arneson/Tactical Studies Rules, 1974). Unlike dungeon maps, however, video game codes were usually made up of numbers and letters in seemingly random combinations.

In the Nintendo Entertainment System and Famicom game *Faxanadu* (Hudson Soft, 1987), players were given a "mantra" after being told not to despair by priests encountered in peaceful towns between dangerous "overworlds." This mantra was a string of letters and numbers that the player could enter later, after writing them down somewhere. *Faxanadu* is noteworthy for calling this code a "mantra" and suggesting that both the player and the avatar had access to it, bridging the diegetic and extradiegetic game experience. (*Metal Gear Solid* [Konami, 1988] used this trope to some acclaim in the battle with Psycho Mantis [Poole 2004, 110]). The near impossibility of remembering such mantras (unlike the famous Konami Code) foregrounds the importance designers and sophisticated players assigned to game-saving practices. The blank pages in old manuals similarly announced the importance of note taking to the gaming experience and suggested that saving and mapping are central to a player's experience of the game, not merely some sort of secondary textual activity. This note-taking play has close relatives in note and map making in text adventure games, choose-your-own-adventure books, and pen-and-paper role-playing games as well. What these manuals and mantras also reveal is that the game save as act, practice, and object exists in multiple places, media, and registers. The save of the console code lived not inside the game or in machine memory but on paper and (briefly) in players' minds and hands. The save code can be conceived as a *paratext* (after Gérard

Genette and, in terms of player practices, Mia Consalvo). Such paratexts seem to oscillate between being peripheral and being central to the game experience. Indeed, it is in the difficulty of writing down these codes and then inputting them accurately that we see them move from tool to plaything. The pleasure a player feels upon successfully inputting the code is often accompanied by an upbeat chime or other audio cue (as in, but not only in, *Faxanadu*): Success! Well played!

In level-based action or platform games, a code usually contains a level name, combined with a certain combination of attributes and equipment to be found there (lives, health, weapons). This is much like how cheat codes have worked in games, from *DOOM* (id Software, 1993) on to the later entries in *Grand Theft Auto* (Rockstar Games), with the difference being that those later games also feature file (to memory) saving options. We see this come together again in the somewhat niche practice of game-state saving, largely limited to emulators and Multiple Arcade Machine Emulator (MAME) gray-market practices. A save state is a snapshot of a set of conditions in a virtual game-playing machine, or emulator. By allowing players to save in games what they otherwise would have had to keep in their head or jot down, save states combine the code save with the cheat.

Carl Therrien has pointed to the possibilities and problems that the emulator's save-state function offers to historians. He shows how these save functions, in combination with video-screen capture, facilitate huge communities of "speedruns" and "long plays" (2012, 16). This loops us back to the high-score-recording aspect of saving. Save-state play pushes these two impulses of saving together and in doing so highlights the fact that saving *is* playing by perversely playing with save files. Save-state players introduce the save where it didn't exist before, bringing the save practice of one kind or era of digital games into another. MAME play delights in the translation of the play practices of one platform, system, or setting (arcade cabinets, personal computers, consoles, etc.) into another. That this translation chiefly takes the form of transformation and importations of different save regimes is telling because it points to the importance of the save in digital play.

Long before the days of online *savescumming* (the manipulation of save states to aid the player in some way) and MAME hacking, the pleasure of this original code play would already be transformed by a new save mode (new for the home console at least): the ability to record data and to save a file as opposed to unlocking a given level state through an alphanumeric code. With the advent of battery-powered memory and then internal and external file-storage systems came new possibilities and challenges for player and designer alike.

Here, I couple the terms *continue* and *load* to the key term *save*. Although they refer to distinct acts and concepts, in gaming practice these acts and concepts are often conjoined. We might think of the strangeness of finding "someone else's game" in the memory of a used cartridge bought at a flea market as an example of how infrequently a player's saves are separated from that same player's continues.

File-based saves, like code-based saves, would allow or even require players to play games of lengths and duration many hundreds of times longer than the quarter-a-minute economy of the arcade, as detailed by Carly Kocurek (2012, 193). That we call these units of time a "sitting" speaks not just to the centrality of the domestic couch and television to this activity but also to the fact that there is always a body, even a docile sitting one, actuating the save decision.

As in the case of the mantra code in *Faxanadu*, saving and checkpoints are very often expressed, triggered, and remediated in the game's diegesis. These saves are performed by player and avatar at once, on the couch and in the game. They may be triggered when certain things are done, reached, or crossed, or they may be selected by players. This save triggering can become an opportunity for particularly tense game play as areas or times between save points are navigated carefully. In all cases, there is a negotiation between player and game state as well as between player and her context of and for play. We might think of how often a player plans to (or promises) to stop playing once she reaches a point in which she can save. This act shows how even when the save is performed in game space, the everyday world that surrounds the player has a huge impact on saving decisions, options, and problems. This means that even in games in which players feel that they can save at any time, where no checkpoint or tool is needed, compromise and contingency still rule. Players save a game not just because of what they have accomplished or experienced within the game but because of the everyday needs of their everyday lives. This being the case, contexts of use and situations and setting in which a players play deeply effect save practices, especially in mobile play (Tobin 2012).

The need to save is of course not only a result of having to go to work or to the bathroom or to answer the phone or doorbell; it also results from players wanting or needing to redo and reset the game. *Braid* (Crystal Dynamics, 2007) and *Prince of Persia* (the Ubisoft rather than Brøderbund iteration, 2003) attempt to bring saving and (re)loading—or, as Chuck Moran (2010) calls it, "diegetic undoing"—into the game space. But a redo is not the same as a save. In both of those games, the management of the ability to redo is distinct from the kind of care and marshalling of both in-game and quotidian time that players engage in anytime they play to save and save to play. Although many other examples abound, the typewriter ribbon in early *Resident Evil* (Capcom, 1996) games, the chests in *Fallout 3* (Bethesda Game Studios, 2009), or the cell phone, cassette tape, and other mobile devices of the later entries in the *Grand Theft Auto* series are in this way clearer examples of the save act being performed and mirrored in the game space, even as that act is performed extradiegetically.

In a reversal of this bringing of the save into game space, a few home systems featured memory-card peripherals (the original Sony PlayStation had the PocketStation and the Sega

Dreamcast had the Visual Memory Unit), which brought play into the saving hardware (Montfort and Consalvo 2012, 86). These devices not only stored files but also acted as (simple) mobile-gaming systems and facilitated playful file manipulation and data management. PocketStations in the PlayStation could save or load some game data on certain arcade systems, establishing an exciting mobile–console–arcade connection that the video game hardware and software company SNK had attempted previously but never seemed to quite work out ("News" 1990, 86). In a stranger twist, the PocketStation was recently remediated in an application for the Sony Vita, allowing a player to play with a virtual memory card.

These devices and re-representations make explicit what we see in all game saving: it both ensures play and is part of it; it is done in play and in order to play. But a paradox or tension lurks in this formation. The possibility of saving both allows and cultivates certain play styles. Saving, although being done in order to play in the future, affects and even foreshortens current play, making it tend toward the careful, the practiced, the composed, and the serious. Play can be a serious, deep, and intense effort. But this is not all play is, can be, or should be.

For an example of another kind of play, we can look to saving's close relative, pausing. Pausing allows players to react to in-game and out-of game issues and needs, just as saving does, but without an injunction to decide that something is *worth* saving. Pausing, at least in single-player contexts, saves play from seriousness, from the future, from instrumentality and work. It returns the "trivial" to the "nontrivial effort" required of players of these ergodic texts (Aarseth 1997, 1). In contrast, when we save, we tend to play for keeps.

When we plan to save, we play in a particular mode. The mere existence of a save feature (of any kind or era) does not force saving on players. If we decide that we won't be saving what we do, we are not merely choosing not to record something or not to mark a place in a text or not to enter something into a possible archive; we are choosing a way to play, an attitude. When we play without intent to save, we play loosely, wildly, without a care to future in-game repercussions. This kind of unsaveworthy play isn't counterplay or radical; it is part of the game's core experience. Players shift from one mode to the other based on various contexts and contingencies. Surely I'm not the only player who blows himself up between saving and switching off the system. This kind of reckless play is explicitly *not* to be saved as a game state or file; when it is saved, that saving is by accident (a mistake we make in play), and we experience it just as that, a mistake that makes things count in ways we didn't want them to.

What we see, then, in the possibilities of contemporary game-save practices are tensions around perfection, labor, effort, and our future. These tensions are the heart of issues that run through the entire history and historiography of game studies. What is play? What is work? Why do we do things that seem difficult or burdensome or pointless or all of the above? Contemporary console save practices push these issues to the forefront, and we see a kind of *care*, as David Sudnow calls it, being pushed to the fore (1983, 54).

But not all play that we save is so studied or so purposeful. We save all the time when we play these games, whether we want to or not, for there is a kind of saving in and through play that resides not in the inside of systems or in memory cards or in the Cloud but rather on the surfaces of game controllers and in and on the bodies of their players. This save is the accrual not of data or progress through game space but of blisters and smudged touch screens. Here is a save that we do not load but only ever continue. In our sore necks and stuck "A" buttons is stored evidence of use, proof of play; through this process, we bend our bodies toward past play as we move into the future and at the same time admit our systems and devices into the category of the well-used thing, the well worn. These patinas and calluses save the experience of the corporeal, the carnal. They save our efforts. They show us in a way that a file or code or even a high score does not that we have really *done* something. It is hard to see ourselves in a screen capture of our own play. In most cases, such a record looks indistinguishable from the game's looping attract-mode screen. In contrast, the legacy of play we save in calluses on our fingers, wear on game keys, and yellowed game pads, like the high score and the quarter on the cabinet, are forms of saving that connect us to our past and future play.

Works Cited

Aarseth, Espen. 1997. *Cybertext: Perspectives on Ergodic Literature*. Baltimore: Johns Hopkins University Press.

Kocurek, Carly. 2012. "Coin-Drop Capitalism: Economic Lessons from the Video Game Arcade." In *Before the Crash: An Anthology of Early Video Game History*, ed. Mark J. P. Wolf, 189–208. Baltimore: Wayne State University Press.

Lowood, Henry. 2009. "*Warcraft* Adventures: Texts, Replay, and Machinima in a Game-Based Story World." In *Third Person: Authoring and Exploring Vast Narratives*, ed. Pat Harrigan and Noah Wardrip-Fruin, 407–428. Cambridge, MA: MIT Press.

Loading ... , North America 6: 82–99. http://journals.sfu.ca/loading/index.php/loading/article/view/104.

Moran, Chuck. 2010. "Playing with Game Time: Auto-Saves and Undoing Despite the 'Magic Circle.'" *Fibreculture Journal* 16. http://sixteen.fibreculturejournal.org/playing-with-game-time-auto-saves-and-undoing-despite-the-magic-circle/.

"News." 1990. *Replay*, March, 83–86.

Poole, Steven. 2004. *Trigger Happy*. New York: Arcade.

Sudnow, David. 1983. *Pilgrim in the Microworld*. New York: Warner Books.

Therrien, Carl. 2012. "Video Games Caught Up in History: Accessibility, Teleological Distortion, and Other Methodological Issues." In *Before the Crash: An Anthology of Early Video Game History*, ed. Mark J. P. Wolf, 9–29. Baltimore: Wayne State University Press.

Tobin, Samuel. 2012. "Time and Space in Play: Saving and Pausing with the Nintendo DS." *Games and Culture* 7 (2): 127–141.

46 SIMULATION

David Myers

The etymology of the term *simulation* is a curious one, simultaneously denoting difference and similarity. The simulation references something other than itself, yet the simulation is also necessarily something other than that reference: an act of *pretense*. The tension between pretense and imitation within the simulation is exacerbated by digital media—for example, within the experientially uncertain "uncanny valley"—and requires increased interpretive effort to resolve.

In this context, what is the difference between a game and a simulation?

In current practice, these two concepts are combined and perhaps, depending on the circumstance, conflated. In common use, there seems an intimate relation between game and simulation, marked by their mutual affinity with the dynamics of digital media. Nevertheless, digital games have rules of different sorts and purposes than do digital simulations, whose rules are perhaps better understood as rule *models*. These rule models are then bound in some important way to that which they would simulate—a restriction not equally binding of game rules.

The Mechanics of Early Simulation Games

Those digital games considered prototypical—among the first of their kind—were clearly designed, in some part, to simulate. For instance, *Spacewar!* (1961), a collaborative construction of MIT graduate students, might be considered to simulate both the physics of space flight and the sociopolitical context of US–Soviet relations that came to prioritize Sputniks, lunar landers, and, eventually, intercontinental ballistic missiles. But figurative references alone are no simulation. The simulation portion of the game is found in the game's code, its embedded algorithms that quantify and formalize cause-and-effect relationships. In *Spacewar!* these relationships were among mass, force, acceleration, and momentum.

Other equally early digital games similarly employed rule models derived directly from first-year physics textbooks—for example, *Lunar Lander* (various versions, circa 1969) and *Artillery* (various versions, circa 1976). The digital medium provided game players the opportunity to experiment with the consequences of algorithms expressed as formulaic equations within the game's code—that is, to vary the output of these equations through manipulation of input variables.

Early interactive simulations of this sort, framed as games, were not restricted to laws of physics. *Hammurabi* (various versions, circa 1968) had the same algorithm-based design as *Lunar Lander*, except that in *Hammurabi* these equations were meant to represent political and economic relationships. Aside from these social and political references, however, *Hammurabi* functioned similarly to simulations of physics: its formulas mechanically transformed player input into output.

Given only slightly more sophisticated programming techniques, simulation games such as *Hammurabi* could be designed to run without any input from the player at all, in recursive loops, using one cycle of output for the next cycle of input. Climate simulations used to make weather predictions (the Navy Operational Global Atmospheric Prediction system and others) run precisely this way: beginning with real-world variables (temperature, humidity, barometric pressure, and the like) and then cycling these initial variables through successive input-output iterations.

Current long-lived digital simulation games of sports leagues—including *Out of the Park Baseball* (OOTP Developments, 1999) and *Championship/Football Manager* (Sports Interactive, 1992)—equally offer options for unattended play, during which the outcomes of either individual competitions or entire seasons can be generated with little to no player intervention. Player intervention and decision making then serve to distinguish the use of simulations, which doesn't necessarily require either of these actions, from the play of games, which does.

This distinction between the manner in which the simulation functions and the manner in which the game is played has consequences in the different ways players gain mastery over the simulation and the game. Mastery of a simulation leads to an increased compliance with its rule models, including its interface and controllers, as in *Microsoft Flight Simulator* (1982), where the player gains an awareness and mastery of fixed-wing flight by conforming her behavior to the requirements of the simulation.

Mastery of a game initially seems similar in that the player gains an awareness and mastery of the game by conforming his behavior to that behavior valued most highly by the game's conditions for winning. However, the master player, once skilled at winning a game, continues to engage and interact with the game rules: pushing and manipulating them to the point of breaking. In the recursive process of digital gameplay and replay, the advanced game player becomes as adept at deconstructing as at complying with game rules, and master gameplay increasingly resembles neophyte game design.

The iterative design of *Civilization* (MicroProse, 1991), one of the more popular (so-called) simulation games in digital-game history, well demonstrates this relationship between game players and game rules, with the game player oscillating between compliance through acceptance and revision through denial. "Sid Meier, who designed *Civilization*, consciously and purposely worked on the *Civilization* series in just this manner: in close collaboration with his co-designer, Bruce Shelley. Shelley tested Meier's designs during repeated play, and Meier adjusted his designs so that those designs more closely conformed to Shelley's optimum play experience" (Myers 2003, 177).

To the extent possible, advanced players tend to adjust the rules of games to evoke valued play experiences more reliably and to satisfy an accompanying desire for further, similar gameplay. This phenomenon has always been a part of advanced gameplay—observed, for instance, in the implementation of "house rules"—but is more commonly provided for and more easily accessed during digital gameplay than during nondigital gameplay.

The representational target for a game is then an ideal—an "optimum play experience"—which remains distinct from the external referent that is the representational target of a simulation. Regardless of the specifics of this external referent (i.e., whether what is simulated is a physical system, an economic system, or some other system), the referent of the simulation is *fixed* within its rule models. In contrast, the optimum play experience of the game is one that necessarily is *unfixed* by the rules of the game. Those games that have the longest sustained player interest tend to have outcomes that are uncertain and, even after extended play, indeterminable—that is, games more like chess and Go (and *Civilization*) and less like Tic Tac Toe.

The Crucible of Player Agency

Focusing on the limits and allowances of player agency as a means of distinguishing digital games and simulations is useful—up to a point. Distinguishing between games and simulations on this basis might assume that each representational form, game and simulation, remains immune to the other's influences. This is not the case.

In its original form (circa 1975), Will Crowther's game *Colossal Cave* was a text-based simulation of spelunking. As *Colossal Cave* was played and replayed, however, it was revised and recoded (similar to how *Civilization* was revised and recoded) in ways that simultaneously reduced its verisimilitude and increased its appeal. The introduction of fantasy objects and characters, treasure hunts, point totals, and other blatantly nonrealistic elements to *Colossal*

Cave turned the original simulation into a gameplay experience that was no longer suggestive of crawling through Kentucky's Mammoth Caves. During this revision, the external target of the original simulation was not retained; over time, that target was revised and replaced according to a desire for more pleasurable gameplay.

In fact, when inspected closely, many representations in early digital simulation games—including representations of spatial and temporal relationships—became less realistic as they were revised and edited during subsequent, recursive play. Iterative game design focused on balancing and, where appropriate, increasing player agency has transformed simulations into something else. The game's own peculiar representational form has become increasingly influential, pervasive, and invasive of the representational form of the simulation.

As an example of this progressively invasive gameplay (and replay) process, consider the well-documented history of wargames, a history that precedes the introduction of digital games in the 1960s (cf., Smith 2010; Dunnigan 1980).

The Transformation of Training

There is a long tradition of using simulations—often in the form of games—in military training. However, it wasn't until the late 1950s that Charles Roberts commercialized the mechanics of simulated warfare in a series of wargames published by his newly formed game company, Avalon Hill.

Avalon Hill was one of several board-game companies—including Simulation Publications, Inc., and Game Design Workshop—that found equal success with a similar design strategy. Although the Avalon Hill wargames employed (roughly) the same mechanics and themes as military training simulations, they were designed to appeal to hobbyists as interested in pleasurable gameplay as in historical accuracy, which required revision of the training simulations. The simulations had to be made less cumbersome to conform to player limitations—for example, by reducing the "breadth of geography, number of icons and complexity of algorithm" (Smith 2010, 11). And they had to be made more gamelike to conform to player desires—for example, by incorporating alternating "turns" of play and hexagon-based playing boards.

During the subsequent period in which board-based wargames were transformed from cardboard to computer, digital media allowed much of the complexities of military training simulations to be restored. These complexities were embedded in the digital game's algorithms, where there was no need, as there had been previously, for player intervention and aid. The rule models of combat simulators receded into the background of gameplay, rising to player awareness and concern only when called upon to adjudicate player decision making.

However, despite the fact that digital media allowed commercial games the same level of realistic detail as military simulations, digital wargames retained their concessions to gameplay. Some of these concessions were obvious ones: points and winning conditions. Others were more subtly hidden in the mechanics of the simulation's rule models.

Physics versus Psychophysics

In games, time is an important resource, and games structure that resource like any other. Turn-based games conventionally allot time in equal portions to all players. If players are judged greedy, then their allotments are strictly controlled—for example, with time clocks in chess.

Board-based wargames—based on the original Avalon Hill designs—were turn-based games. Soon after these war games were translated to digital media, "real-time" wargames began to appear in the video game marketplace. In these real-time games, all players "play" (e.g., move their playing pieces) whenever they like, with the computer constantly monitoring and updating consequences of play.

Real-time strategy games proved popular. The genre generated its own well-known acronym (RTS) and a familiar design template. Currently popular digital wargame series with strong simulative components—for example, *Men of War* (1C Company, 2009) and *Empire: Total War* (Sega, 2009)—are published with either RTS or real-time tactics components (or with both).

Based on their name alone, we might expect RTS games to include a more accurate and realistic simulation of time than similarly themed turn-based games, but this is not the case. A battle in the *Empire: Total War* series—the Battle of Waterloo, for instance—might involve hundreds, if not thousands, of historically accurate units, and each of these units may be controlled at some level during gameplay. However, when such games are played in a real-time mode, it is impractical to micromanage and exert this control during battle. Once the game battle begins, it continues, diminishing the extent players can assert agency and control—and therein ostensibly simulating a "real" battle. Nevertheless, even in this unattended and automated state, the "real time" of the simulation is distorted. The Battle of Waterloo, for instance, took place over the course of several days, with the main engagement lasting from noon until evening. The "real time" of the game's version of this battle is more compact and paced and sequenced in ways that appeal to its players.

RTS game designs are best understood to simulate, if anything, a player's *sense* of real time—or, equally, a pleasurable *illusion* of real time (including options to "pause"). The "real time" in strategy wargames is therein very similar to other representations of time in digital games—"bullet time," "quick time," and so on—that are also designed and implemented for gameplay purposes. Each of these representations is an anthropomorphic *misrepresentation* of time in which some more objective time is edited and transformed into a game resource. This resource is then allotted and doled out according to the rules of the game. During this allotment process, the game waxes as the simulation wanes.

Spatial representations in war games have likewise been transformed and distorted according to the needs of both gameplay and players. The hexagonal overlay of the Avalon Hill board games is now a ubiquitous system for transforming simulation map boundaries and borders into a jagged series of 120-degree angles. Coordinate systems of this sort—using area-to-area movement schemes—necessarily compromise more realistic landscapes, creating abstractions that conform only to a topology of play.

Logical relationships represented within a simulation's rule models—in particular those associated with laws of probability—can be equally compromised. In popular versions of *Civilization*, for instance, a phalanx might destroy a bomber, and an archer might sink a battleship. These individual units within the game—representations of ancient and modern military forces—are governed by a rudimentary set of algorithms assigning all such units some probability of defeating their opposites. Although the odds are greatly stacked against the archer fighting the battleship, at some moment during repeated and extended play David occasionally slays Goliath.

Maintaining desirable possibilities of this sort in simulations of combat and conflict—the possibility that tables can be turned and underdogs can win—assigns the game's most critical functions to player agency and intervention at the expense of realism.

In games that offer no pretense of simulation, this pervasive influence on game design is obvious: game space and game time are calibrated according to the capabilities and limitations of human hand–eye coordination and related cognitive functions. They govern the rate at which *Tetris* (ELORG, 1984) blocks fall, the screen size in *Bejeweled* (PopCap, 2001), and the movement of opponents in *Galaga* (Namco, 1981). In simulation games, this governance is less obvious but still critical: sequencing and pacing and systemic logic are assigned game-determined representations based on player capabilities, expectations, and desires.

Representations of Misrepresentations

War games are not the only digital games that gain a significant portion of their appeal from being classified and marketed as simulation games. Will Wright's game *SimCity* (Maxis, 1989) was the first of a series of games designed using a common "sandbox" template that have since, like RTS games, established their own market genre. With the growing stature of the video game industry, educators have more actively promoted the use of digital games in educational contexts, where there is a great deal invested in demonstrating how game-specific skills might be used to accomplish more practical, outside-the-game goals. The effectiveness of the simulation game as a learning device—its "realness"—is then a claim that educational game advocates find useful. This claim has found popular appeal, even in game-based contexts in which the "realness" of the simulation is often subverted.

Indeed, a steadfast faith in the representational verisimilitude of digital simulations sometimes prompts a belief that these simulations, when representing real-world processes, might become *too* real and in their bludgeoning reality go beyond any possibility of human agency and control. Such simulations then evoke a primitive fear that signs and symbols might rob us of our sense of what is and what is not—the same fear that drives a *Body Snatchers* (Finney 1955) scenario: simulations (the reference or *pretense*) might replace us (the referent or *actual*).

However, these more frightening simulations—alongside those touted as teaching and learning tools—are founded in an exactitude of representation. In the *simulation game's* assertion of player agency, this exactitude is neither required nor preferred nor implemented. The simulation portion of a simulation game must provide a representation of reality only insofar as player agency remains applicable and, when applied, effective. If the simulation interferes with or diminishes player agency while conforming to an external target of representation, then that conformation needn't be enforced.

Given this subservient role, in games and elsewhere, digital simulations are more likely to conform to psychophysics than to physics and consequently to appear less often as objects of mistrust and fear.

Prior to their successful incorporation into game form, for instance, digital-based human simulations were often portrayed in films as well intentioned but, Frankenstein style, quite berserk (e.g., HAL in *2001: A Space Odyssey* [Stanley Kubrik, 1968] and Joshua in *Wargames* [John Badham, 1983]). When properly designed to empower human agency, however, simulations can gain the ability to right reality's wrongs—as does, prototypically, Jake in *Avatar* (James Cameron, 2009). In such revisionist simulations, the fear of the simulation replacing reality is allayed by a desire for the simulation to do just that: to provide an "alternate" reality in substitution for an otherwise inferior and "broken" version (see also McGonigal 2012).

Positioning the simulation game as an alternate reality of this more friendly and recognizable sort allows us to exploit our gameplay; we are simultaneously self-centered and socially constructive. Our pleasures need no longer be so guilty, and any harsher reality that might subsequently intrude is no longer quite so persuasive.

Works Cited

Dunnigan, James F. 1980. *The Complete Wargames Handbook*. New York: William Morrow.

Finney, Jack. 1955. *The Body Snatchers*. New York: Dell.

McGonigal, Jane. 2012. *Reality Is Broken: Why Games Make Us Better and How They Can Change the World*. New York: Penguin.

Myers, David. 2003. *The Nature of Computer Games*. New York: Peter Lang.

Smith, Roger. 2010. "The Long History of Gaming in Military Training." *Simulation & Gaming* 41 (1): 6–19.

47 TOYS

Jon-Paul C. Dyson

In February 1975, Al Alcorn, creator of Atari's game *Pong* (1972), traveled with colleague Gene Lipkin to the 72nd Annual American Toy Fair in New York City to market a home version of the arcade game he had made. The trip was a failure. Alcorn and Lipkin failed to secure a single buyer interested in selling the new television game; Atari ultimately made a deal with Tom Quinn of Sears to market the product through the Sporting Goods division. In retrospect, the reason Atari failed to successfully land a toy distributor had much to do with the company's lack of understanding of how deals were made and constructed in the toy industry, but the toy industry's initial lack of interest in picking up *Pong* also symbolized what would be a long-standing and uncertain relationship between video games and toys. Indeed, a consideration of the video game *as* a toy has largely been absent from most scholarship on video games. Video game scholars have instead tended to emphasize video games' descent from computers, to discuss video games' relationship to television technology, or to treat the video game within a formal study of the properties of games. Even today, a fundamental question remains: Are video games toys?

Rather than answering this question by trying to parse the definitions of the terms *video game* and *toy*—after all, scholars have trouble even defining the basic concept of play—let us instead unpack Atari's decision to go to Toy Fair. What seems most clear is that Atari founder Nolan Bushnell and other company executives recognized that *Pong* was first and foremost an object of play, an object meant to promote an activity that is fun, voluntary, and engaged in for its own sake. Atari was already in the play business with its growing line of coin-operated arcade games, but arcade games were public forms of play for which there existed a mature industry of manufacturers and distributors who made pinballs, slot machines, and redemption amusement devices for locations such as bars and boardwalks. When it came to a plaything for the home, Atari looked first to the toy industry.

Toys are a product of the market economy. For centuries, there have been well-established toy manufacturers and a thriving toy industry. This toy industry grew enormously during the nineteenth and twentieth centuries with the rise of a prosperous middle class and consumption-based economies in Europe, North America, and Asia. Toy makers churned out an ever-increasing number of new toys, often by incorporating new technologies into old play forms. Wooden blocks gave way to metal construction kits such as the Erector set, and these kits in turn were supplanted by plastic building components such as the LEGO brick. Doll manufacturers continually introduced new materials and mechanics to dolls to make their appearance more lifelike and to give them the ability to speak, move, and even use the bathroom. Since at least the 1920s, toy makers had created circuit-based electric questioner games. When Atari put a game on a chip, it was only logical that the company should look to the toy industry to market the home version of *Pong*, as had Magnavox in 1972 when it announced Odyssey in the pages of *Playthings*, the main publication of the toy industry.

Throughout the 1970s and 1980s, toy companies manufactured a significant number of home electronic games of all types. Mattel's Intellivision (1979), the ColecoVision (1982), Milton Bradley's Vectrex (1982), and the Nintendo Entertainment System (NES, 1985) all came from companies with a background in toys. When Nintendo launched the NES, it advertised it clearly as a toy, marketed it to toy stores, and packaged it with R.O.B. (Robotic Operating Buddy), a robot in a year when *Playthings* cited robots as one of the top toy trends of the year. This relationship between toys and video games during the 1970s and 1980s was even stronger in the handheld market. Ralph Baer, the man who patented the idea of playing a video game on a television and invented the Magnavox Odyssey, also created one of the best-selling toys of the 1970s and 1980s, Simon. Coleco and Mattel introduced many sports handhelds, including popular football and baseball handheld games. Milton Bradley released the Microvision, the first handheld gaming device with interchangeable games, in 1979. Ten years later Nintendo built on the success of its popular Game and Watch systems with the Game Boy.

Coleco offers a case study of how one prominent toy company became a major player in the video game business. Founded in 1932 as Connecticut Leather Company, Coleco began manufacturing toys in the 1950s. By the 1960s, its line of playthings ranged from backyard pools to plastic soldiers, but it specialized in equipment themed around sports. In its *TOYS/GAMES/SPORTING GOODS* catalog for 1975, for example, it heavily advertised its numerous electric, vibrating football games; *T.V. Football* (a nonelectronic, gambling game that players engaged in while watching televised football); and a wide-variety of mechanical basketball, soccer, and hockey games. The company was adept at translating sports into toys and games for the home, so it is not surprising that within a year of the debut of the home version of *Pong*, Coleco had created its own TV game. TELSTAR, which used General Instrument's AY-3-8500 chip to play tennis, hockey, and handball, led the Coleco catalog in 1976, followed by the usual assortment of air hockey, electric football, and shaft-driven hockey games. Coleco followed with many handheld electronic games, including the iconic *Electronic Quarterback*,

Head-to-Head Baseball, and Mini-Arcade versions of *Pac-Man* and *Donkey Kong* (among others). The company eventually got into the full-fledged console market with the ColecoVision in 1982. When Coleco introduced the Adam in 1983, it symbolized the transformation of the computer itself into a toy.

In many ways, the computer had always been a toy for a category of players known as hackers. The computers they used may have been paid for with defense dollars or business funds, but early hackers such as Steve Russell and other members of MIT's Tech Model Railroad Club turned these machines to playful purposes by creating games such as *Spacewar!* (1962). As the game spread across the country, computer devotees' willingness to play with the computer grew. In his *Rolling Stone* profile of programmers at Stanford in 1972, "Spacewar: Fanatic Life and Symbolic Death among the Computer Bums," Stewart Brand labeled hackers as "fanatics with a potent new toy." *Creative Computing*, the first general-interest computer magazine founded by David Ahl, published articles and programs to help users draw pictures, play music, or generate poetry. When Ahl compiled programs for his book *BASIC Computer Games* (1973), he included many programs such as *BUNNY* (draws a Playboy bunny), *BUZZWD* (generates sets of random, three-word computer jargon), and *ZOOP* (a program that responded to the system commands of a BASIC [Beginner's All-Purpose Symbolic Instruction Code] compiler with various silly and insulting phrases) that were not games in any traditional sense. Steven Levy's book *Hackers* (1984) described how this ethos penetrated throughout the computer industry. The Scandinavian demoscene in the 1980s similarly celebrated the use of computers as playthings to create art and music that showed off the programmer's creativity and programming powers.

And yet only rarely has the computer ever been identified as a toy, perhaps because it was originally used exclusively by adults, and toys have traditionally been associated with children. Indeed, in portraiture through the nineteenth century the presence of a toy often identified the person holding it as a child, and the type of toy might signal the gender. Brian Sutton-Smith pointedly asks in his book *Toys as Culture*, "Aren't children nowadays those creatures who are known by their toys, plays and games, who are classified, according to age and sex, by their toys?" (1986, 87). If a computer or a video game is a toy, by implication is the person using it a child? This is indeed the argument made by cultural historian Gary Cross, who in his critique of modern masculinity, *Men to Boys: The Making of Modern Immaturity* (2008), cites statistics from the Entertainment Software Association showing that people continue to play video games into adulthood to support his argument that men are failing to put off childish things and take on the responsibilities of manhood.

But the strict historical association between toys and children has long been breaking down. The roots of this breakdown lie in changes in understanding the family. During the course of the nineteenth century, the toy remained an object of childhood, even as Western society experienced profound shifts in the family's role and function. Whereas once the family had largely been a locus of economic production or civic stability within a patriarchal social order, it increasingly became a center of emotional security ruled by bonds of mutual affection. In the phrase of the historian and social critic Christopher Lasch (1977), it became a "haven in a heartless world." In the nineteenth century, cultural authorities had claimed religion was the glue that kept the family together, but by the early twentieth century toy manufacturers began to offer their products as key means of building bonds between generations. For the toy industry, it was not so much the family that prayed together that stayed together, but the family that played together.

Toy advertisements exemplified this notion. A. C. Gilbert, creator of the Erector set, promised parents to "solve the boy problem" and turn a boy into a man through play with these construction sets. Lionel similarly advertised its trains as a means to bring together father and son. Covers of Milton Bradley's *Battleship* (1967) pictured a father and son happily playing the game, while the mother and daughter smilingly watched from the next room as they washed the dishes. Manufacturers of video games for the home quickly adopted this visual advertising language from the toy industry. Atari marketed its early coin-operated games with pictures of adults, and the first advertisement for *Pong* (by Sears Telegames) in the Sears *Wish Book* for 1975 similarly featured a man and woman, but the company quickly introduced entire families into the advertisements. Other advertisements from other companies likewise featured the video game as a toy that everyone in the family could enjoy. By the 1980s, however, the toy industry had become much more segmented in large part because of the growth of Toys 'R' Us and other specialized retailers, so when Nintendo released the NES, it squarely went after the youth market.

Since then the video game has been variously advertised as a device for children or for teenagers or for adults. As the market for video games has grown, as players have persisted in their play even as they have grown up, and as game makers have found more channels with which to reach customers, game makers have been able to segment and specialize. When Xbox debuted in 2001, for example, Microsoft clearly aimed it for an older audience, with images of young adults, not kids, playing the game; the repetitive use of the word *power* in the advertisements; and the promotion of the violent shooter *Halo* (Bungie) as a launch title. As gaming audiences aged, they still wanted to hang onto their toys, even if these objects weren't called that anymore. The Wii would successfully revive the image of the console as a toy that could bring entire families together, but by that time the video game clearly knew no age restrictions.

If the toy was traditionally the province of childhood, it is no surprise that there is a strong connection between toys and the rituals of gift giving. As anthropologists have pointed out, the gift is a key device for maintaining societal relations, both inside and outside the family.

In modern culture, the gift has its locus at Christmastime, when toys make up a significant percentage of the presents. As scholars such as Stephen Nissenbaum (1997) have pointed out, in America Christmas went from being forbidden to being celebrated as a means of promoting family bonds, and toy manufacturers encouraged these trends. The toy is the ultimate Christmas gift because it serves no other purpose than to be enjoyable for the user—therefore, the act of giving cements loyalty and love between the gift giver and the gift recipient. Video game companies have long tapped into this association between toys and gift giving. Manufacturers have released almost every new major home video game system at Christmastime, from *Pong* in the Sears Christmas *Wish Book* in 1975 to the Xbox One and PlayStation 4 in 2013. Although the growth in the number of adult players means that people have more disposable income throughout the year to buy games, the holiday season is still by far the time of greatest video game sales.

Thus, in manufacturing, advertising, audience, and sales, video games and toys have been intertwined. This historical perspective on the relationship between toys and video games is perhaps unsatisfying to some who want to emphasize theoretical differences, but, as noted earlier, when we try to define toys, just as when we try to define play or games, we enter a rabbit warren of definitional complexity. Many ways of differentiating between toys and games seem arbitrary or not useful. For example, the distinction that games have rules, although true, does not obviate the fact that play of all kinds has rules, though often those rules are unarticulated and developed iteratively through the play. When children play with toys, they usually follow rules, though they generally make them up. Or perhaps it might be asserted that games have an end state, that a game is bound by time, whereas toys are not. And yet plenty of modern games such as *Dungeons & Dragons* (Gygax and Arneson/Tactical Studies Rules, 1974) and *World of Warcraft* (Blizzard Entertainment, 2004) have no fixed end state. Are they better described as toys, then?

Two of Will Wright's games exemplify the problems arrived at by trying to reach a theoretical separation between games and toys. Wright has compared his two most successful games, *SimCity* (Maxis, 1989) and *The Sims* (EA Maxis, 2000), to playthings that are not games: *SimCity* to model railroad building and *The Sims* to dollhouse play. Most people would consider model railroads and dollhouses as toys, and yet *SimCity* and *The Sims* were widely considered games. Indeed, when Wright and Maxis brought *SimCity* to Brøderbund for potential distribution, Brøderbund encouraged the design team to add more gamelike elements such as reacting to a nuclear meltdown or dealing with a rampaging monster. As it turned out, players generally ignored these scenarios in favor of the pure pleasure of running their cities.

Some years later, when Wright began developing *The Sims*, he thought consciously about the relationship between games, models, and toys. In his design papers for the game, he wrote down some of his thinking. He noted that whereas there were plenty of good computer games and models (i.e., simulations) at the time, there were fewer computer toys. He wrote:

> COMPUTER → TOYS
>
> THIS IS WHERE COMP. IS MOST LACKING
>
> SCREENSAVERS — HOME DESIGN — INTERNET
>
> NEED THINGS WHICH CAN BE SUBVERTED CREATIVELY BY USER
>
> ENABLE "CREATIVITY"
>
> WHAT I'M STRUGGLING TO DESIGN & BUILD (n.d.)

He noted that the best toys have a wide range of possibilities and encourage exploring and experimenting. Psychologists have long noted that good toys have many affordances that give children the opportunity to play in many different ways. So Wright set about creating a program that would have the affordances of toys but also partake of some of the qualities of a game or a model. The result, *The Sims*, was a spectacular product. However, although Wright may have stressed its toy elements, its publishers marketed it as a game. Electronic Arts (EA) proudly advertised *The Sims* as "Game of the Year" and the "#1 Selling PC Game of All-Time." With *The Sims*, as with *Pong*, it proved difficult to make hard-and-fast delineations between what was a toy and what was a game.

That decision to market *The Sims* as a game, not a toy, was at its core a simple business decision. Just as Atari brought *Pong* to the toy industry because there was a well-developed and self-identified toy industry, so it made sense for EA to market *The Sims* through the now mature electronic games industry. And yet this may also hint at deeper cultural attitudes toward games and toys. Games, especially in recent years, have earned a measure of esteem and cultural elevation, from the articulation of mathematical game theory to the development of robust associations for adult players of games such as *Scrabble*, chess, and poker. Although it's OK for an adult to play games, it's still suspect for an adult to play with a toy.

In recent years, the toy industry itself has become increasingly digital. Processors have powered products such as Alphie, Furby, Tickle Me Elmo, LEGO Mindstorms, LeapFrog, and increasing numbers of other products. Activision's *Skylanders* (2011) and Disney's *Infinity* (2013) blur the line between toy and game as well as between the physical and virtual worlds as players play both on screen and with the plastic figures in real life. The popularity of these cyborg toys raises the question of whether toys necessarily need to have a physical property. In an age in which much of our play is purely virtual, can a toy be purely digital?

What does seem most clear is that a toy is an agent of transformation that lets the user play, imagine, daydream in ways that help him or her grow mentally, imaginatively, emotionally—even to grow bigger in his or her own mind. Toys touch his or her life forever.

How toys do this has changed dramatically over the centuries, and video games are only the latest variation. It's not always possible to untangle the relationship between video games and toys. But perhaps that doesn't matter. Ultimately, the benchmark for determining whether something is a good video game or a good toy is the same: Does it promote good play? And over time that metric has never changed and never will.

Works Cited

Ahl, David, ed. 1973. *BASIC Computer Games*. Morris Plains, NJ: Creative Computing.

Brand, Stewart. 1972. "Spacewar: Fanatic Life and Symbolic Death among the Computer Bums." *Rolling Stone*, December 7. http://www.wheels.org/spacewar/stone/rolling_stone.html.

Cross, Gary. 2008. *Men to Boys: The Making of Modern Immaturity*. New York: Columbia University Press.

Lasch, Christopher. 1977. *Haven in a Heartless World: The Family Besieged*. New York: Basic Books.

Levy, Steven. 1984. *Hackers: Heroes of the Computer Revolution*. Garden City, NY: Nerraw Manijaime and Doubleday.

Nissenbaum, Stephen. 1997. *The Battle for Christmas*. New York: Vintage Books.

Sutton-Smith, Brian. 1986. *Toys as Culture*. New York: Gardner Press.

Wright, Will. n.d. Game Design Notebook—Robot Wars, Maxis acquisitions, Sims. Will Wright Collection. Brian Sutton-Smith Library and Archives of Play at The Strong, Rochester, NY, Series I, Box 1.

48 WALKTHROUGH

James Newman

There can be no doubt that the walkthrough is an integral part of video game culture. Most gamers will be familiar with the opportunity to purchase a strategy guide as part of a special offer with a new game. Many gamers, whether or not they care to admit it, have consulted an online guide in search of a solution to a particularly difficult sequence or to learn more about an in-game encounter. The sheer number, not to mention the comprehensiveness, of walkthroughs circulating in print and online already marks them out as an important video-gaming phenomenon. Moreover, although we might think of them as uniquely modern creations whose existence relies on the Internet or coincides with the expansion of ancillary markets for merchandising and ephemera, walkthroughs have a history that dates back almost as far as video games themselves.

However, despite their prevalence and heritage, walkthroughs remain among the most controversial and divisive of all contemporary video-gaming texts. For some, a walkthrough's instructions on how to solve puzzles or navigate environments provide the solution to problems that potentially bar any further progress through the game and, as such, allow access to levels and sequences that might otherwise remain forever unplayed. For others, the presentation of solutions to puzzles undermines the point of gameplay. The very use of a walkthrough is thus viewed as the ultimate expression of cheating, an uncomplicated admission of a player's failure and thus his or her inexpert status. Being more conciliatory to the player, we might see the existence of walkthroughs as indicators of failures in game design where linearity and excessive demands on skill combine to shut down progress for all but the most adept, committed, or lucky. Yet some walkthroughs are easier to read as testaments to meticulous and expert play that seeks to eke out every last feature, secret, and facet of a game, whether intended or known by its developers. In this way, the walkthrough might be seen to breathe life into a game by offering new opportunities for play that go beyond what we might think of as the normal strategies or standard objectives. These walkthroughs are created by conspicuously expert players, often combining their efforts and sharing their works-in-progress online as they explore games and gameplay in self-consciously investigative ways. Here, the walkthrough is the formal record of a collective knowledge passed on to readers, whose future play and approach to the game is framed by these prefigurative texts. By contrast, other "walkthrough" publications, available in book and game stores and lavishly bound with full-color images and glossy pages, might be seen as a part—perhaps even a cynically exploitative part—of the processes of production, tightly enmeshed within the regimes of development, marketing, and retail and generating additional revenue as they are bundled with new games. Interestingly, if we look back to the earliest days of publishing on video game strategy and gameplay, however, we see a different picture.

Defining Our Terms

In the early 1980s, riding the popularity of the new medium of video games, a new genre of book emerged. With titles such as *The Video Master's Guide to Defender* (Broomis 1982) and *The Winners Book of Video Games* (Kubey 1982), these pulpy paperbacks, often with hand-annotated screen layouts, promised to reveal winning players' secrets and strategies or, as the front cover of *How to Beat the Video Games* (Blanchet 1982) unblushingly put it, the "surefire techniques to outsmart all the most popular games." The content of these books focused on boosting scores, beating the game "at its own game," and in some cases even making money from expert play, as described in the section "Hustling *Pac-Man*" in Ken Uston's book *Mastering* Pac-Man (1982, 123). Importantly, these books were written *for* players *by* players rather than being tie-in products associated with development or retail or the games they addressed. However, their concern with revealing the patterns and algorithms underlying gameplay so as to translate this systemic knowledge into successful and playful tactics and strategies makes these books a clear ancestor of the modern walkthrough, although nowhere in these volumes is that term used explicitly. By 1988, however, the term *walkthrough* took center stage in Shay Addams's collection of adventure-game solutions, *Quest for Clues* (1988). Here, each title is introduced with a brief backstory before "the Walkthrough" guides the player, step by step, clue by clue, through the game. The spartan formatting of each required text-input string and the representation of the game world in flowchartlike diagrams that forgo allusions to geographical contiguity in favor of nodes and flows of logic are particularly striking and foreshadow the reductive, codelike presentation of contemporary player-produced walkthrough texts.

However, where Addams's books used the word *walkthrough* in an unambiguous, almost triumphant, manner, today the situation is made altogether more complex by numerous overlapping and competing terms. The Help pages at GameFAQs, one of the largest online repositories of player-authored texts, highlight the problem in positing the synonymity of the terms *guide*, *file*, *walkthrough*, *strategy guide*, and *FAQ*. Indeed, if we add IGN's "Wiki Guides" to the list and try to distinguish between "unofficial" and "official" strategy guides published by Prima and Brady, for instance, we get a sense of the terminological imprecision. Although we might broadly concur with the fundamental assertion that the various-titled documents are united by their aim to "provide help for gamers," we might take issue with GameFAQs' claim that the terms are synonymous. Certainly, they are often used interchangeably, but there are some important distinctions to be made.

Fundamentally, it is useful to distinguish between "official" and "unofficial" texts. As veteran writer Alan Emrich notes, a key difference between the two centers on the point of access to the game (n.d.). In exchange for a cut of royalties, a game's developer or publisher grants access to pre-release code and to members of the development team who can provide "inside information." The resulting guide may be released simultaneously with the game, very often as part of a reduced-price bundle. For some commentators, the tie between strategy guide, game design, and commercial opportunity has become even closer: "In certain video games there will be secrets, bonuses, treasures or other things that are unknown to the player. No information regarding these things is placed inside the game. No amount of deductive logic and reasoning will result in the player discovering these things. ... Finding these secrets is only a matter of knowing or not knowing. The easiest, and sometimes only way, to become someone who knows is to read the strategy guide" (Rubin 2004, 260–261).

As Emrich reveals, although unofficial guides *might* be written by authors with access to the game's programmers or play testers, there is no guarantee that they are, and because these guides are based on the game's retail release, they rely on a combination of firsthand play and insights gleaned from the wider community of players. *How to Win Video Games* assures the reader that its information is obtained from only "the best video-gamesters" (Editors of *Consumer Guide* 1982, 3), and its colophon is similarly at pains to note the absence of any affiliation or contractual agreement between the book's publishers or the manufacturer of the games to which its pages refer.

In this respect, the writers of the *"Unofficial" Strategy Guide* and the "Beat the Game" texts of the early 1980s have much in common with the myriad gamers producing the kinds of walkthrough texts distributed via sites such as GameFAQs, IGN Wikis, and countless other fan sites across the web, though IGN's efforts are typically made available for free.

As we see, *walkthrough* has become something of an umbrella term for texts and publications of considerable diversity. These texts and publications range from collectible products of the commercial publishing industry created in collaboration with a game's developers to plaintext documents full of deduced techniques uploaded by players to fansites.

Characteristics

Although forms and formats vary, walkthrough texts share some common characteristics. They typically present relational spaces that bear much in common with the caricature of the pirate's treasure map: "Take three paces forward, go left five paces," and so on. And just like the pirate's map, guidance is offered in the form of imperatives. The opening paragraph of Alex's guide for the game *Shadow of the Colossus* (Team Ico, 2005) is a case in point: "Pass through a small valley with some trees and once you are at the foot of the mountains you are right on track as the first colossus is waiting for you up there and ready to smash you into the ground. So let's get into the fight" (2013). We can note also in this brief excerpt the use of the second-person address that positions the gamer-reader and his or her character as one and the same. However, there are other qualities to the mode of address. Diane Carr, Diarmid Campbell, and Katie Ellwood (2006), for instance, identify what they refer to as a cautionary tone in walk-through writing in that it frequently warns the player about the shocks or surprises that might be forthcoming. For example, in a walkthrough for *Final Fantasy VII* (Square/Eidos Interactive, 1997) the writer advises, "Once you leave the train, check the body of the closest guard twice to get two Potions. Then head north. You'll be attacked by some guards. Take them out with your sword (you may win a Potion for killing them) and then move left to go outside" (Megura 2000). This "regulatory mode" (Burn 2006) in the walkthrough author's language might lead us to view the text as one that imposes restrictions on the player, confining him or her to certain paths or styles of play. However, Daniel Ashton and James Newman (2010) have noted that the walkthrough acts as a lens through which to view the game as a place of threat, albeit a threat abated and guarded against by means of the walkthrough. In this way, we should note the differential performances of expert status that are played out and reinforced as experts enlighten the nonadept. As Mia Consalvo (2003, 273) notes, where walkthroughs are posted online to sites such as GameFAQs, they may be reviewed and rated by readers, thus offering yet another way by which expert status is identified and performed.

One notable feature of the texts distributed through sites such as GameFAQs is their minimalist appearance. Where strategy guides published from the 1990s on use screengrabs and bespoke annotated maps and illustrations, player-produced walkthroughs typically eschew typographical elegance, color, and image in favor of apparently outdated fixed-pitch plaintext and occasional ASCII (American Standard Code for Information Interchange) artwork. Like Addams's lists of text inputs and flowcharts in *Quest for Clues*, the plaintext document has the effect of reducing the game's representational system to its barest and most stark. Where cutscenes, high-definition graphics, animation, and multichannel sound might define the video game in certain arenas, in the player-authored walkthrough the game is nought but a simulation that requires a series of inputs to achieve the desired outputs. As such, there is little if any discussion of the graphics, sound, or plot, and in this way we see yet another connection between the modern player-authored walkthrough and the books of the 1980s with their sparing but seductively wobbly freehand drawings and serious, focused text. These walkthroughs are the codification of successful gameplay technique packaged and delivered as what Burn has termed "technical, dispassionate text" (2006, 91). However, there is perhaps more to the focus on logic, procedure, and process. As James Newman notes, "[The walkthrough's] mimicry of computer printouts, console terminals and the procedures of debugging ... create[s] a text in which emotional response is suppressed and subsumed by the machine-like sequence of commands that, if successfully executed, will bring success. In a sense, we can view the walkthrough as a program in itself, or perhaps more accurately a decompiled account of the game program. ... [T]he walkthrough author is essentially a reverse engineer" (2008, 103).

To Completion and Beyond

If there were nothing more to walkthroughs than delivering solutions and aiding the stuck gamer, we might perhaps accept the charges that their use constitutes nothing more than cheating. However, further exploration of the range of player-produced documents as well as the materials that extend beyond the sequential, virtual tour begins to add some intriguing complexity to the notion of how one *should* play a game and who decides.

Player-produced texts such as Super Nova's "Mission Failure FAQ" (2003) for *Goldeneye 007* (Rare, 1997) focus attention on precisely those aspects of the game that walkthroughs usually ensure players never see. Here, failure is constituted as a gameplay objective like any other. Both the production and use of such texts reveal an approach to play that is not motivated by a desire to move through the game's narrative to its conclusion or to solve its trouble spots but rather by an interest in apprehending the full extent of what the game has to offer. In some ways, we should not find this motivation surprising. The desire to document every branch of the game's narrative tree or every combination of weapon arises from the same interest in comprehensively knowing the game's potential. We see the beginnings of this interest in Uston's guide *Mastering* Pac-Man (1982). This 156-page treatise on *Pac-Man* (Namco, 1980) seeks to zero in on the system that underpins the game. Uston's analysis is concerned with the four monsters' artificial intelligence, the way in which Pac-Man moves more quickly along empty channels when eating through dots, and the way the tunnels function. Most importantly, the book's purpose is to translate this knowledge into exploitable strategy. Understanding the differences in their pattern-finding behavior allows measured responses to the four monsters; knowing that Pac-Man can outrun and outmaneuver certain monsters by moving in open space or repeatedly cornering them is key to evading their attacks, thus unveiling the counterintuitive and harnessable fact that Pac-Man can influence the movement of the monsters by hesitating or even momentarily reversing and moving toward them while in the tunnels. Ultimately, Uston's work sets out a number of distilled patterns that may be used to tackle the various mazes. Each is carefully explained, illustrated, and mapped, making use of advanced techniques such as "hesitation," the "reverse flick," and "tunneling" (while allowing for "conservative improvisation," again balancing the exploratory and regulatory), and is entirely deduced from systematic, investigative play. What Uston presents is essentially a form of reverse engineering that seeks to apprehend *Pac-Man* at the systemic, algorithmic level rather than at the representational level (Newman 2008).

Playing with the Game

Many player-authors propose renegotiations of the player–designer relationship that both encourage and normalize self-consciously investigative, resistant, and even deviant gameplay. Most obviously, such renegotiation may include the creation of new gameplay objectives whose parameters are set out in the same imperative tone and that work to establish expert performances and aspirations to mastery. These new objectives might take the form of speedruns, low-ammo runs (using fewer than usual weapons), or pacifist modes (dispatching as few enemies as possible), for instance. Indeed, in a knowing nod to the latter practice that further normalizes it, Bizarre Creations added a "Pacifist Mode" achievement to *Geometry Wars* (2003), and Nintendo's *Mario Galaxy* and *New Super Mario Bros.* series routinely require speedrun or technique-restricted replays of levels.

In addition to finding new ways to replay games, exploring ways to break the game's logic or perform beyond what its simulation appears to have been designed to allow occupies the attentions of many of the walkthrough authors whose work circulates online. In contrast to official strategy guides, whose walkthroughs codify a sanctioned repertoire of performances, approaches, and techniques, texts such as Chris Schultze's "*Goldeneye 007* Glitches FAQ" (2000) revel in the identification of more than 200 inconsistencies, glitches, and failures in the simulation. Importantly, although there is clear interest in unveiling and documenting glitches in all their variety, ranging from comparatively innocuous graphical aberrations to glitches that effectively power-up the player and game-crashing bugs, there is no sense of the author's disappointment or animosity regarding the presence of such glitches. The glitches are not revealed to accuse the game's developers of sloppy work, and the game remains one that is revered in the pages of the walkthrough. Rather, like the cataloging of ways in which missions can be failed, these glitches represent part of the totality of the game that can be known, played, and played with.

These texts crystallize issues that cut to the very heart of game studies. Whose performance should be legitimized and recorded? How *should* a player play the game, and how might we account for the variation in performances? There are clear contradictions at work here in that walkthroughs are at once exploratory and regulatory, simultaneously pushing at the boundaries of what is possible while effectively policing those same boundaries. However, it is by raising these questions that walkthroughs reveal themselves as key documents not just for the stuck gamer but for game studies scholars. For those concerned with game history, walkthroughs are among the most valuable texts and represent perhaps the most comprehensive records presently available. What is key to their utility is their focus on mapping not just the extent of the game as designed but also, in the case of those documents produced by players, the gameplay opportunity revealed through systematic, investigative play, which might well involve the exploitation of inconsistencies in the code, new forms of play, or new challenges shared among constituencies of players—many of which run counter to the apparently normalized vision of the game. Moreover, by documenting the differences in versions, conversions, and the way games and gameplay evolve over time, walkthroughs remind us that games are unstable objects brought to life through the ongoing interactions between designers and players. At the least, they provide essential context for any game historian, but perhaps we might argue that they are the ultimate archival documents of play and gameplay.

Summary

As we have seen, the sheer variety of walkthroughs is quite astonishing, ranging from authorized accounts of the game as designed to reverse-engineered investigations of exploits and glitches. Although there is no doubt that assisting the player in difficulty remains a defining feature of walkthroughs, we must surely appreciate walkthroughs as more than quick-fix solutions or "cheat sheets." In their various forms, they reveal much about how each game was made, how its creators envisioned it being played, as well as how it is actually played and performed by constituencies of players. Considered en masse, walkthroughs represent the most detailed and comprehensive documents of gameplay opportunity that currently exist. As archival documents of video games and the potentials and strategies for play, they are unrivalled in their detail by any formal scholarship to emerge from the academy or professional games criticism. They offer invaluable insights into the pleasures of play and, most crucially, replay. That walkthroughs remain misunderstood is both a product of their mislabeling as part of the paraphernalia of cheating and the lack of study they have been granted. That they remain unstudied is one of game studies' most serious oversights.

Works Cited

Addams, Shay. 1988. *Quest for Clues.* Londonderry, NH: Origin Systems.

Alex. 2013. "Shadow of the Colossus." GameFAQs. http://www.gamefaqs.com/ps2/924364-shadow-of-the-colossus/faqs/39500.

Ashton, Daniel, and James Newman. 2010. "Relations of Control: Walkthroughs and the Structuring of Player Agency." *The Fibreculture Journal* 16. http://sixteen.fibreculturejournal.org/relations-of-control-walkthroughs-and-the-structuring-of-player-agency/.

Blanchet, Michael. 1982. *How to Beat the Video Games.* New York: Fireside.

Broomis, Nick. 1982. *The Video Master's Guide to Defender.* New York: Bantam Books.

Burn, Andrew. 2006. "Reworking the Text: Online Fandom." In *Computer Games: Text, Narrative and Play,* ed. Diane Carr, David Buckingham, Andrew Burn, and Gareth Schott, 88–102. Cambridge, UK: Polity Press.

Carr, Diane, Diarmid Campbell, and Katie Ellwood. 2006. "Film, Adaptation and Computer Games." In *Computer Games: Text, Narrative and Play,* ed. Diane Carr, David Buckingham, Andrew Burn, and Gareth Schott, 149–161. Cambridge, UK: Polity Press.

Consalvo, Mia. 2003. "*Zelda 64* and Video Game Fans." *Television & New Media* 4 (3): 321–334.

Editors of *Consumer Guide.* 1982. *How to Win Video Games.* New York: Pocket Books.

Emrich, Alan. n.d. "What I Observed as the Barbarians Stormed the Gate." Decline of Guides. http://www.alanemrich.com/Writing_Archive_pages/decline.htm.

Kubey, Craig. 1982. *The Winners' Book of Video Games.* New York: Warner Books.

Megura, Kao. 2000. "*Final Fantasy VII* FAQ v2.2." GameFAQs. http://www.gamefaqs.com/ps/197341-final-fantasy-vii/faqs/2376.

Newman, James. 2008. "*Playing with Videogames.*" Abingdon, UK: Routledge.

Rubin, Scott. 2004. "Response to: Who Are Walkthroughs and FAQs for?" In *Difficult Questions about Videogames*, ed. James Newman and Iain Simons, 260–261. Nottingham, UK: Suppose Partners.

Schultze, Chris. 2000. "*Goldeneye 007* Glitches FAQ." GameFAQs. http://www.gamefaqs.com/n64/197462-goldeneye-007/faqs/9961.

Super Nova. 2003. "*Goldeneye*—Mission Failure FAQ." GameFAQs. http://www.gamefaqs.com/n64/197462-goldeneye-007/faqs/23475.

Uston, Ken. 1982. *Mastering* Pac-Man. Rev. ed. New York: Signet.

49 WORLD BUILDING

Marcelo Alejandro Aranda

Historical narratives, fiction, and video games have a shared need to construct worlds. The earliest of Western historians, Herodotus, skillfully wove together histories, folklore, and his own observations to make other lands and times vivid to his audience. Early twentieth-century authors such as H. P. Lovecraft and J. R. R. Tolkien created detailed cultural, scientific, and cosmological backgrounds for their stories, in effect creating fictional worlds. By dropping fragments of a deep historical past and other aspects of a broader universe within their narratives, these authors imbued their fictional worlds with a sense of depth and complexity. Video games similarly use world building to add depth to their interactive environments. The integration of a historical background into a game serves the same function as perspective in a landscape painting; both perspective and historical world building add depth and complexity to a composition. With a few pieces of information left here and there, the player's imagination assumes the existence of a larger, more complete background for his or her characters and actions.

World building was originally found in games with elaborate narratives, such as role-playing games, but has increasingly entered into mainstream action-adventure games. Not all games use world building to immerse players. Some game genres, such as platformers or puzzle games, do not use this method because they engage players through immediate gameplay rather than through narrative and context. In games with a complex narrative structure and developed world background, players explore the world built by a designer, slowly expanding their geographic and cultural knowledge of the game's setting and eventually coming to an understanding of the game's internal rationale. In effect, by engaging with historical background in a game, the player interacts with it as a historian interprets source materials.

Historians use a similar process of world building but generally approach this task from a different starting point. As a historian, I gather source material from the past to construct a narrative that will allow me to understand the causality, contingency, and consequences of historical figures and their actions. Having constructed my narrative, I can then use it to explain the changing social temporality of particular historical contexts. Here I am drawing on ideas from William Sewell's *Logics of History* (2005), wherein he discusses the implicit social theories that historians use to understand the past. I take it as an implicit given that any of my sources, no matter how limited in scope, emerge from a much larger and more complex context than is immediately evident. For example, a source that discusses how the occurrence of a solar eclipse causes a panic in the market stalls in front of a cathedral reveals the relationship between commerce and religion, the financial and political clout of religious authorities, and popular understanding of extraordinary astronomical events.

Over the course of a game, a player gathers fragments of the internal historical narrative, learning about the setting in a piecemeal manner. As she gathers aspects of information that are not in accord with her previous knowledge of the past, she is forced to reconstruct her understanding of the causes for her present circumstances. By finding new sources of information about the historical past, the player changes and enhances her comprehension of the game narrative, and thus the game world expands and becomes more concrete. It is precisely because of this similarity between the process of historical research and exploratory gameplay that insights from the philosophy of history can be useful to game designers.

I am using the term *history* here to refer not only to the past of human activity on earth but also to the constructed narratives we tell each other about the past (Kapell and Elliot 2013). Games that interact with history as a narrative do so in a very different manner from that of a simulation or a wargame. The latter two genres strive for a certain level of accuracy and verisimilitude, balanced with the needs of gameplay. In contrast, the purpose of games such as *Final Fantasy Tactics* (Square, 1997), *Soul Reaver 2* (Crystal Dynamics, 2001), and *Assassin's Creed III* (Ubisoft Montreal, 2012) is to introduce the player to a different world.

We can identify three broad categories of historical world building in video games. In the first, the game uses a fictional history in another world, and the player explores it. In the second, the player travels to different historical periods. Finally, in the third, the historical past is used as a setting for gameplay, and knowledge of our world adds depth to the game. The three examples I use for these categories are *Final Fantasy Tactics*, the *Legacy of Kain* series (Crystal Dynamics), and the *Assassin's Creed* series, respectively. My choice of case histories is not meant to suggest that these games are the definitive ones to use historical world building. Rather, I use them as examples because they all share a self-reflexive consciousness about the nature of narrative and history.

Released in 1997, *Final Fantasy Tactics* (*FFT*) was based on *Tactics Ogre* (Quest, 1995), an earlier game by Yasumi Matsuni, the lead game designer. In the 1990s, Matsuni was profoundly disturbed by the ethnic cleansing in the Balkans (Parish 2011), which led him to write *Tactics Ogre*, a tactical role-playing game with a complex storyline and morality system. For *FFT*, Square wanted Matsuni to tone down the gameplay difficulty level to make the game accessible to a wider audience. Even with these mechanical changes, many of the same themes of war, choice, and history are present in both games. The narrative of *FFT* is quite complex, dealing with the aftermath of a fifty-year-long war, a succession crisis, and the subsequent struggle between the most powerful dukes of the kingdom of Ivalice.

FFT begins with a message from Arazlam Durai, a historian of Ivalice preparing a revisionist history of the War of the Lions, using documents left by one of his ancestors. Arazlam mentions he will focus on the exploits of Ramza Beoulve, a man denounced as a heretic by the clergy of Ivalice. In the framing narrative, the historian Arazlam reveals the problem that compelled him to write his revisionist account: "There is no official record of the role [Ramza] played on history's stage. However, according to the Durai Papers, the existence of which became known to the public only this last year—they had long lain concealed in Church archives—this forgotten young man is in fact the true hero. The Church maintains he was a heretic, an inciter of unrest and disturber of the peace. Which account is to be believed? Join me in my search to uncover the answer" (*FFT*, prologue).

Thus, the game is ostensibly the true account of the War of the Lions as told in the Durai papers, and Ramza, Arazlam, and the player share the process of interacting with historical sources to rewrite their understanding of Ivalice's past. Within the game, Ramza comes across not only the written sources that reveal the secret history of the Glabados Church but also technomagical artifacts from prior civilizations that allow him to complete his quest. Both Ramza and the player ultimately come to understand that politics and religion are intertwined in Ivalice and that the stories we tell about the past are contingent on who is writing them. As is evident, *FFT* has multiple narrative levels at work, in effect a nested history, and is quite consciously aware of the constructed nature of history.

In the second category of history as world building in video games, the player travels to different historical periods. The consequences of time travel on history in the world of Nosgoth is the subject of the *Legacy of Kain* series. Amy Hennig, the lead developer, used her background in English literature and film studies to introduce a rich thematic and narrative complexity. The second game in the series, *Soul Reaver* (1999), has Raziel, a vampire clan leader, as the protagonist. Raziel returns, seemingly for revenge, as a spectral harvester of souls to kill the other vampire clan leaders and to defeat Kain, their ruler. At the game's end, Raziel confronts Kain at a time machine, and the player learns that there is more to the narrative than simple revenge. In the midst of their fight and his activation of the time machine, Kain reveals that "our futures are predestined. … We each play out the parts fate has written for us. Free will is an illusion" (*Soul Reaver,* finale). Kain travels into Nosgoth's past with Raziel on his heels, and determinism, agency, and teleology become the central themes in the next installment.

Soul Reaver 2 (2001) has Raziel confront his human past and the contingencies that led to the future from which he came. He begins to understand that the Sarafan priesthood of whom he was part was not "noble and altruistic" but instead a fanatical religious order intent on genocide and that not all vampires are like the monsters he fought in Nosgoth's future. As happens to Ramza in *FFT*, Raziel's encounter with the past changes his perspective on his circumstances in the present, something that Hennig intended because she believes "that time travel is ultimately a journey of epiphanies, where the protagonist realizes the role that he already played in history" (quoted in Davison and Rybicki 2000). Kain ultimately arranges for Raziel to encounter the historical version of the Soul Reaver, the magical blade he carries. Two iterations of the same object in the same place at the same time creates a paradox, making it possible to rewrite history.

Kain orchestrates events to allow a potential moment where history can be rewritten, as he explains to Raziel in a cutscene that elaborates the historical rules at work in Nosgoth:

> I have seen the beginning and the end of our story, however—and the tale is crude and ill conceived. We must rewrite the ending of it, you and I. ... Thirty years hence, I am presented with a dilemma—let's call it a two-sided coin. If the coin falls one way, I sacrifice myself and thus restore the Pillars. But as the last surviving vampire in Nosgoth, this would mean the annihilation of our species. ... If the coin lands on the reverse, I refuse the sacrifice and thus doom the Pillars to an eternity of collapse. Either way, the game is rigged. ... Drop a stone into a rushing river—the current simple courses around it and flows on as if the obstruction were never there. You and I are pebbles, Raziel, and have even less hope of disrupting the time stream. The continuum of history is simply too strong, too resilient. (*Soul Reaver 2*, William's Chapel)

The philosophy of history Kain explains is that history is cyclical and teleological. It is cyclical in that similar situations and circumstances recur throughout the history of Nosgoth, and many of the characters refer to the Wheel of Destiny as "the inexorable cycle of death and rebirth to which all men are compelled" (*Soul Reaver 2*, prologue). Such a teleological and determined condition removes mystery from the past and precludes possibility in the future. The history of Nosgoth from beginning to end has essentially already been written, and the only way that Kain can reintroduce individual agency and free will is by causing paradoxes that force history to rewrite itself. At the beginning of *Legacy of Kain: Defiance* (2003), Kain states the rationale for his actions: "Given the choice whether to rule a corrupt and failing Empire or to challenge the fates for another throw, a better throw against one's destiny[,] ... [o]ne can only match, move by move, the machinations of fate and thus defy the tyrannous stars." Raziel comes to realize that he, like the younger Kain, is being used as an instrument, so he spends the rest of the series trying to escape the deterministic shackles of Nosgoth's history and become the arbiter of his own fate.

Some of the strongest criticisms of the *Legacy of Kain* games when they were released were that the player had little opportunity to explore Nosgoth or to move off the main narrative track. In contrast, one of the main attractions of the *Assassin's Creed* game series is the freedom to explore one's own historical past. Within the game world, the device allowing the player to explore the past is the Animus, a virtual-reality device that gives the player access to ancestral memories from different historical periods. In the main arc of the series, players take on the role of Desmond Miles and experience the Second Crusade, Renaissance Italy, and the American Revolution through ancestral memories. Thus, like *FFT*, the *Assassin's Creed* series is a nested history, with the framing narrative about Desmond set in the present day and each new historical setting forming another chapter within the larger narrative.

In the first game in the series, *Assassin's Creed* (2007), Desmond explores the memories of Altair, a member of the historical Assassin order in the twelfth-century Levant. Vidic, the Templar scientist leading the Animus project, considers Desmond as nothing more than an instrument for knowledge gathering. In his interactions with Vidic, Desmond expresses some of his anxieties about what he is seeing:

Desmond: Some of the stuff I'm seeing in the Animus ... sometimes it seems ... wrong. ...

Vidic: It doesn't ... match up with what you read on an online encyclopedia? What your high-school history teacher taught you? Let me ask you something: do these supposed experts have access to secret knowledge, kept hidden from the rest of us?

Desmond: There are books, letters, documents ... all sorts of source material from back then. Some of it seems to contradict what the Animus is showing me. ...

Vidic: It's part of what makes the Animus so spectacular. There's no room for misinterpretation—

Desmond: There's always room. (*Assassin's Creed*, ellipses used to indicate pauses)

Interestingly, whereas Vidic regards the Animus as the ultimate primary source allowing history to become a positivistic science, Desmond recognizes that there is still a possibility of misinterpreting what they see in his ancestral memories.

In *Assassin's Creed III* (2012), the final installment of Desmond's storyline, he explores the memories of Connor, a half-British/half–Native American ancestor caught up in the American Revolution. For this game, Ubisoft wanted to depict the Revolutionary War in a nuanced manner, so it consulted the historian François Furstenberg, a specialist on that period, who commented: "Anything that complicates the narrative is a good thing" (quoted in Barakat 2012). Examples of such complication include the fact that Connor has adversaries in both the rebel and the loyalist camps.

Perhaps some of the most interesting exchanges about the role of history occur between Desmond and Shaun, an Assassin scholar who helps Desmond to understand the historical settings he explores. Shaun highlights how history is constantly rewritten to suit the needs and interests of the moment but unfortunately at the same time also makes past events and texts vulnerable to being hijacked for partisan ends: "Your politicians are constantly referencing the founding fathers—and insisting they must have been in support of one thing or another. I have never seen such a blatant disregard for history. ... I don't think most of your presidents and senators and judges care what the founders thought. They just want to know how they can bend old words to achieve modern goals" (*Assassin's Creed III*, Precursor Cave). Although somewhat cynically stated, Shaun's observation is an insight that many young historians realize as they go through the process of conducting historical research and creating knowledge. In the process of exploring the past, you also learn how it has been constructed.

I have highlighted instances of sophisticated use of historical narrative by game designers, but historians can also learn much from games such as *Legacy of Kain* and *Assassin's Creed*. These games can simulate and consider the role of paradox, contingent influences, and counterfactual narratives—issues that can be difficult for historians to explore using limited archival sources. Although historians do not often study past magical or technical civilizations or the presence of time-traveling historical figures, the scholarly problems stemming from subaltern sources, contradictory historical accounts, and revised historical narratives are all ones they grapple with in their research.

Works Cited

Barakat, Matthew. 2012. "Founding Father Featured in Popular New Video Game." *Washington Post*, October 30. http://p.washingtontimes.com/news/2012/oct/30/founding-father-featured-in-popular-new-video-game/. Accessed June 19, 2014.

Davison, John, and John Rybicki. 2000. "*Legacy of Kain*: Funk Soul Brother." *Official US Playstation Magazine* 36 (September). Excerpted at http://nosgoth.yuku.com/reply/50800/GLoK-Interviews-with-Amy-Hennig#reply-50800. Accessed June 17, 2014.

Kapell, Matthew Wilhelm, and Andrew B. R. Elliot, eds. 2013. *Playing with the Past: Digital Games and the Simulation of History*. New York: Bloomsbury.

Parish, Jeremy. 2011. "Let Us Remember Together: A *Tactics Ogre* Retrospective." 1UP, February 8. http://www.1up.com/features/tactics-ogre-retrospective. Accessed June 17, 2014.

Sewell, William H. 2005. *Logics of History: Social Theory and Social Transformation*. Chicago: University of Chicago Press.

CONTRIBUTORS

Marcelo Alejandro Aranda is a lecturer in Science in the Making: Integrated Learning Environment (SIMILE), a new year-long residential program in the history of science at Stanford University. He recently completed his dissertation in the History Department at Stanford and is currently revising it into a book tentatively titled "The Scientific Culture of the Seventeenth-Century Spanish Empire." Aranda has also participated in the Mapping the Republic of Letters project for a number of years and is interested in how new technologies are integrated into existing intellectual cultures, both in the early modern period and in the present.

Brooke Belisle is assistant professor of visual culture at Stony Brook University in the Department of Art and the Consortium for Digital Art, Culture, and Technology. She is also an editor of the *Journal of Visual Culture.* Her research explores emergent, experimental, and expanded aesthetic formats that cross multiple media technologies and historical moments. She wrote the contribution for this volume while she was a 2013–2015 New Faculty Fellow of the American Council of Learned Societies.

Caetlin Benson-Allott is director and associate professor of film and media studies at the University of Oklahoma. She is the author of *Killer Tapes and Shattered Screens: Video Spectatorship from VHS to File Sharing* (University of California Press, 2013) and *Remote Control* (Bloomsbury Press, 2015). Her work on platform studies, digital media, and contemporary film cultures has appeared in journals such as *Film Quarterly*, where she wrote a regular column from 2011 to 2013; *The Atlantic*; *Cinema Journal*; the *Journal of Visual Culture*; *Jump Cut*; and *Feminist Media Histories*, as well as multiple anthologies.

Stephanie Boluk is assistant professor in the Humanities and Media Studies Program at Pratt Institute. Her research and teaching incorporate game studies, media studies, utopian studies, and critical theory to explore video games, electronic literature, alternative currencies, finance capitalism, and the convergence of leisure and labor in contemporary information economies.

She is currently coauthoring a book with Patrick LeMieux titled *Metagaming: Videogames and the Practice of Play* (Duke University Press, forthcoming). For more information, see stephanieboluk.com.

Jennifer deWinter is associate professor of rhetoric and faculty in the Interactive Media and Game Development Program at Worcester Polytechnic Institute. She teaches courses on game studies, game design, and game production and management. She has published on the convergence of anime, manga, and computer games both in their Japanese contexts and in global markets. Her work has appeared in numerous journals and edited collections, and she is coeditor of *Computer Games and Technical Communication: Critical Methods and Applications at the Intersection* (Ashgate, 2014, with Ryan Moeller) and *Video Game Policy: Production, Circulation, Consumption* (Routledge, 2016 with Steven Conway) as well as editor for the textbook *Videogames* (Fountainhead, forthcoming). In collaboration with Carly A. Kocurek, she is the series editor for the Influential Game Designer book series, for which she wrote the inaugural book, *Shigeru Miyamoto* (Bloomsbury, 2015).

Jon-Paul C. Dyson is director of the International Center for the History of Electronic Games and vice president for exhibits at the Strong Museum in Rochester, New York, which has the world's most comprehensive collection of playthings. Dyson has supervised the development of many exhibits, including *Reading Adventureland*, *American Comic Book Heroes*, *eGameRevolution*, *Pinball Playfields*, *Atari by Design*, and others on the history and impact of play and video games. He also has written and spoken widely on the history of play and video games. He has a PhD in US history and a lifetime of gaming experience going back to mainframes.

Kate Edwards is the executive director of the International Game Developers Association (IGDA), appointed in December 2012. She is also the founder and principal consultant of Geogrify, a Seattle-based consultancy for content culturalization, and a unique hybrid of applied geographer, writer, and corporate strategist. As Microsoft's first geopolitical strategist on the Geopolitical Strategy team she created and managed, Edwards was responsible for protecting against political and cultural content risks across all products and locales. Since leaving Microsoft, she has provided guidance to many companies on a wide range of geopolitical and cultural issues. She is also the founder and former chair of the IGDA's Localization Special Interest Group, a former board member of IGDA Seattle, the co-organizer of the Game Localization Summit at the Game Developers Conference, and a regular columnist for *MultiLingual Computing* magazine. In October 2013, *Fortune* magazine named Edwards one of the "10 most powerful women" in the game industry.

Mary Flanagan founded and directs the internationally acclaimed game research laboratory Tiltfactor. She has written more than 20 critical essays and chapters on digital culture and play, and her recent books in English include *Critical Play* (MIT Press, 2009) and *Values at Play in Digital Games* (MIT Press, 2014, with Helen Nissenbaum). Flanagan has broken ground with collaborator Helen Nissenbaum by investigating how games, interactive systems, and online activities can be redesigned to prioritize human values. In this work, they have proven that using humanist principles to shape software and product development offers a profoundly important strategy for innovation. Flanagan's work has been supported by commissions including the British Arts Council, the National Endowment for the Humanities, the American Council of Learned Societies, the Institute of Museum and Library Services, and the National Science Foundation. She has served on the faculty of the Salzburg Global Seminar and the White House Office of Science and Technology Policy Academic Consortium on Games for Impact. Flanagan is the Sherman Fairchild Distinguished Professor in Digital Humanities at Dartmouth College.

Jacob Gaboury is assistant professor of digital media and visual culture in the Department of Cultural Analysis and Theory at Stony Brook University. His research engages the history and critical theory of digital media through the fields of visual culture, queer theory, and media archaeology. He is currently finishing a monograph entitled "Image Objects," which offers an archaeology of early three-dimensional computer graphics in the 1960s and 1970s, with a focus on the critical but neglected history of the University of Utah's computer graphics program. Gaboury has held fellowships through the Max Planck Institute for the History of Science, the Charles Babbage Institute, the Institute for Electrical and Electronics Engineers, the Association of Computing Machinery, and the Smithsonian Institute's National Museum of American History. His work has been published or is forthcoming in the *Journal of Visual Culture*, *Media-N*, and *Camera Obscura*.

William Gibbons is assistant professor of musicology at Texas Christian University, where he teaches courses in music history and musical multimedia. He is the author of *Building the Operatic Museum: Eighteenth-Century Opera in Fin-de-Siècle Paris* (University of Rochester Press, 2013) and coeditor *of Music in Video Games: Studying Play* (Routledge, 2014, with K. J. Donnelly and Neil Lerner), and his articles have appeared in a number of journals, including *Game Studies*, *Music and the Moving Image*, the *Journal for the Society for American Music*, and *Opera Quarterly*. His current book project is "Unlimited Replays: The Art of Classical Music in Video Games," under contract with Oxford University Press.

Raiford Guins is associate professor of culture and technology at Stony Brook University. He is also founding curator of the William A. Higinbotham Game Studies Collection at Stony Brook University and principal editor of the *Journal of Visual Culture*. He has recently published *Game After: A Cultural Study of Video Game After* (MIT Press, 2014) and is currently researching his next book, "Atari Modern: A Design History of Atari's Coin-Op Arcade Video Game Cabinets, 1972–1979." His writings on game history also appear in many journals and magazines, including the *Atlantic*, *Cabinet*, *Design and Culture*, *Design Issues*, *Game Studies*, *Journal of Design History*, *Journal of Visual Culture*, and *Reconstruction: Studies in Contemporary Culture*. Guins coedits the MIT Press book series Game Histories with Henry Lowood.

Erkki Huhtamo is known as a founding figure of media archaeology. He has published extensively on media culture and media arts, lectured worldwide, given stage performances, curated exhibitions, and directed television programs. He is a professor in the Department of Design Media Arts and the Department of Film, Television, and Digital Media at the University of California, Los Angeles. His most recent book is *Illusions in Motion: Media Archaeology of the Moving Panorama and Related Spectacles* (MIT Press, 2013).

Don Ihde is Distinguished Professor of Philosophy, Emeritus, at Stony Brook University. His work on the philosophy of technology is well known internationally, and his work on the notion of embodiment and technologies appears in a number of his books, including *Bodies In Technology* (Minnesota University Press, 2002), *Embodied Technics* (Automatic Press, 2010), *Experimental Phenomenology: Multistabilities* (2nd edition, State University of New York Press, 2012), and *Listening and Voice: Phenomenologies of Sound* (2nd edition, State University of New York Press, 2007), as well as in many articles. He is currently working on acoustic technologies and is a founding philosopher of the postphenomenological style of science-technology analysis.

Jon Ippolito is an artist, writer, and curator born in Berkeley, California, in 1962 who turned to making art after failing as an astrophysicist. When he applied for what he thought was a position as a museum guard at the Guggenheim, New York, he was hired in the Curatorial Department, where in 1993 he curated the exhibit *Virtual Reality: An Emerging Medium* and subsequent exhibitions that explore the intersection of contemporary art and new media. In 2002, Ippolito joined the faculty of the University of Maine's New Media Department, where with Joline Blais he cofounded Still Water, a lab devoted to studying and building creative networks. His writing on the cultural and aesthetic implications of new media has appeared in the *Washington Post*, *Art Journal*, and in his coauthored books *At the Edge of Art* (Thames and Hudson, 2006, with Joline Blais) and *Re-collection: Art, New Media, and Social Memory* (MIT Press, 2014, with Richard Rinehart). More information about him can be found at three.org/ippolito.

Katherine Isbister is professor of computational media at the University of California, Santa Cruz. She was the founding director of New York University's Game Innovation Lab. Isbister's research focuses on designing games that heighten social and emotional connections for players, with a goal toward innovating design theory and practice. Her lab's games have been featured in such venues as IndieCade (Yamove! Finalist in 2012), the World Science Festival, and museums, including the Liberty Science Center. Isbister's book on game character design, *Better Game Characters by Design: A Psychological Approach* (CRC Press, 2005), was nominated for a Game Developer Magazine Frontline Award. Her edited volume *Game Usability: Advancing the Player Experience* (CRC Press, 2008, with Noah Schaffer) brings together best practices in gameplay testing and user research.

Mikael Jakobsson is research scientist and lecturer in the Comparative Media Studies Program at MIT and research coordinator for the MIT Game Lab. He conducts research at the intersection of game design and game culture. With a foundation in interaction design, he investigates how gaming activities fit into social and cultural practices as well as how this knowledge can inform the design and development process. He has twenty years of experience teaching and advising all levels of students in higher education. Recent areas of interest include game studies, game design, and game criticism. He also reviewed video games for ten years before leaving his native Sweden.

Steven E. Jones is professor of English and codirector of the Center for Textual Studies and Digital Humanities at Loyola University, Chicago. He is the author of a number of books and articles on technology and culture, digital humanities, and video games, including *Against Technology: From the Luddites to Neo-Luddism* (Routledge, 2006), *The Meaning of Video Games* (Routledge, 2008), *Codename Revolution: The Nintendo Wii Platform* (MIT Press, 2012, with George K. Thiruvathukal), and *The Emergence of the Digital Humanities* (Routledge, 2013).

Jesper Juul is associate professor at the Royal Danish Academy of Art, School of Design. Since the late 1990s, he has been working with the development of video game theory at the IT University of Copenhagen, MIT, and the New York University Game Center. His publications include *Half-Real: Video Games between Real Rules and Fictional Worlds* (MIT Press, 2005), on video game theory, and *A Casual Revolution: Reinventing Video Games and Their Players* (MIT Press, 2009), on how puzzle games, music games, and the Nintendo Wii brought video games to a new audience. He maintains the *Ludologist*, a blog on "game research and other important things." His latest book, *The Art of Failure,* was published by MIT Press in 2013.

Eric Kaltman is a computer science graduate student at the Expressive Intelligence Studio at the University of California, Santa Cruz. His research focuses on the technical history of computer games and strategies for their long-term preservation. He previously worked on educational research games, networked air-quality monitoring systems, and the archiving of Stanford's computer game collections.

Matthew Kirschenbaum is associate professor in the Department of English at the University of Maryland and associate director of the Maryland Institute for Technology in the Humanities. He is coeditor of *Zones of Control: Perspectives on Wargaming* from the MIT Press (2016).

Carly A. Kocurek is assistant professor of digital humanities and media studies and director of digital humanities at the Illinois Institute of Technology. Her research covers the culture of video gaming. She holds a PhD in American studies from the University of Texas. Her first book, *Coin-Operated Americans: Rebooting Boyhood at the Video Game Arcade* (University of Minnesota Press, 2015), is a history of the early video game arcade in the U.S. Her work has appeared in the journals *Game Studies*, *Visual Studies*, the *Journal of Gaming and Virtual Worlds*, *In Media Res*, *Flow*, and the *New Everyday* as well as in anthologies, including *Before the Crash* (Wayne State University Press, 2012) and *Gaming Globally* (Palgrave McMillan, 2012). She is coeditor, with Jennifer deWinter, of the Influential Game Designers book series (Bloomsbury).

Peter Krapp is professor and chair of film and media studies at the University of California, Irvine, where he also contributes to the Departments of Informatics and English. He is the immediate past chair of the Irvine Academic Senate. Among his publications are *Deja Vu: Aberrations of Cultural Memory* (University of Minnesota Press, 2004) and *Noise Channels: Glitch and Error in Digital Culture* (University of Minnesota Press, 2011).

Patrick LeMieux is an artist, game designer, and assistant professor teaching game studies and critical game making at the University of California, Davis. He is currently coauthoring a book with Stephanie Boluk titled *Metagaming: Videogames and the Practice of Play* (Duke University Press, forthcoming). For more information, visit patrick-lemieux.com.

Henry Lowood is curator for the history of science and technology collections as well as for the film and media collections at Stanford University. He is also a lecturer in the Science and Technology Studies Program and the History and Philosophy of Science Program at Stanford as well as in the School of Library and Information Science at San Jose State University and the Art Department at the University of California, Santa Cruz. His most recent book is *The Machinima Reader* (MIT Press, 2011, coedited with Michael Nitsche). Lowood coedits the MIT Press book series Game Histories with Raiford Guins.

Esther MacCallum-Stewart is research fellow at the Digital Cultures Research Centre, University of the West of England. Her work examines the ways in which players understand game narratives, and throughout her research career her work has continued to interrogate the relationship between players and warfare—for example, through live-action role-playing groups or re-creations of combat and warfare in gaming narratives. She has also written widely on love, sex, and sexuality in games as well as on the ways in which players interpret games for their own ends as fans and producers. She can be contacted at neveah@gmail.com or @neveahfs at Twitter.

Ken S. McAllister is professor of rhetoric at the University of Arizona (UA), codirector of the Learning Games Initiative Research Archive, and planning director of the UA School of Information. Specializing in critical new media studies and transdisciplinary collaboration, McAllister lectures and publishes widely on the digital humanities, information studies, high-performance computing, digital-archive development, and computer games.

Nick Montfort develops computational art and poetry, often collaboratively. Montfort, who is on the MIT faculty, is the author of the books of poems *#!* and *Riddle & Bind*, the coauthor of *2002: A Palindrome Story*, and has developed more than fifty digital projects including the collaborations The Deletionist and Sea and Spar Between. The MIT Press has published four of his collaborative and individual books; *Exploratory Programming for the Arts and Humanities* is coming soon.

David Myers is Distinguished Professor of Communication in the School of Mass Communication at Loyola University, New Orleans. With an article published in *Simulation & Gaming* in 1984, he was one of the first scholars to extend the study of games and play to include analyses of video games. He is the author *of The Nature of Computer Games: Play as Semiosis* (Peter Lang, 2003) and *Play Redux: The Form of Computer Games* (University of Michigan Press, 2010). He can be contacted by email at dmyers@loyno.edu.

James Newman is professor of digital media, university teaching fellow, and subject leader for film, media, and creative computing at Bath Spa University. He is the author of numerous books on video games and gaming cultures, including *Videogames* (Routledge, 2004; 2nd edition, 2013), *Playing with Videogames* (Routledge, 2008), *100 Videogames: Teaching Videogames* (British Film Institute, 2008, with Iain Simons), and *Best Before: Videogames, Supersession, and Obsolescence* (Routledge, 2012). Newman is cofounder of the UK National Videogame Archive, which is a partnership with the Science Museum, and he is a contributor to the GameCity family of projects.

Jenna Ng is Anniversary Research Lecturer in Film and Interactive Media at the University of York. She works primarily on cultural analyses of digital media and visualization, with particular interests in the imaging technologies of digital video, mobile media, haptic devices, and motion- and virtual-capture systems. She is the editor of *Understanding Machinima: Essays on Filmmaking in Virtual Worlds* (Bloomsbury, 2013), a collection of essays on machinima that also incorporates Quick Response codes and mobile digital content (http://m.understandingmachinima.com/), and has published work in various essay collections and journals. She was previously a Newton Trust/Leverhulme Early Career Fellow at CRASSH, University of Cambridge, and a postdoctoral fellow at HUMlab, Umeå University, Sweden.

Michael Nitsche is the director of graduate studies for the Digital Media Program at the Georgia Institute of Technology, where he teaches mainly on issues of hybrid spaces and what we do in them. He uses performance studies, craft research, human–computer interaction, and media studies as critical approaches and applies them to interaction design for digital media, mobile technology, and digital performances. He directs the Digital World and Image Group, which has received funding from the National Science Foundation, Alcatel Lucent, Turner Broadcasting, and GCATT, among other institutions and organizations. Nitsche's publications include the books *Video Game Spaces* (MIT Press, 2009) and *The Machinima Reader* (MIT Press, 2011, coedited with Henry Lowood).

Laine Nooney is a cultural historian of video games and computing whose research interests include media archaeology, critical/feminist materialism, and technology and inclusivity. Her most recent work, on Roberta Williams and the problem of gender in video game history, appears in *Game Studies*. She is presently preparing a book manuscript on the corporate and cultural history of the home-entertainment software company Sierra On-Line, titled "Before We Were Gamers: Sierra On-Line and the Archaeology of Video Game History." Nooney is on the editorial group for the *Journal of Visual Culture*, an adviser to the Softalk Apple Preservation Project, and co-organizer of the first Different Games Conference. She is an assistant professor of digital media in the School of Literature, Media and Communication at Georgia Tech. Nooney tweets as @Sierra_Offline.

Hector Postigo is associate professor in the Department of Broadcasting, Telecommunications, and Mass Media in the School of Communications and Theater at Temple University. He is the author of *The Digital Rights Movement: The Role of Technology in Subverting Digital Copyright* (MIT Press, 2012).

Jas Purewal is a digital-entertainment lawyer and founder of Purewal & Partners LLP, a digital-entertainment and tech legal and business consultancy based in London. He is an expert in video games and European digital entertainment law and advises a wide range of developers, publishers, platforms, and other businesses in the industry. He writes the video game law blog *Gamer/Law* (www.gamerlaw.co.uk), one of the leading blogs on interactive entertainment law. He also teaches interactive entertainment law and business at several universities and the UK National Film and TV School. Purewal was educated at the University of Nottingham and is admitted to practice as a solicitor in England and Wales. He has been a lifelong gamer and technology enthusiast since his father bought him his first computer when he was eight.

Reneé H. Reynolds is a PhD candidate at the University of Arizona in the Rhetoric, Composition, and Teaching of English Program. Her dissertation will focus on the rhetoric of intellectual, cultural, legal, and linguistic ownership within the comics and graphic narrative industry.

Judd Ethan Ruggill is associate professor of communication at Arizona State University. He is also codirector of the Learning Games Initiative, a transdisciplinary, interinstitutional research group that studies, teaches with, and builds computer games.

Marie-Laure Ryan is an independent scholar based in Colorado. She is the author of *Possible Worlds: Artificial Intelligence and Narrative Theory* (Indiana University Press, 1991), *Narrative as Virtual Reality: Immersion and Interactivity in Literature and Electronic Media* (Johns Hopkins University Press, 2001), and *Avatars of Story* (University of Minnesota Press, 2006). She has also edited several collections, including *The Routledge Encyclopedia of Narrative* (Routledge, 2005, with David Herman and Manfred Jahn), *The Johns Hopkins Guidebook to Digital Humanities* (Johns Hopkins University Press, 2014, with Lori Emerson and Ben Robertson), and *Storyworlds across Media* (University of Nebraska Press, 2014, with Jan-Noël Thon). She has been scholar in residence at the University of Colorado, Boulder, and Johannes Gutenberg Fellow at the University of Mainz, Germany. Her website is at users.frii.com/mlryan/, and she can be reached at marilaur@gmail.com.

Katie Salen Tekinbaş is a game designer at heart and the founding executive director of Institute of Play, a nonprofit doing work in games and learning. She is also professor of games and digital media at DePaul University and once codesigned a karaoke ice cream truck driven by a squirrel. Salen Tekinbaş led the team that founded Quest to Learn in 2009, a public school for grades 6–12 in New York City. The school uses a game-based learning model and supports students in an inquiry-based curriculum with questing to learn at its core. She is coauthor of *Rules of Play* (2003, with Eric Zimmerman), coeditor of *The Game Design Reader* (2005, with Eric Zimmerman), author of *Quest to Learn: Growing a School for Digital Kids* (2011), and editor of *The Ecology of Games: Connecting Youth, Games, and Learning* (2007), all from MIT Press. She has worked as a game designer for more than 12 years and is a former coeditor of the *International Journal of Learning and Media*. Salen Tekinbaş has been involved in the design of slow games, online games, mobile games, and big games in both the commercial and independent games sectors. She was an early advocate of machinima and continues to be interested in connections between game design, learning, and transformative modes of play.

Anastasia Salter is assistant professor of digital media at the University of Central Florida. She is the author of *What Is Your Quest? From Adventure Games to Interactive Books* (University of Iowa Press, 2014) and the coauthor of *Flash: Building the Interactive Web* (MIT Press, 2014, with John Murray). Her research focuses on positioning games and digital narratives as media artifacts with consequences for learning, social engagement, and participatory culture.

Mark Sample is associate professor of digital studies at Davidson College, where he studies and teaches computational media and digital culture. His work has appeared in *Game Studies, Digital Humanities Quarterly*, and *Journal of Digital Humanities*. He is also a coauthor of *10 PRINT CHR$(205.5+RND(1));:GOTO 10* (MIT Press, 2013).

Bobby Schweizer is lecturer in the College of Computing and Digital Media at DePaul University. His research considers how architecture and the urban form are mediated through video games. He also studies close reading/playing methods of games using live-streaming and recorded footage. Schweizer has coauthored *Newsgames: Journalism at Play* (MIT Press, 2010) and coedited *Meet Me at the Fair: A World's Fair Reader* (ETC Press, 2014).

John Sharp is associate professor of games and learning in the School of Art, Media, and Technology at Parsons The New School for Design, where he is codirector of the Prototyping, Education, and Technology Lab. He is a game designer, graphic designer, art historian, and educator. His research is focused on game design curriculum, video game aesthetics, and the creative process. Sharp is a member of the game design collective Local No. 12, which makes games out of cultural practices. He is also a partner in Supercosm LLC, a consultancy for nonprofits and organizations in the arts, education, and entertainment fields.

Miguel Sicart is associate professor at the Center for Computer Game Research at IT University Copenhagen. He is the author of *The Ethics of Computer Games* (2009), *Beyond Choices: The Design of Ethical Gameplay* (2013), and *Play Matters* (2014), all published by the MIT Press. Sicart researches and teaches play and game design from a multidisciplinary perspective, combining philosophy of technology and interaction design to explore playful forms of expression. His work can be found on his personal website, miguelsicart.net, or at playmatters.cc.

David Sirlin, founder of Sirlin Games, is a tabletop game designer and publisher as well as a consultant to video game developers. He was lead designer of Capcom's games *Puzzle Fighter HD Remix* (1996) and *Street Fighter HD Remix* (2008), and he has been a longtime tournament player in *Street Fighter*. He's author of *Playing to Win* (Lulu, 2005), a book about competitive gaming.

Rebecca E. Skinner is a visiting scholar at the University of California, Berkeley, Center for Science, Technology, Medicine, and Society; a graduate of Berkeley's Department of City and

Regional Planning; and a former research associate and adjunct lecturer at Stanford University. Her scholarship specializes in the history of artificial intelligence (*Building the Second Mind: 1956 and the Origins of Artificial Intelligence Computing* [Self-Published, 2012]) and the Cold War (*Out of the Closed World: How the Computer Revolution Helped to End the Cold War* [Self-Published, 2013]).

Melanie Swalwell is an ARC Future Fellow and associate professor in the Screen and Media Department at Flinders University. She is the author of many chapters and articles on the histories of digital games and coeditor of *The Pleasures of Computer Gaming: Essays on Cultural History, Theory, and Aesthetics* (McFarland, 2008, with Jason Wilson). Her game history research has also appeared in nontraditional formats, such as the scholarly interactive Cast-offs from the Golden Age (Vectors, 2006) and the photographic exhibition *More Than a Craze* (2010). Swalwell is project leader of Play It Again, which in conjunction with its partners is researching the history and preservation of digital games in New Zealand and Australia in the 1980s. She is currently writing a book on homebrew gaming in the 1980s.

David Thomas is assistant professor, attendant, in the Department of Architecture at the University of Colorado, Denver. He also is the director of academic technology in the university's online division. His research focuses on the theory and design of fun. His search for "fun objects" ranges from games to architecture. For more than 20 years, he has worked as a professional video game journalist, tying many of these interests into an inquiry of fun as a native aesthetic of games.

Samuel Tobin is assistant professor of communications media and game design at Fitchburg State College in Massachusetts who studies play, media, and everyday life. He holds a doctorate in sociology from the New School for Social Research in New York City. He is the author of *Portable Play in Everyday Life: The Nintendo DS* (Palgrave Macmillan, 2013).

Emma Witkowski is assistant professor in the School of Media and Communication, RMIT University, Melbourne, Australia, where she teaches and researches socio-phenomenological aspects of game cultures. She has published on e-sports, gender and LAN play, masculinities in high performance competitions, and running with playful exercise-apps.

Mark J. P. Wolf is professor and chair of the Communication Department at Concordia University Wisconsin. He has a B.A. (1990) in film production and an M.A. (1992) and Ph.D. (1995) in critical studies from the School of Cinema/Television (now renamed the School of Cinematic Arts) at the University of Southern California. His books include *Abstracting Reality: Art, Communication, and Cognition in the Digital Age* (University of America Press, 2000), *The Medium of the Video Game* (University of Texas Press, 2001), *Virtual Morality: Morals, Ethics, and New Media* (Peter Lang Publishing, 2003), *The Video Game Theory Reader* (Routledge, 2003), *The World of the*

D'ni: Myst *and* Riven (University of Michigan Press, 2006), *The Video Game Explosion: A History from* PONG *to PlayStation and Beyond* (ABC-CLIO/Greenwood Press, 2007), *The Video Game Theory Reader 2* (Routledge, 2008), *Before the Crash: An Anthology of Early Video Game History* (Wayne State University Press, 2012), the two-volume *Encyclopedia of Video Games: The Culture, Technology, and Art of Gaming* (ABC-CLIO/Greenwood Press, 2012), *Building Imaginary Worlds: The Theory and History of Subcreation* (Routledge, 2012), *The Routledge Companion to Video Game Studies* (Routledge, 2014), *LEGO Studies: Examining the Building Blocks of a Transmedial Phenomenon* (Routledge, 2015), *Video Games around the World* (MIT Press, 2015), the four-volume *Video Games and Gaming Culture* (Routledge 2016), and two novels for which his agent is looking for a publisher. He is also founder and coeditor of the Landmark Video Game book series. He has been invited to speak in North America, Europe, and Asia as well as on *Second Life*. He is on the advisory boards of *Videotopia* and the *International Journal of Gaming and Computer-Mediated Simulations* as well as on several editorial boards, including those for *Games and Culture*, the *Journal of E-media Studies*, and *Mechademia: An Annual Forum for Anime, Manga, and the Fan Arts*. He lives in Wisconsin with his wife, Diane, and his sons, Michael, Christian, and Francis.

INDEX

2048 (game), 53
2D (two-dimensional), 83, 128–129, 132, 180, 182, 230, 262, 360–363, 365, 366n9
3D (three-dimensional), 26, 83–85, 129, 131–132, 177–181, 184, 205, 207, 231, 247, 287, 293, 359, 361–363, 365, 366n9

Achievement, 1–4, 13, 94, 104, 206, 238, 244, 293, 297, 320, 322, 414
 casuals, 4–5
 completists, 4, 6
 hunters, 5
Acorn, Al, 58, 90, 400
ActionScript, 54, 56
Activision, 7–8, 49, 90, 111, 213, 259, 271, 273–274, 406
Adventure games, 13–18, 49, 121, 229, 383, 386, 434
Aesthetic choice, 109, 114, 214
Agency, 30, 39, 153, 155, 212, 228n2, 246, 298, 299–300, 343, 345, 381, 383, 395–399, 421–422
Albaugh, Mike, 94
American Toy Fair, 401
Android phone, 86
Anna (machinima), 293
Arcade
 cabinet, 64, 90, 92, 95, 111, 238
 as game type, 37, 41, 47, 49–50, 54, 59, 64–65, 68–69, 71, 82–83, 110, 127, 134, 160–161, 174, 199, 215, 229–231, 240, 308–309, 336, 361, 363, 387, 389, 403
 industry and institution, 21–26, 82, 91–94, 110, 191, 230, 238, 242, 244, 385, 388, 401
Archives, 14, 50, 120, 139n2, 421
Aristotle, 192n1, 230, 321n4
Arneson, Dave, 39, 131, 191, 227, 284, 318, 335, 378–380, 382, 386, 405. *See also* Gygax, Gary
Artificial intelligence, 29–34, 54, 127, 203–204, 224–225, 248, 338, 413
Assassin's Creed (game), 241, 420, 423–424
Asymmetric game, 169–170, 172–175, 279
Atari
 company, 16, 34, 64–65, 90–94, 110–111, 134, 136, 138, 140n6, 199, 203, 231, 273, 401–402, 404, 406
 2600 (also, VCS), 7, 14, 47, 49, 58–59, 68, 74, 82, 83, 199, 234, 308, 370
Atari Football (game), 93–95
Automata, 29–30, 32
Avatar, 37–41, 57, 75, 85, 107, 129, 130, 184, 213–214, 242, 244, 294, 299, 309, 339, 344, 386, 388

Badge, 1, 3–4, 6–7
Baer, Ralph H., 66, 89, 231, 273, 402
Baio, Andy, 319
Balance (game design), 91, 169–175, 301
Barad, Karen, 345–347
Barenaked Ladies (music group), 292

Bell Telephone Laboratories, 54
Bennett, Jane, 345
Blow, Jonathan, 264–265, 319
Bogost, Ian, 6, 14, 55, 59, 83, 123, 140n5, 147, 322n7, 343–344, 373
Boss (game-level), 41–42
Braid (game), 261–262, 265, 319–320, 388. *See also* Blow, Jonathan
Brain-Computer Interface, 249
Brand, Stewart, 81, 403
Breeze, Mez, 56
Bristow, Steven D., 67
Buchsbaum, Walter H., 65, 68
Buckwalter, Len, 68
Bullet hell (game type), 114
Bushnell, Nolan, 34, 64, 93, 110, 401

Caillois, Roger, 231, 337, 354, 377
Calhamer, Allan B., 317
Call of Cthuhlu (game), 381, 383
Cao Fei, 394
Carmack, John, 204–205. *See also DOOM*; Romero, John
Carter, Marcus, 320
Casual games, 114, 229
Cayley, John, 56
Cheat
  cheating, 409, 413, 416
  cheats, 106, 288, 305, 307
  code, 212
Checkers, 30–31, 34, 153
Chess, 29–33, 34n5, 153–154, 170–171, 174, 224, 226, 279–282, 337, 352–353, 385, 395, 397, 406
Chun, Wendy Hui Kyong, 53
Coin-operated games, 23–25, 48, 64–65, 67, 109, 237, 401, 404. *See also* Arcade, cabinet
Cold War, 30, 196, 224, 226, 282, 315–316
Colecovision, 47, 68, 74, 83, 161–162, 402–403
Collaboration (in-game), 76, 91, 105
Collecting
  in-game item, 182, 234, 314 (*see also* Achievement)
  institutions, 133, 137
  practice of, 6, 48
Collectors, 6, 46, 49–50, 137
Combat (game), 110, 112
Commander Keen, 205
Completion logic, 2, 8
Complexity (rules), 78, 90, 111, 113–114, 383, 396
Console (game), 1–4, 14, 21, 26, 45–47, 49, 53, 63–71, 74, 81–86, 89, 109–111, 119, 135, 138–139, 145, 156, 160–163, 165, 174, 177, 179, 182–185, 190, 195, 199, 204–205, 215, 221, 229, 240–242, 263, 271, 273, 290, 308, 325, 327, 330, 333n3, 336, 345, 347, 361–362, 370, 386–387, 389, 403–404, 413
Constraint, 78, 84, 115, 195, 264, 299, 318, 319, 355
Controller (game), 40–41, 49, 57, 69, 73–75, 81–86, 87n1, 89, 91, 94, 113, 115, 129, 177–180, 182, 184, 244, 271, 309–311, 330, 390, 394
Copyright, 59, 269–274, 292, 330, 332
CRT (cathode ray tube), 49, 86–87, 134–136
Customization, 40, 310
Cybernetics, 30, 81

Dadaist, 288
Dames Making Games, 243, 261
Dance Dance Revolution (game), 14, 115, 337
Dark Souls (game), 113
Darwin (game), 54–55
Degenerate, 170–171
Dexterity, 113–114, 244, 379
Diary of a Camper (machinima), 106, 289
Die Gute Fabrik, 91, 95
Difficulty settings, 111, 115
Diplomacy (game), 317
Directional pad (d-pad), 74, 83
Disney Infinity (game), 406
Donkey Kong (game), 74, 160, 178, 319, 344, 403
DOOM (game), 14, 76, 103, 105, 111, 136, 138, 183, 203–208, 287–288, 310, 362, 385, 387
Dresher, Melvin, 315
Dual in-line package switches, 110

Duchamp, Marcel, 154, 226, 289
Dungeons & Dragons (game), 39, 131, 191, 227, 232, 284, 318, 335, 378–379, 383, 386, 405. *See also* Arneson, Dave; Gygax, Gary
Dynamic audio, 159, 163, 165
Dynamic difficulty, 112

Edutainment, 121–122, 124
Embodied play, 153, 364
End user license agreement (EULA), 59, 60n3
Engine
  game, 54, 59, 76, 103, 105–107, 203–208, 225, 276n21, 287–288, 290–291, 293, 326, 362
  rendering, 287
Entertainment Software Association (ESA), 242, 245n2, 403
Erector set, 402, 404
Error, xv, 73–74, 78, 99, 110, 113, 211–212, 215–217
Escher, M. C., 361, 365
E-sports, 93, 261, 265, 322
EVE Online (game), 91, 319

Fairchild Channel F, 65–66, 68
Fairness, 109, 151, 169, 172
Famicom (NES), 83, 85, 386
Fans (gaming), 48, 50, 134, 136, 144, 157, 237–238, 291–292, 381, 430
Faxanadu (game), 386–388
Feedback, 73–74, 76, 84, 113, 198, 300–301, 307, 371
Final Fantasy Tactics (game), 420
Flash, 54, 56–57, 60n1, 262
Flood, Merrill, 315
Frag Dolls, 241

Gaiman, Neil, 321n1
Galland, Antoine, 313
Game & Watch (Nintendo), 83
Game boy, 45, 49, 136, 160, 183, 344, 346, 402
Game design, xiv, xvii, 7, 47, 89–90, 93, 95, 101, 109, 112, 114, 121, 124, 147, 154, 197, 199, 203–204, 206–208, 213, 232, 235, 280, 288, 298, 302–303, 322, 325, 327, 329, 332, 333nn1–2, 343, 346, 352–355, 395–398, 409, 411
Game designer, 15, 37, 40–42, 53, 76, 90, 124, 156, 161, 172, 175, 178–179, 187, 195–196, 198–199, 243, 259–260, 263–264, 322, 347, 363, 373, 420
Game development, xvii, 18, 37, 41, 74, 84, 90, 100, 104, 114, 195–199, 203, 207, 243, 265, 329, 343
GameFAQs, 410–412
Game guidebook, 386
Gamepad, 75, 83–86
Gameplay, 5, 37, 39–40, 46–47, 53–54, 57, 59, 70, 84–85, 87, 89–91, 94, 103, 105–106, 109–110, 112–113, 115, 121, 128, 138, 156, 158, 159, 161, 164–165, 169–175, 177–179, 181, 195, 203, 207, 211–214, 216, 227, 231, 233–234, 270, 288–292, 297, 300, 306–307, 309–310, 311n2, 326–327, 333n1, 337, 343–344, 354–355, 364, 383, 394–399, 409–410, 413–415, 419–421
Game production, xvii, 207–208, 242, 426
Gamer, 2, 5, 7, 25–26, 39, 41–42, 97–101, 104, 106, 134, 138, 147, 175, 184, 191, 206, 214, 237–238, 240–245, 249, 261, 284, 291, 330, 409, 411–415
Gamercard, 2–3
Gamer identity, 237, 241–242, 244
Gamerscore, 2–3
Game studies, xi, xiii, xiv, xv, xvi, xvii, 15, 49, 73, 143, 147, 153, 185, 192, 198, 213, 216, 229, 315–317, 346, 353, 372, 389, 415–416
Game technology, xv, 203–207, 216
Game theory, 30, 87n1, 314–316, 343, 406
Game world, 5, 37, 40, 83, 85, 106, 128–129, 132, 188, 197, 204, 206, 212, 214, 247, 288, 290, 298–300, 338–340, 364, 366, 410, 420, 423
Gamification, 7, 78, 123–124, 144, 300
Garfield, Richard, 313–314, 317, 322
Gauntlet (game), 91–92
GaymerX, 243–244
Gender, 24, 242–244, 245n1, 347, 403

Genre, 13–15, 17–18, 38–39, 42, 43n4, 91, 109, 114, 120–121, 153, 156–157, 159, 164–165, 215, 221–222, 229–235, 259–261, 265, 279, 288, 301, 307, 309, 313, 319–320, 330, 332, 339–340, 361–364, 377–379, 381, 383, 397–398, 410, 419–420
Gianturco, Alex, 319
Gibbs, Martin, 320
Glitch, 56, 104, 138, 211–216, 216n1, 217n3, 307, 333n1, 346, 414–415
Graphical user interface (GUI), 307
Graphics, 13, 15, 18, 39, 71, 75–76, 119, 164, 177, 181, 203–206, 212, 215, 240, 261, 270–271, 287, 310, 336, 359–360, 362–365, 382, 412
Guins, Raiford, 48
Gygax, Gary, 191, 378–379. *See also* Arneson, Dave; *Dungeons & Dragons*

Hacker, 83, 103–105, 207, 403
Halo 3 (game), 115, 292
Hamlet on the Holodeck, 248, 371–372
Hansen, Mark B. N., 248, 251
Haptic force-feedback, 84
Hardware, 33, 45, 49, 50, 57, 69, 77, 78, 89, 103, 110–111, 133–136, 140n8, 160–162, 164, 178, 188, 199, 205, 211–214, 226, 229, 232, 243, 271, 287, 307, 325, 330–332, 344–347, 362, 389
Harrop, Mitchell, 320
Hawkin, William J., 68
Historical narrative, vxii, 419–420, 424
History of computing, xiii, xv, xvi, 214, 223–224
History of games, xiv, xvi, 66, 103, 307
Howard, Nigel, 316–317
Human-computer interaction, 73, 216, 307
Human-computer interface (HCI), 73, 76, 84

Iconography, 230, 234, 240, 244
Identity politics, 237
id Software, 59, 76, 103, 105–106, 111, 136, 170, 183, 203, 205–208, 228n2, 259, 287–288, 310, 329, 362, 387. *See also* Carmack, John; *DOOM*; *Quake*; Romero, John
Imagination, 13, 15, 21–22, 83, 132, 238, 248–249, 265, 336, 419
Immersive, 26, 75, 84, 157, 180, 192, 212, 247–250, 255, 383
Indy 800 (game), 90, 92–93
Infinity Ward, 111, 326
Information architecture, 310
Input, 30, 39–40, 64–65, 73–76, 82–84, 105, 113, 119, 159, 162, 164–165, 177, 182–185, 211–212, 248, 299–300, 306–310, 311n1, 330, 338, 370–371, 377, 387, 394, 410, 412
Intellectual property, xvii, 59, 138, 269–274, 292, 322n7, 332
Interactive audio, 159
Interactive fiction, 15–17, 41
Interactivity, 25, 230, 234, 248, 251
International Game Developers Association, 240
Internet Archive, 55, 134, 217n3
iPhone (Apple), 86, 140n6, 274, 345–346
Isometric, 361–362, 364–365, 366n3

Jaguar, 84
Joystick, 14, 47, 49, 74, 81–83, 94, 128, 136, 308
J.S. Joust (game), 91

Kay, Alan, 60, 81
King, Geoff, 382
Kittler, Friedrich, 285
Konami (cheat code), 57, 386
Kriegsspiel, xvii, 120, 279–285, 377
Krzywinska, Tanya, 382

Lantz, Frank, 322
LARP (live action role-play), 378
La sortie des usines (film), 288
Leeroy Jenkins (machinima), 289–290
Left 4 Dead (game), 2, 90, 112, 152
Legacy of Kain (game series), 420–422, 424
Legend of Zelda: Ocarina of Time, The (game), 178, 319
LEGO, 130–131, 402, 406
Levy, Steven, 403

Lipkin, Gene, 401
Livingstone, Ian, 380
Logg, Ed, 91
LOGO (computer language), 370–371
Lowood, Henry, 288, 385
Luck (games of), 25, 114

Machinima, 106–107, 207, 215, 287–294
Macklin, Colleen, 264, 322
Magic: The Gathering (game), 169, 174, 313–314, 321n1
Magnavox Odyssey, 65–66, 82–83, 89, 273, 402
MAME (Multiple Arcade Machine Emulator), 50, 134, 136, 387
Manipulation, 73, 77, 84, 104, 106–107, 113, 160, 177, 179, 297, 309, 319–320, 371, 387, 389, 394
Marino, Mark, 57
Masocore, 114
Matchups, 172–174
Mateas, Michael, 372, 374
Materialism, 345–346
Mattel Intellivision, 59, 68, 74, 198, 402
Mauro, Robert, 68
Meaningful difficulty, 115
Mechner, Jordan, 53
Metagame, 300, 313–321
Metagame, The (game), 322
Metal Gear Solid (game), 178, 386
Microsoft Flight Simulator, 130, 363, 394
Microsoft Kinect, 74, 85–86, 184
Microvision, 402
MIDI (Musical Instrument Digital Interface), 162–163
Miller, Alan, 90
Milton Bradley, 120, 234, 402, 404
Minecraft (game), 54, 276n20, 294, 327
Minsky, Marvin, 29–30, 32–33, 35n9
MMORPG (massively multiplayer online role-playing game), 378–379, 383
Molotov Alva (machinima), 293
Montfort, Nick, 59, 83, 86, 140n5, 343–344, 389
Morgenstern, Oskar, 30, 315
Motion control, 84–86
Multiplayer gaming, 206, 288
Multi-User Dungeon (MUD), 5, 191, 379
Murray, Janet, 248, 371
Mystery House (game), 136, 347

Naked Game, The (game), 54–57
Narrative, 15–16, 40, 46, 59, 92, 105, 121–123, 130, 153, 159, 164, 204, 227, 230, 233–234, 287, 289, 291–292, 325, 327, 329, 335–340, 371–372, 383, 413, 419–421, 423–424
Newell, Allen, 31–33
New games movement, 95, 264, 355
Nintendo Entertainment System (NES), 47, 57, 68, 74, 83, 134, 160, 162, 205, 215, 386, 402
Nintendo Wii, 40, 69–70, 113, 127, 215, 429
Nonplayer character, 37–38, 308, 366n9
Nostalgia, 13, 45, 47, 49, 136, 165, 215
Note-taking, 386

Oculus Rift, 182, 184, 247, 249–251
One Thousand and One Nights, 313
Operating system, 57, 137–138, 343–345, 347

Panorama, 21–22, 248, 251–254
Panoramic, 250, 254
Patch (computer), 174, 207–208, 329, 333n1
Patches (Activision achievements), 6–8
Patents, 269–274
Pause (game function), 164, 184, 308, 311n1, 397, 423
PC gaming, 204, 208, 326
Penny Arcade (amusement arcade), 24, 45
Penny Arcade (comic book), 238
Permadeath, 113–115
Perspective (visual), 128, 148, 164, 183, 229, 253, 260, 289, 359–365, 366n3, 419
Pervasive play, 177
Peterson, Jon, 280, 377
Photography, 151–152, 177, 181, 185, 250, 255, 359, 362
Photorealism, 177, 181

Platform, 70, 74, 83, 86, 105, 121, 137, 140, 145, 153, 156, 180, 185, 195, 205–208, 211, 224, 233, 249, 262–265, 274, 306, 325, 330–331, 333n3, 343–348, 363, 382, 387
Platform game (platformer), 83, 145, 233, 261, 265, 320, 363, 365, 387, 419
Player-character, 16–18, 39, 41, 83, 90, 289, 330, 338, 361, 379
Players, 74, 81–82, 84, 86, 89–95, 99–100, 103, 105–107, 109–112, 114–115, 119, 121, 123, 128, 131, 151, 153–154, 156, 159–165, 169–175, 178–179, 183–184, 188, 190–191, 198, 206–208, 211–214, 216, 221, 227, 229, 234, 237–238, 240–243, 247, 264–265, 281, 284, 288–290, 298–302, 308–310, 313, 316–322, 327, 330, 332, 335–337, 339, 345, 347, 351–356, 362–364, 366, 372–373, 378–383, 385–390, 394–395, 397–398, 402–406, 409–411, 413, 415, 419, 423
PlayStation, 84, 140, 163, 184, 330, 388–389
PlayStation 2, 69, 74, 165, 183
PlayStation 3, 69–70, 213
PlayStation 4, 69, 240, 243, 405
PlayStation Move, 74, 85, 184
Pong (game), 34, 47, 54, 57–58, 64, 82, 90, 93, 128, 156, 160, 232, 273, 401–402, 404–406
Popular music, 160, 165
Prince of Persia (game), 53, 388
Prisoner's dilemma, 315–318
Procedural literacy, 371–373
Procedural programming, 369
Procedural rhetoric, 373–374
Puzzles, 15–18, 114, 120, 122, 224, 409

Quake (engine), 59, 362
Quake (game), 105–106, 170, 183, 207–208, 228n2, 287–289, 293, 329
Quake movies, 207–208, 288

RAND, 32
Red vs. Blue (machinima series), 280, 292
Retro (style), 45, 48–49, 54, 261, 365
Retro-games, 48–49, 50, 140n6, 215, 365
Reward, 1–2, 5–7, 25, 110–111, 151, 159, 333n2, 338
Robotron 2084 (game), 84
Role-playing games (RPGs), 39, 86, 87n1, 123, 227, 265, 308, 315, 318, 335, 379, 382, 386, 419
Romero, John, 205, 209n1. *See also* Carmack, John; *DOOM*
Russell, Steve, 26, 81, 403

Sandman, The (comic book series), 321n1
Scenario (in-game), 34n3, 75, 115, 279, 281, 285, 308, 335, 337, 399, 405
Score, 3, 6, 24, 31, 41, 110, 113, 242, 385, 387, 390, 410
Screen essentialism, 86
Sears, 90, 401, 404–405
Second Life, 136, 293–294, 337
Sega Dreamcast, 69, 74
Selection (screen), 115, 307–309, 310–311
Serenity Now (*World of Warcraft* guild), 290
Serious games, 76, 120, 144–145, 147, 229, 264, 337, 373
Sewell, William, 420
Shahrazad (*Magic: The Gathering* card), 313–314, 317, 319, 321n2
Shanghai World Expo, 294
Shannon, Claude, 29, 31
Sharp, John, 322n5
Sicart, Miguel, 113–114, 373
SimCity (game), 60, 122, 372, 398, 405
Simon (game), 402. *See also* Milton Bradley
Simon, Herbert, 31–33, 34n4, 34n6
Sims, The (game series), 40, 291, 337, 340, 405–406
Simulation, 39, 67, 75, 89, 93, 119–121, 129–131, 183, 190, 196, 198–199, 225, 242, 281, 293, 298, 302, 311, 359–361, 363, 365, 366n7, 373–375, 380, 393–399, 406, 412, 414, 420
Singularity, 248–249
Skill (player), 24, 94, 104–107, 109–114, 120, 121, 123, 127–130, 132, 143, 172, 178, 197, 205, 212, 241–242, 274, 288, 293, 299, 301, 337, 370, 385, 394, 399, 409, 419

Skill games, 231–232
Skylanders (game), 406
Softdisk, 205
Software development, 57, 195, 198
Sony Vita, 389
Sound, 3, 23, 57, 84, 105, 136, 159–165, 195, 206, 211, 214–216, 287, 293, 309, 325–326, 412–413
Spacewar! (game), 26, 33–34, 59, 64, 74, 81–82, 86, 136, 198, 225–226, 230, 307, 393, 403. *See also* Russell, Steve
SpecialEffect, 243
Speedrun, 115, 138, 319, 387, 414
Sportsfriends (game), 91, 95
Sports video games, 93
StarCraft (game), 114, 169–172, 174, 318, 320, 322
StarCraft II (game), 107, 319
Steam (Valve), 1, 86, 104, 263, 319, 327
Stereoscope, 250–252, 254–255
Stereoscopic, 22–23, 183, 247–248, 250, 254–255
Strategy guide, 409, 411–412, 414
Super Mario Bros. (game), 113, 153, 159, 233, 309, 319, 344, 414
Surrealist artists, 153–154, 226, 288
Symmetric games, 169–175

Tabletop games, 191
  gaming, 195, 322, 382
  tabletop role-playing game (TRPG), 378–380
Team play, 94, 362
Team sports, 89–90, 93
Tech Model Railroad Club (TMRC), 82. *See also* Russell, Steve; *Spacewar!*
Tekinbas, Katie Salen, 112, 318, 353
Tennis for Two (game), 35n8, 74, 82, 103–104, 198, 225, 307
Terrell, Richard, 319
Text adventure, 16, 18, 227, 382, 386
"Thing-power," 345
Tier list, 173
Toys, 90, 401–407
Trademarks, 269–271, 273
Trak ball (controllers), 94–95
Trophies, 1
Turing, Alan, 32
TV-Game, 67
Twitch, 274, 319

Ubisoft, 60, 184, 241, 388, 420, 423
Unreal (game engine), 59, 276n21
User-generated content, 187, 208, 331
User interface, 74, 76, 99, 107, 113, 212, 270, 307, 310, 311n1, 370

Variable difficulty, 110
Video game history, 90, 330
Virtual reality (VR), 131–132, 177, 182–183, 184, 247, 256, 423
von Neumann, John, 30, 279, 315–316
von Reisswitz, Georg Leopold, 279

Wallis, James, 382
Wargames, 29, 34, 120, 172, 232, 279–286, 335, 376–377, 379, 396–397. *See also* Kriegsspiel
Warthogs (*Halo*), 291–292
WASD keys, 82, 136
Wiimote, 74, 85
Wired magazine, 322
Women in Games International, 243
Wright, Cosmo, 319
Wright, Will, 322, 398, 405

Xbox, 4, 74, 330, 404
Xbox Live, 2–3, 103, 262
Xbox One, 69–71, 184, 405
Xbox 360, 1–4, 69–71, 77, 85, 174, 290, 344

Yokoi, Gunpei, 83
YouTube, 274, 319, 330, 385

Zimmerman, Eric, 112, 260–261, 318, 353